Arms Control and Disarmament

Paolo Foradori • Giampiero Giacomello
Alessandro Pascolini
Editors

Arms Control and Disarmament

50 Years of Experience in Nuclear Education

palgrave
macmillan

Editors
Paolo Foradori
School of International Studies
University of Trento
Trento, Italy

Giampiero Giacomello
Department of Political and Social Sciences
University of Bologna
Bologna, Italy

Alessandro Pascolini
Department of Physics and Astronomy "Galileo Galilei"
University of Padua
Padua, Italy

ISBN 978-3-319-62258-3 ISBN 978-3-319-62259-0 (eBook)
https://doi.org/10.1007/978-3-319-62259-0

Library of Congress Control Number: 2017955412

Cover illustration: Enzo Brandi / EyeEm - Getty

Printed on acid-free paper

This Palgrave Macmillan imprint is published by Springer Nature
The registered company is Springer International Publishing AG
The registered company address is: Gewerbestrasse 11, 6330 Cham, Switzerland

To all the friends of ISODARCO

Contents

List of Acronyms

ABM	Anti-ballistic Missile
ACDA	Arms Control and Disarmament Agency
AEC	Atomic Energy Commission
ARPA	Advanced Research Project Agency
ASAT	Anti-Satellite Weapon
ASW	Anti-Submarine Warfare
BAMBI	Ballistic Anti-Missile Boost Interceptor
BMD	Ballistic Missile Defence
BWC	Biological Weapons Convention
CD	Conference on Disarmament
CERN	European Organization for Nuclear Research
CFE	Conventional Forces in Europe
CIA	Central Intelligence Agency
CIS	Commonwealth of Independent States
ComSat	Communication Satellite
CPSU	Central Committee of the Communist Party of the Soviet Union
CSCE	Conference on Security and Cooperation in Europe
CTBT	Comprehensive Nuclear-Test-Ban Treaty
CWC	Chemical Weapons Convention
DNPE	Disarmament and Non-Proliferation Education
DPRK	Democratic People's Republic of Korea
DSAT	Defence Against ASAT
DSP	Defense Support Program
EC	European Community

EFTA	European Free Trade Area
eif	Entry Into Force
EMP	Electromagnetic Pulse
ESA	European Space Agency
EURATOM	European Atomic Energy Community
FBS	Forward-based System
FMCT	Fissile Material Cut-off Treaty
FOBS	Fractional Orbital Ballistic System
FRP	Fire-resistant Pit
FSU	Former Soviet Union
GCD	General and Complete Disarmament
GEO	Geostationary Orbit
GLCM	Ground-launched Cruise Missiles
GPS	Global Positioning System
HEU	Highly-enriched Uranium
IAEA	International Atomic Energy Agency
ICBM	Intercontinental Ballistic Missile
IDCSP	Initial Defense Communication Satellite Program
IGO	Inter-governmental Organization
IMS	International Monitoring System
IMU	Inertial Measurement Unit
INF	Intermediate-Range Nuclear Forces
INS	Inertial Navigation System
IO	International Organization
IPP	Industrial Partnering Program
IRBM	Intermediate-Range Ballistic Missiles
ISODARCO	International School on Disarmament and Research on Conflicts
ISTC	International Science and Technology Centre
LEO	Low Earth Orbit
LEU	Low-Enrichment Uranium
LOW	Launch-on-Warning
LRTNF	Long-Range Theatre Nuclear Forces
LUA	Launch-under-Attack
MAD	Mutually Assured Destruction
MAR	Multi-Function Array Radar
MBFR	Mutual and Balanced Force Reductions
MDA	Missile Defense Act
MinAtom	Russia's Ministry of Atomic Energy

MIRV	Multiple Independently Targetable Re-entry Vehicles
MLF	Multilateral Force
MOX	Mixed-Oxide
MRBM	Medium-Range Ballistic Missiles
MRV	Multiple Re-entry Vehicle
MSR	Missile Site Radar
NASA	National Aeronautics and Space Administration
NATO	North Atlantic Treaty Organization
NGO	Non-governmental Organization
NPT	Non-Proliferation Treaty
NSA	Non-State Actors
NTM	National Technical Means
OSCE	Organization for Security and Cooperation in Europe
P5	The Five Permanent Members of the United Nations Security Council
PAL	Permissive Action Link
PAR	Perimeter Acquisition Radar
PBCS	Post Boost Control System
PLO	Palestine Liberation Organisation
PTBT	Partial Test Ban Treaty
RAND	Research and Development Corporation
RECM	Radio-electronic Countermeasure
RV	Re-entry Vehicles
R&D	Research and Development
SACEUR	Supreme Allied Commander Europe
SALT	Strategic Arms Limitation Talks
SAR	Synthetic Aperture Radar
SBI	Space-based Interceptors
SBL	Space-based Laser
SDI	Strategic Defense Initiative
SLBM	Submarine-launched-ballistic Missile
SORT	Strategic Offensive Reductions Treaty
SPD	Sequential Payload Delivery System
SRAM	Short-Range Attack Missile
SSAC	State Systems of Accounting and Control
START	Strategic Arms Reduction Treaty
SWU	Separative Work Unit
TACMAR	Tactical Multi-Function Array Radar
TBM	Theatre Ballistic Missile

TBT	Test Ban Treaty
TEL	Transporter–Erector–Launcher
TNF	Theatre Nuclear Force
TNW	Tactical Nuclear System
UAV	Unmanned Air Vehicle
UN	United Nations
UNGA	United Nations General Assembly
UNSCEAR	United Nations Scientific Committee on the Effects of Atomic Radiation
UNSCOM	United Nations Special Commission on Iraq
USAF	United States Air Force
US DoD	United States Department of Defense
US DoE	United States Department of Energy
USEC	United States Enrichment Corporation
WMD	Weapons of Mass Destruction

List of Tables

ACKNOWLEDGMENTS

Like all books, this volume too is the product of the work of many people and, as editors, we think it is fundamental to express our appreciation for all who have contributed.

First and foremost, we would like to thank all the authors and editors of the previous volumes from which chapters included here have been drawn. In this volume, we have kept all the contributions the way they were when first published, adding just a few notes where necessary and re-editing the texts for a homogeneous presentation.

Our gratitude also goes to publishers of the original volumes who have graciously granted us permission to re-publish those chapters here.

The very idea of celebrating the 50th anniversary of the International School on Disarmament and Research on Conflicts (ISODARCO) with this book maturated in discussions with Carlo Schaerf, ISODARCO's President, David Carlton, Catherine McArdle Kelleher and Judith Reppy, and we are deeply grateful for their assistance and suggestions.

Introduction

Paolo Foradori, Giampiero Giacomello and Alessandro Pascolini

Seventy years after the bombing of Hiroshima and Nagasaki and 45 years after the Non-Proliferation Treaty (NPT) entered into force, nuclear weapons continue to pose a serious threat to global security. Disarmament and non-proliferation education (DNPE) has long played a crucial role in international efforts to combat the spread of nuclear weapons and ideally to eliminate this unique category of weapons of mass destruction. As the United Nations states, the key goal of DNPE is

> to impart knowledge and skills to individuals to empower them to make their contribution, as national and world citizens, to the achievement of concrete disarmament and non-proliferation measures and the ultimate goal of general and complete disarmament under effective international control.[1]

P. Foradori
School of International Studies,
University of Trento, Trento, Italy

G. Giacomello
Department of Political and Social Sciences,
University of Bologna, Bologna, Italy

A. Pascolini
Department of Physics and Astronomy "Galileo Galilei",
University of Padua, Padova, Italy

This volume is a collection of contributions by some of the world's leading experts in the nuclear field who have participated in the educational activities of the International School on Disarmament and Research on Conflicts (ISODARCO) during its half-century long existence. It features work by people who contributed in fundamental ways to shaping policies, strategies, theories, scholarly analyses and debates in the study of non-proliferation and disarmament. Among others, contributors include two Nobel laureates (Thomas Schelling and Joseph Rotblat); one of the founding fathers of the academic discipline of International Relations (Hans Morgenthau); a world pioneer and leading figure in systems analysis, game theory and conflict resolution (Anatol Rapoport); outstanding scientists who directly participated in the development of nuclear weapons and later in efforts to control them (Joseph Rotblat, Herbert York, Richard Garwin, Bernard Feld); diplomats/policymakers who were key in creating the current international non-proliferation regime (George Bunn) and prominent scholars who authored some of the classic works on arms control and non-proliferation issues.

All contributors were lecturers in ISODARCO courses and their lectures became book chapters that originally appeared in some of the 22 volumes published based on ISODARCO courses since the first volume in 1967. On the occasion of ISODARCO's 50th anniversary, these valuable contributions are now being republished to celebrate the organization after five decades of continuous engagement in non-proliferation education and training and, above all, to provide the younger generation with a set of remarkable readings that have retained, largely unaltered over time, their highly valuable and enlightening insights into nuclear issues. In this volume, we explicitly intended to present the contributions in their original form and language, the way they were when first published, adding just a few notes where necessary and re-editing the texts for a homogeneous presentation. The purpose of such a choice is to offer the reader a path through the evolution of the debate on arms control and nuclear disarmament as well as to show the many hurdles and drawbacks those scholars working on nuclear education have had to overcome in all those years.

For guidance in selecting those chapters to be included in the present book from among almost 300 essays, the editors considered 6 main general criteria. First, the contributions had to directly address nuclear weapons-related matters, thus excluding a long list of excellent writings that addressed other categories of security threat. The primary decision was therefore to restrict the subject matter of the book to the traditional "core business" of ISODARCO. As a second criterion, the editors considered it important to include some genuine "classics" that could not be missing from a nuclear arms control and non-proliferation "must read" list; for example, the works by Hans Morgenthau and Thomas Schelling. Third, it was considered important for the chapter selection to reflect the broad and diversified range of viewpoints that have characterized ISODARCO's non-ideological and ecumenical approach, whereby all opinions and perspectives are legitimate and welcomed as long as they are well argued, reasoned and contribute to enhancing understanding, analysis and critical thinking. As a consequence, the volume purposefully includes chapters with a clear progressive stance on disarmament as well as more conservative assessments, passionate calls for nuclear abolition as well as technical and cold studies, and policy-oriented examinations suggesting practical recommendations as well as theory-grounded evaluations. Similarly, and fourth, the selection aimed to provide some geographical balance, with contributions from as many nationals as possible, although an overrepresentation of American authors proved inevitable. A fine mix of scholars and practitioners was also deemed important, although efforts at gender balancing miserably failed, with only two female contributions. Fifth, a chronological/historical criterion guided the selection in order to cover as much and as evenly as possible the 50 years of ISODARCO's lifetime, with a view to offering a comprehensive overview of the numerous issues featured during different periods of the nuclear age.[2] It goes without saying that in addition to the above criteria, a final but decisive selection element has been the inherent value of each piece of work, both as a telling testament of a specific moment in the history of nuclear weaponry and policies and as embodying continuous relevance for our understanding of current disarmament and non-proliferation issues.

Having selected the contributions according to those criteria, the book is divided into three parts and presents the contributions in chronological order to reconstruct the historical evolution of nuclear weapons and the attempts to control, limit and possibly eliminate them. It thus begins with the sixties/seventies (Part I, Early Attempts to Arms Control), moving to the critical years of the East/West confrontation of the seventies and eighties (Part II, The Hard Times, 1978–1989) and ends in the current post-bipolar international context (Part III, After the Cold War). This history largely overlaps with the list of courses, workshops and seminars that

ISODARCO has organized since its establishment in 1966 and the resulting publications.

The first chapter by Bernard Feld originated out of the very first ISODARCO course organized in Frascati, near Rome, in 1966. The lecture by Feld, and the resulting chapter re-published here, emphasizes the unprecedented danger represented by nuclear weapons in the history of humanity. It analyses the complex relationship between technology and the arms race, making a strong case in favour of involving the independent scientific community in the "rational" study of and engagement with issues so potently related to world security. In the second chapter, Francesco Calogero analyses the technical and political problems connected to the introduction of anti-ballistic missiles (ABM), an event that is presented as one of the destabilizing developments of the arms race and an obstacle to arms control. His contribution, originally presented in 1968, contains many observations that continue to be relevant in today's international scenario characterized by the revival of ABM defence systems supported by the United States. Anatol Rapoport's third chapter discusses the application of game theory to international relations, where decision makers are called upon to make difficult decisions and the lives of fellow citizens are at stake. Those decisions are even more difficult to make in a nuclear-armed world, where mistakes and miscalculations can have catastrophic consequences.

In the following chapter, Herbert York recounts the complex history of the development of multiple warheads (MIRV) for missiles and its relationship to ABM systems. MIRV technology made it possible to deliver multiple warheads with very high accuracy to several different targets with a single missile, greatly enhancing the destructive power of nuclear weapons. The complex relationship between war and technological innovation is at the core of York's analysis. Arms control agreements are then examined in Chap. 5: Thomas Schelling identifies the motivational structures that underlie understandings and negotiations in such a context. With a special focus on the ABM negotiations, the author argues that the key issues are figuring out what the purpose of an agreement is, what we would want to call an agreement, what types of agreements there are and how to evaluate them. In Chap. 6, Hans Morgenthau passionately argues in favour of nuclear non-proliferation and disarmament as vital objectives to save humanity from certain catastrophe. The fact is that a nuclear weapon is not a weapon in the conventional semantic sense. It is not a rational means to a rational end; it is an instrument of unlimited, universal destruction. Hence, the threat or the actuality of a nuclear war is not a rational instrument of national policy (as can be argued in the tradition of Machiavelli, Lipsius and Clausewitz) but rather an instrument of suicide and genocide.

The East-West confrontation of the Cold War is considered in the final chapter of the First Part. Michail Milstein analyses some fundamental military strategic concepts and addresses the superpowers' negotiations for strategic arms limitation. Arms control talks between the Soviet Union and the United States on the limitation of, and even more so on the reduction of, strategic arms helped build trust between the two countries, strengthened the stability of relations between them and, in doing so, favourably influenced the entire international situation by reducing the threat of war.

Part II of the book covers the "hard times" of the late seventies and eighties, when the rivalry between the superpowers intensified to critical levels (the highest number of deployed nuclear weapons was reached in 1986) and risked escalating out of control to a point of no return. Mutual destruction became a very realistic scenario. The section opens with a study by Joseph Rotblat on the need to keep nuclear energy and technology under strict control, dwelling in particular on the radiation hazards of civilian (and military) fission fuel cycles. The role of theatre nuclear weapons in Europe in the dynamics of East-West confrontation is then the central focus of Chap. 9 by Lawrence Freedman, who discusses in great detail the need to, but also the difficulties in, negotiating an arms control agreement for this specific category of nuclear weapons.

In a time of increasing tensions between the countries of the North Atlantic Treaty Organization (NATO) and the Warsaw Pact, Chap. 10 by George Bunn analyses some of the key bilateral and multilateral arms control and non-proliferation agreements. He suggests practical recommendations on how to overcome distrust among the parties, which is regarded as the major obstacle to a successful accord. In Chap. 11 David Carlton offers his sober reflection of disarmament, assessing the many international systemic features inhibiting the elimination of nuclear weapons. While the maximalist goal of complete abolition is described as unattainable under the existing circumstances, he advocates for more minimalistic but achievable objectives such as gradual and limited measures of arms control covering only some countries and specific categories of weapons.[3] Chapter 12 by Cui Liru is a very interesting examination of the implications of the nuclear first-use option from a Chinese perspective and of the possible ways to overcome the potentially catastrophic consequences of this high-risk strategy.

The strategy of deterrence is thoroughly discussed in the next two chapters. In Chap. 13, Jane Sharp addresses the issue of extended deterrence to NATO countries, tackling the delicate balance between the assurance needs of the *protégées* and the inherent problems of credibility within such a strategy. To prevent an "overkill" scenario of a nuclear exchange between the two Cold War superpowers, in Chap. 14 Richard Ullman analyses in detail the concept of minimum deterrence with its doctrinal and very practical implications.

Part III of the volume is dedicated to the post-Cold War international context. The end of the bipolar system, the dissolution of the Soviet Union and the political changes in Eastern Europe deeply modified the landscape of international relations, introducing both new opportunities and new challenges for arms control and nuclear disarmament. Among the latter, in Chap. 15 Richard Garwin considers the issues of advanced weapons on earth and in space and their implications for arms control, and assesses both the technical and political dimensions of a possible space arms control regime. Harald Müller examines the challenges to non-proliferation in the new international environment in Chap. 16. Discussing the notion of a "regime" and the motivations (as well as the barriers) for nuclear acquisition (or restraint), the author investigates the health of the NPT through the case of Germany and its decision not to proliferate. Chapter 17 by Frank von Hippel and Oleg Bukharin is concerned with the legacy of the Cold War in terms of securing the abundance of fissile material produced by the two superpowers. The massive reduction of nuclear arsenals has created the challenge of storing and disposing of, in a secure and safe fashion, huge quantities of weapon-quality plutonium and highly enriched uranium. The authors present the results and perspectives of US-Russian cooperation on this new and often neglected security threat.

The post-Cold War opportunities for non-proliferation and multilateral arms control negotiations are addressed by Patricia Lewis in Chap. 18. Again, encompassing a fine combination of technical and political considerations, the author analyses the new verification technologies, mechanisms and procedures at our disposal to ensure compliance with the obligations undertaken by sovereign states. Special emphasis is also placed on the "democratization" of verification, particularly the role for academia, non-governmental organizations and the media—in other words, epistemic communities. The first decade of the new millennium ended with very positive and promising perspectives towards nuclear disarmament, after the London meeting of presidents Dmitry Medvedev and Barack Obama, the advanced work on a new Strategic Arms Reduction Treaty (START) agreement, the growing cooperation between Russia and the United States and Obama's explicit pledge for a "world without nuclear weapons" in his Prague speech of 2009.[4]

The 2010 ISODARCO course was devoted to a critical appraisal of these new perspectives on the road to nuclear zero and their consequences for arms control. The resulting book includes these two contributions. Chapter 19 by Alexei Arbatov looks at the Russian view on deterrence and makes the strong argument that a future nuclear-free world cannot be just the present world minus nuclear weapons but must be characterized by a very different system of international security and governance. The section ends with the chapter by Matthew Evangelista, who discusses military strategy in a world beyond deterrence, analysing the choices and contradictions in US proposals about nuclear abolition and its commitments to extended deterrence.

The book ends with a Conclusion by the editors focusing on the role of epistemic communities and of education in promoting non-proliferation and disarmament. It is followed by an Afterword by Carlo Schaerf, founder and president of ISODARCO, sketching in a personal recall the origins, motivations, objectives, working ethos and habits of the organisation after 50 years of continuous engagement in the fight against nuclear weapons.

We believe that this volume can provide useful insights on nuclear weapons and policies, thereby contributing to the efforts of the international community to combat the presence and further spread of these weapons with a view to their complete elimination; a difficult, but not impossible, long-term objective. At the end of the day, each of us bears some responsibility for what the future of humanity will look like. On that note, we would like to conclude this introduction by recalling the final words of the Russell-Einstein Manifesto, which led directly to a conference of scientists held in Pugwash, Nova Scotia, in 1957 and inspired the formation of a unique and innovative transnational organization, the Pugwash Conferences on Science and World Affairs (of which ISODARCO represents the Italian group):

> There lies before us, if we choose, continual progress in happiness, knowledge, and wisdom. Shall we, instead, choose death, because we cannot forget our quarrels? We appeal as human beings to human beings: Remember your humanity, and forget the rest. If you can do so, the way lies open to a new Paradise; if you cannot, there lies before you the risk of universal death.[5]

Paolo Foradori is an Associate Professor at the University of Trento, School of International Studies. From 2009 to 2011 he was Marie Curie Fellow at the Center for Non-proliferation Studies, Monterey Institute of International Studies, California, and until 2014 an associate at the Managing the Atom Program, Belfer Center for Science and International Affairs at Harvard University's Kennedy School of Government.

Giampiero Giacomello is Associate Professor of Political Science in the Department of Social and Political Sciences, University of Bologna, where he has taught Strategic Studies for several years. His research interests include strategic theory, cyber-security and war-gaming and simulation. He has published extensively on these themes and has been a visiting professor at several European and American universities. His association with ISODARCO dates back to 1988.

Alessandro Pascolini is a theoretical physicist and senior scholar at the University of Padua, where he also taught courses on arms and disarmament and science communication. He is the vice-president of ISODARCO.

Notes

1. United Nations, *United Nations Study on Disarmament and Non-Proliferation Education* (Ney York: UN General Assembly Report A/57/124, 2002).
2. It should be noted that for approximately a decade starting in the mid-1990s, in addition to nuclear non-proliferation and disarmament, ISODARCO focused on a range of very different security issues (ethnic conflicts, international terrorism, revolution of military affairs, climate change, new military technology, etc.) that emerged in those years as being particularly challenging. This explains the wide temporal gap between the first chapters and the last two of Part III.
3. David Carlton played a crucial role in the development of ISODARCO providing his expertise as editor of most of the books published as proceedings of the courses.
4. Obama's landmark Prague speech is available at The White House Office of the Press Secretary, "Remarks by the President Barack Obama in Prague as Delivered," Hradcany Square, Prague, Czech Republic, 5 April 2009, http://prague.usembassy.gov/obama.html.
5. The text of the Manifesto is available in Joseph Rotblat, ed., *Scientists, the Arms Race and Disarmament* (London: Taylor and Francis, 1982): 301–303.

PART I

Early Attempts to Arms Control

Introduction to Part I

The first part of the book reflects the contribution of the International School on Disarmament and Research on Conflicts (ISODARCO) to two of the most preeminent issues characterizing the first decades of the nuclear age: the need of proper understanding of the new weapon and its impact on the conduct of international relations; and the first attempts at the regulation and control of nuclear arsenals after the disarmament failures of the late forties.

While some considered nuclear weapons to be simply another type of weapon, just bigger and more powerful, more attentive and lucid observers immediately grasped the nature of the bomb as a weapon quantitatively and qualitatively different from anything invented before. Indeed, the catastrophic, disproportionate, and indiscriminate levels of death and destruction that nuclear weapons could instantly deliver qualify them as something truly unprecedented in human history. The subsequent invention of thermonuclear weapons and ballistic missiles exacerbated this scenario even further.

Many ISODARCO scholars and practitioners included in this volume share a similar view and urged the elaboration of innovative thinking to come to terms with the extraordinary novelties of the nuclear era. Among them, Hans Morgenthau—widely praised as the father of the modern discipline of International Relations—convincingly argues that a nuclear weapon is not a weapon in the conventional semantic sense, since it fundamentally contradicts the Clausewitzian rationality of being usable as an instrument of national policy due to its inherent suicidal and genocidal character. In the same vein, the need for new analytical categories and tools to make sense of the changed security environment is central in the work of Anatol Rapoport, who considers the application of game theory to the complex process of decision-making in a nuclear armed world where mistakes and miscalculations can have catastrophic consequences.

A proper understanding of the new weapon cannot avoid an in-depth comprehension of the technological dimensions underpinning nuclear weapons and policies. This conviction has always been a distinctive feature of ISODARCO's program of non-proliferation and disarmament education, with continuous communication and interchange between science and world affairs. The contributions by Bernard Feld, Francesco Calogero, and Herbert York perfectly reflect this approach. How could one possibly appreciate the counter-intuitive notion that missile-defence in a nuclear context might be a destabilizing factor without knowing how missile technology and multiple independently targetable re-entry vehicles (MIRV) systems operate?

As the mounting tensions of the Cold War made the aspiration to nuclear disarmament increasingly unrealistic, attempts at regulating and controlling nuclear weapons became imperative. Nobel Prize winner Thomas Schelling has produced the most acute analyses of arms control, conceiving of it as a process of cooperative security aiming at reducing the likelihood of war, diminishing the political and economic costs of preparing for war, and minimizing the scope and violence of war, if it were to occur. With that same priority, the lessening of the threat of war and the promotion of international peace through trust-building arms control talks between the United States and the Soviet Union is also central to the analysis by Michail Milstein.

Almost half a century since these themes were first discussed in ISODARCO courses and later turned into publications, their value remains surprisingly relevant in today's nuclear context. Their insight continues to shed light and offers important lessons for our understanding of contemporary nuclear weapons and policies; as for instance is the case of the old, and today still super-topical, issue of missile defence.

The invitation to develop original, innovative, "out of the box thinking" is all the more relevant as new actors (new nuclear armed states, proliferators, and non-state actors), new risks and threats (nuclear terrorism, nuclear safety and security), new doctrines and strategies (limited nuclear war and complex deterrence) have emerged and complicate today's security landscape. Similarly, deep knowledge and an understanding of science and technology continue to be crucial to appropriately grasp, for instance, the significance of the programs of nuclear weapons stewardship and modernisation, the rapid advancement in missiles and other delivery vehicles, as well as the intricate technicalities of non-proliferation efforts as the recent deal between Iran and the P5+1 countries has shown.

Finally, the "Importance of Agreements," to quote the title of Thomas Schelling's chapter, remains critical. Despite the deep reductions since the peak of the Cold War, several thousand nuclear weapons remain in the world's arsenals today, some of which are mounted on ballistic missiles ready to be launched, a potent reminder of the immediate capability for world annihilation. Negotiating and agreeing on further reductions at small numbers of nuclear weapons might become even more challenging. If this is the case, the insight and the lessons from the past might be all the more useful.

CHAPTER 1

Technological Aspects of World Security

Bernard T. Feld

It is commonplace to say that war has become so horrible that it is no longer possible to contemplate it. Yet it is being contemplated today, not only by the military, whose profession is so to do, but also by the so-called civilian strategists. And the more they contemplate war, the more possible war seems to become; and as the opportunities for manipulation of force for the achievement of political objectives become more apparent, so does the horror recede and the talk of mega-deaths take on a kind of cocktail-party unreality.

That wars should be regarded as intolerable is not new. This happened after every major war, but the horror slowly faded as the world picked up the pieces and proceeded to politics as usual. Although each revolutionary new military invention had led to the prediction that war has now become too horrible to contemplate, the world has always grown accustomed to the new horror and after a while has not known how to avert it.

The situation since the first atomic bombs were exploded differs from that of the past for quantitative rather than qualitative reasons. First, there

Originally published in Edoardo Amaldi and Carlo Schaerf, eds., *International School on Disarmament and Arms Control: Proceedings of the First Course* (Rome: ISODARCO, 1967): 33–41.

B.T. Feld (✉)
Massachusetts Institute of Technology, Cambridge, MA, USA

P. Foradori et al. (eds.), *Arms Control and Disarmament*,
https://doi.org/10.1007/978-3-319-62259-0_1

is a new scale of destructiveness: a major city and its entire population can now be destroyed in minutes, rather than after long months of massive bombardment. Second, there is the now universal recognition by governments that science and technology are essential for the maintenance of military strength, so the pace of military technology since the end of the war has remained essentially at the wartime level. Coupled with the general explosive growth of science in the last 20 years, this has meant that the time between radical new innovations in methods of mass destruction is now measured in a few brief years rather than generations. However, there is a new aspect to the situation, the beginning of a recognition—though it has not yet penetrated to the furthest reaches of our governmental structures—that the national security of all nations may be better served by restraints, controls and agreements to limit armaments and their use than by the race for the improvement of such weapons.

But arms races have a life of their own. They are not turned off by intellectual recognition of their futility. Nor is it simple to stop governmental activity in the weapons field once started. The time must be right, the political and psychological situation—both internal and external—must be favourable. Such favourable constellations of circumstances have not occurred very often in the past.

One such opportunity occurred at the end of World War II. At that time the American Government prepared a wise and far-sighted plan for the international control of such weapons, the so-called Acheson-Lilienthal plan. There are many reasons why this plan was never accepted, and not all of these have to do with the intransigence of the Russians.

Probably the main reason for the failure of these proposals as presented by Mr Baruch was the fact that the Soviet Union, at the conclusion of the last war, was neither interested in nor ready for the freezing of the status quo with respect either to her armed strength or to the possibility for its further expansion. Stalin was not interested in any kind of agreement that would have prevented the Soviet Union from independently acquiring nuclear weapons. On the other hand, although the Acheson-Lilienthal proposals were both far-sighted and magnanimous, there was in their presentation a large element of hypocrisy—of appearing to offer the moon in the full certainty that the offer could not possibly be accepted. Certainly, there was little excuse, beyond the immediate provocations of the cold war, for the cynical dragging on of the disarmament proceedings in the United Nations until well into the fifties, in which the Western Allies piously advocated a world nuclear government and the Soviet Union equally piously proposed a purely verbal "ban the bomb" agreement, without either side having the slightest expectation of any progress. Aside from a missed opportunity, when it might have been possible to place some controls over the development of nuclear weapons before they had been produced in large numbers, or when some sort of agreement might have been reached to limit their numbers and types, the net effect of the United Nations disarmament discussions of the late forties and early fifties was to develop among politicians and people the most profound cynicism concerning the intention of the major powers regarding any possible limitation of their armaments.

An opportunity of a different kind, less spectacular but possibly as significant, was missed in the early fifties when the hydrogen bomb was first achieved by the United States, followed in about one year by the Soviet Union. At that time an important segment of American scientific opinion held that another initiative should be taken to halt the growing arms race and, by mutual and binding agreement with the Soviet Union, to limit the new development that was clearly and spectacularly on the horizon. The resulting internal struggle in the United States was not only lost by the scientists but it ended in a vicious vendetta against their spokesman, J. Robert Oppenheimer, and flagrant intimidation against any independent scientific initiative. In this case, not only was an opportunity missed, but also unreason, in the name of McCarthyism, prevailed instead.

Another missed opportunity for significant armament controls began with the launching of the first Soviet Sputnik. Unfortunately, following the initial Russian successes in the launching of satellites and owing to their obvious ability to use the same types of rockets to deliver nuclear weapons from great distances, the United States passed through a period of panic induced by the belief that our capabilities in the field of rocketry were lagging hopelessly behind those of the Russians. The scientific community was called upon for emergency aid (it was in this period that the President's Scientific Advisory Committee was activated) and a massive and rapid programme of rocket development and missile construction was initiated to overcome the so-called missile gap. As is now well known, this missile gap never existed to any appreciable degree and, in fact, now exists in the reverse sense. American missile capabilities for long-range delivery of nuclear weapons now exceed those of the Soviet Union by a large factor.

During the late fifties and early sixties, when the potentialities for intercontinental ballistic missile (ICBM) development were being explored and understood, it became clear that such developments could provide a new opportunity for halting the arms race and for reducing nuclear armaments. The need for large numbers of nuclear weapons had previously been justified by the military, not because of the need to use large numbers in any conceivable conflict, but rather because American and Soviet defensive developments, in particular anti-aircraft defences, had become sufficiently effective so that a large number of bombers were required for even a few to be sure of penetrating these defences. However the same is not true of ICBMs, since there is no effective defence against these; therefore the number of missile weapons required for any possible military application is very much less than the number of bomber weapons required for the same application, most especially if the rockets themselves can be rendered invulnerable against an enemy attack.

This situation was clearly recognized by many scientists in the United States in the late fifties and a number of proposals were put forward for limiting such forces according to a doctrine now known as "minimum deterrence." Apparently Soviet military strategists understood these arguments much better than their American counterparts, for the Soviet Union has limited her missiles to a number sufficient to ensure that under all circumstances any American nuclear attack on the Soviet Union would result in unacceptable retaliatory damage to the United States. Probably, Russian acceptance of the concept of nuclear deterrence was made easier by the desire to economize on military expenditures. American affluence made such economic considerations less important and allowed us to assemble a nuclear force, which is evidently a number of times larger than any conceivably needed to deter an enemy attack.

However, this opportunity for nuclear arms limitation is not completely gone, since even conservative military spokesmen recognize that both the United States and Soviet Union are now in a position to halt further procurement of missiles and even to get rid of some obsolete delivery systems, such as heavy bombers. Recently the American government has proposed a freeze on further missile procurement, as well as an agreement for both sides to destroy obsolete bombers. Although this proposal has been supported on the grounds that its effect would be to limit armaments to levels much lower than those that would be reached without such an agreement (thus constituting, in effect, a measure of disarmament) the freeze proposal has nevertheless been regarded by the Soviet Union and other nations as an attempt to maintain American nuclear superiority. Hence, it has not been very enthusiastically received and, I believe, it will not have much chance of success unless the governments are willing to go a step further and reduce armaments to lower levels.

Nevertheless, despite the lack of agreement on freezing or reducing missile stocks, the current missile situation has achieved a sort of quasi-stability, in which the number of American and Soviet ICBMs appears to be approaching a plateau. Although the present unbalanced situation cannot continue over a long period, there is a kind of short-range stability arising from purely economic considerations and from the recognition of sufficiency. But even such temporary stability is likely to be destroyed by projected technological developments in both the United States and the Soviet Union; that is the development of anti-ballistic-missile (ABM) systems. Should such systems be deployed, the number of existing missiles would soon be regarded as insufficient to ensure capability of retaliation against attack. But what is worse, even if such programmes are only partially successful (that is to say, if it should be possible to develop a system that would shoot down a certain fraction of all attacking missiles), as long as work on such systems goes on, and as long as neither side is fully cognizant of the status of such work on the other side, the tendency will be to assume the worst and to build many more missiles than are needed. Thus, just as soon as any significant progress is apparent towards the development of an ABM, the armaments race will start off again and current stockpiles of missiles will be greatly increased.

Unfortunately, all attempts to convince the Soviet Union that it would be to our mutual advantage to forego the deployment of an ABM system have fallen upon deaf ears. In part, I suppose, the incentives which we have offered have not been sufficient; in part, we are up against a very profound aspect of Russian military psychology, which has always emphasized the defence and which therefore finds it extremely difficult to conceive of and to agree on a plan in which the deliberate suppression of purely defensive measures is required. Still, it would appear to be to our mutual advantage at least to agree on a moratorium on ABM deployment at this time.

Clearly, any diminution or reversal of the arms race will require much more than just all governments refraining from inflaming it—it will require agreements for the limitation and eventual reduction of armaments and of weapons systems.

It is obviously not possible to predict in detail how future scientific and technological developments will change the character of the arms race. However it does not take a great deal of foresight to observe that in at least one field, that of biology and biochemistry, the probabilities of the development of new weapons of mass destruction are very frightening indeed. Biological and chemical weapons have, until now, been relatively ineffective and are therefore seldom used in combat. But, considering the revolutionary developments which have taken place in this field in this last decade, it is clearly unrealistic to expect that this situation will remain static (most especially the pace of biological and chemical weapons development programmes). The field has fortunately, until now, remained mainly in the hands of second-rate technologists.

The fields of biological and chemical warfare offer a striking example of an aspect of the arms race in which restraint and self-control exercised by the major nations at this time would be profoundly in their self-interest. Such weapons, if and when developed, will be relatively cheap as compared to nuclear weapons and will therefore be of extreme interest to small countries with limited resources and aggressive intent. However, at this time the most effective development programmes in these fields are being carried out by the major powers. When successful, the main effect of such developments will be to provide new and cheap weapons possibilities to many small countries although, at the same time, there will be no improvement in the relative positions of the major powers. In this circumstance, the only obstacle preventing the major powers from mutual agreement not to develop such weapons would seem to be a stubborn non-recognition of self-interest, or else a short-sighted politically induced inability to talk sense to each other on matters of common concern.

Finally, perhaps the most important aspect of the nuclear arms race, which might have been controlled at any time during the past decade but which is now rapidly approaching the point of no return, concerns the spreading of nuclear weapons to other countries. For many years only three nations possessed nuclear weapons. Then France joined the club, and now China. Quite clearly, each addition to the list of nations possessing nuclear weapons increased the incentives of and the pressures upon non-nuclear nations to reconsider their positions. We are now witnessing a heated internal debate in India, in which the forces of restraint have only barely and temporarily managed to remain in control. Similar debates are going on in countries like Sweden, Switzerland, Japan, the United Arab Republic, Israel, et cetera. Once the dam breaks, it will be impossible to contain the flood. The question is how to convince those non-nuclear nations that are capable of producing nuclear weapons that they should continue in their present course of restraint.

A number of suggested agreements have been proposed to prevent or at least impede further proliferation of nuclear weapons. The present partial test ban is one such measure, insofar as it is difficult for participating nations to develop nuclear weapons. It does not prevent this, however, as long as tests are permitted underground. Extension of the test ban to include underground testing would be a major advance. Other proposals include, for example, an agreement among the nuclear powers not to give nuclear weapons, or the materials or information required for their construction, to nations not yet possessing them and for the non-nuclear nations to agree not to make weapons or to obtain weapons materials or information.

There is no reason why such an agreement among the nuclear powers could not be signed today, at least by the three major nuclear powers, if the political atmosphere were more favourable.

As for the agreement among non-nuclear nations, this would be easier to achieve if these nations were provided with guarantees by the major nuclear powers, preferably through the United Nations, against the possibility that lack of nuclear weapons would jeopardize their national security or leave them open to nuclear attack or to blackmail resulting from the threat thereof. There has been the suggestion that such guarantees might take the form of assurance from the nuclear powers that a nuclear attack on the non-nuclear nations would be followed by immediate retaliation in kind. However, there are very great political difficulties in providing such guarantees, most especially in view of the current Sino-Soviet difficulties. Perhaps the most effective guarantee at this time would be an agreement on the part of all the nuclear powers that they would not use nuclear weapons against any nation not possessing them.

Other measures could be and should be undertaken to impede the spread of nuclear weapons; for example, the universal application of the controls over nuclear materials now in effect for projects under the International Atomic Energy Agency (IAEA). At present there are many purely national programmes in addition to a large number of programmes resulting from bilateral agreements, as well as regional arrangements such as the European Atomic Energy Community (EURATOM) for the peaceful exploitation of nuclear energy. Although the United States has announced that she will put her future bilateral agreements under the same inspection system as is used in IAEA, this procedure now covers only a small fraction of present-day nuclear energy development programmes. As a result, we are now rapidly approaching a situation where many nations, through perfectly legitimate nuclear energy programmes, will soon have a sufficient accumulation of fissionable materials and capabilities to be able to embark on independent nuclear weapons programmes.

Clearly, the adoption of all such controls and their acceptance by the non-nuclear nations will be possible only when these nations have agreed to forego independent nuclear weapons development. Equally clearly, in the present atmosphere such a resignation of potential nuclear status would only be possible if the major powers would demonstrate to the non-nuclear nations, by adoption of effective arms limitation measures, that they are willing to exhibit restraint on their own development and use of nuclear weapons.

The key lies in the world "restraint." In a world in which naked power is admired, and in which the recognition of status flows from the exhibition of and willingness to use such power, it is not likely that nations capable of achieving nuclear weapons would voluntarily forego this possibility. It is not obvious, that, in the present political context, either the United States or the Soviet Union is capable of showing sufficient recognition of self- and mutual interest to provide the needed guarantees and incentives to other nations to convince them to restrain themselves. On the other hand, the stakes are exceedingly high—imagine a world in which most nations possess nuclear weapons and their means of delivery and the incentives to the nuclear power are also great.

The problem is whether we can recognize these stakes and incentives clearly enough to make the necessary political moves, and to forego the wrong political moves, in order to bring a halt to the arms race before it runs completely out of hand; whether we can start to limit arms and control the uses of force in order to establish the stable world order that will avoid the otherwise inevitable catastrophe.

Bernard T. Feld (1919–1993) was Professor of Physics at the Massachusetts Institute of Technology. During World War II he worked on the Manhattan Project for the development of the atomic bomb. He was Editor of the Bulletin of the Atomic Scientists and served as Secretary-General of Pugwash Conferences on Science and World Affairs from 1973 to 1978.

CHAPTER 2

Anti-ballistic Missiles

Francesco Calogero

Introduction

In this chapter I will discuss the new perspectives opened by the most recent technological developments relevant to the strategic and political confrontation between the Superpowers. I will focus attention mainly on the anti-ballistic missile (ABM) system, but will also introduce in the picture other relevant gadgets, such as the multiple independently targetable re-entry vehicles (MIRV) and the fractional orbital ballistic system (FOBS).[1]

It may be wise to preface my remarks sketching briefly the outline of the contemporary strategic framework. This is based upon *deterrence*, which is assumed to be the guarantor of stability on the strategic level. Deterrence is implied by the second-strike *assured destruction* capability of each Superpower, namely the capability to inflict unacceptable damage to the other side, by means of a retaliatory strike, even after absorbing a full counterforce first strike. Because stability is based on the

Originally published in Edoardo Amaldi and Carlo Schaerf, eds., *International School on Disarmament and Arms Control: Proceedings of the Second Course* (Rome: ISODARCO, 1969): 73–95.

F. Calogero (✉)
Physics Department, University of Rome "La Sapienza", Rome, Italy

P. Foradori et al. (eds.), *Arms Control and Disarmament*,
https://doi.org/10.1007/978-3-319-62259-0_2

effectiveness of this assured destruction threat to deter either side from launching a strategic surprise attack, its preservation is deemed fundamental both by decision makers and strategic analysts. A categorization of new weapon developments is therefore suggested, in terms of their effect on the equilibrium based on deterrence; developments which reinforce second-strike countervalue capability are termed stabilizing, and developments which increase first-strike counterforce capability are termed destabilizing.

This is of course a very simplified picture, but it does nonetheless represent a zero-approximation assessment, which is significant. Its paradoxical nature should be immediately emphasized, for it implies that weapons aimed at the population are considered, in some sense, less dangerous than arms which are primarily targeted at the enemy's strategic forces. A similarly paradoxical assessment is associated with defensive measures, those directed to the preservation of one's own strategic forces being considered stabilizing, those aimed at protecting the civilian population being instead perceived as having a destabilizing character, in as much as they tend to negate the retaliatory second-strike capability to the other party in the strategic game. As a matter of fact, these so-called destabilizing developments, rather than leading to a dangerously unstable situation, are likely to result in a reinvigoration of the arms race, because as a rule they force the other party to react, by an increase in number and/or performance of its strategic weapons, or by the development of new ones, so as to re-establish the deterrence balance.

The developments of ABMs and of new missile capabilities like MIRVs and FOBS should be evaluated within this framework. In the zero-approximation just defined, the introduction of ABMs must be considered a destabilizing development and therefore such to induce a reinvigoration of the arms race; the same conclusion applies to MIRVs. To be sure, the actual situation is more complex than outlined up to now; in this chapter I will try to analyse it in more detail, but the final conclusion will not be too far off from this zero-approximation estimate.

Discussion of ABMs from a Technical Point of View

The ABM system is designed to detect, identify, intercept and destroy the warheads of incoming hostile missiles. At present the only feasible method to achieve this aim is by shooting a missile at the incoming vehicle.

Kill Method

Of the several possible methods to destroy or incapacitate the incoming warheads three appear at present as more relevant: blast, neutrons, x-rays.

Blast. This is the most conventional method; the warhead of the intercepting missile is detonated close to the incoming missile and the shock wave from the explosion destroys it. This kill technique only works within the atmosphere, that is in the terminal part of the trajectory of the incoming missile, because the shock wave is transmitted through the atmosphere. The warhead of the intercepting missile must of course be itself a nuclear weapon, presumably with a yield in the tens or hundreds of kilotons, and must be detonated rather close to the incoming missile (presumably within hundreds of metres).

Neutrons. The explosion of a nuclear weapon releases an intense neutron flux. The neutrons released may penetrate an incoming nuclear weapon and induce a sufficient number of fissions within its fissile material to modify its properties (for instance, by melting) so as to prevent the subsequent explosion of the nuclear warhead. Because the neutron flux must be rather intense, this kill mechanism also requires a close "hit" and is therefore effective only for intercepts occurring at relatively short range from the defensive missile launch site (less than 35 km). A disadvantage of this method is that the defender does not know for sure whether the incoming warhead has in fact been incapacitated until the weapon strikes its target and does, or fails to, explode.

X-rays. Most of the energy released by a nuclear explosion in the megaton range detonated above the atmosphere is in the form of x-ray. This electromagnetic radiation can travel over large distances with very little attenuation (above the atmosphere); on the other hand, if it reaches in sufficient quantity the incoming missile, it may damage its heat shield (for instance, by evaporating it), so that the vehicle will burn up upon its reentry in the atmosphere. The kill radius by this mechanism is quite large (many kilometres), so that the intercept does not require a close hit and may therefore be accomplished at ranges of several hundred kilometres from the ABM launch site. This kill mechanism is sometime referred to as the ZAPP effect, ZAPP being the comics-jargon symbol for a very large detonation.

These kill techniques open up the possibility to carry out both a *point defence* and an *area defence*. The point defence concept refers to a system designed to protect a small area (radius: few tens of kilometres); this aim is achieved by a defensive missile designed to destroy or incapacitate the incoming warhead in the terminal stage of its trajectory, after it has re-entered the atmosphere, say at a distance of less than 30 km from its eventual target. Either one of the first two kill mechanisms described above could be used for point defence. The missile to be used for terminal defence needs to have a relatively short range, but very high acceleration. Of course a large area can be defended by a large number of point defence sites.

The area defence concept refers to a system designed to protect a large area (radius: several hundreds of kilometres): this aim is achieved by a relatively long-range defensive missile designed to intercept the incoming missile at a distance of several hundred kilometres away from the defensive missile launch site. The intercept would of course occur well above the atmosphere and would utilize the ZAPP effect as its kill mechanism. A small number of defensive missile sites of this kind could cover a very large area, for instance the whole continental territory of the United States.

Other types of defence against the threat of offensive nuclear missiles have been sometimes mentioned ("screen" type defences, temporary modifications of extra-atmospheric space to impede missile travel, temporary modifications of the space within the atmosphere above one's own country to impede missile re-entry, kill of enemy's offensive missiles in boost phase, etc.), but none of these appear feasible at present.

The Attacker's Response

As soon as party A in the strategic confrontation develops some means to downgrade the effectiveness of party B's offensive missiles, it evokes B's response aimed at maintaining the previously existing offensive capability. The prospective "attacker" has several means to achieve this objective: it appears in fact that in this game the advantage is largely with him rather than with the "defender."

The countermeasures by the attacker can be divided into two categories: technological devices to maintain the capability of his missiles to deliver their weapons on target, in spite of the ABM system; attack strategies designed to by-pass or overcome the ABM system.

Penetration Aids

A number of penetration aids have been discussed in the open literature; these are technical devices to increase the capability of offensive missiles to penetrate an ABM system and deliver their nuclear weapons on target. A list of those which appear more practical include:

Harden the warhead by appropriate shielding, or by improvement in its design (including the choice of materials), to make it more able to withstand blasts, neutrons and/or x-rays.

Shoot a large number of *decoys* along with the actual missile, to saturate the identification capability of the ABM system. The booster itself may be designed to fragment after the separation of the warhead-carrying missile, in such a way that several fragments simulate, on the ABM radar screens, incoming missiles. Another very cheap possibility (in terms of booster payload, and therefore also in terms of cost) is to use balloons which are ejected and inflated after the missile has emerged from the atmosphere, so that they follow a similar trajectory in the vacuum of extra-atmospheric space; it is easy to design their shapes and reflecting properties to mimic real missiles on the ABM radar screens. The discrimination would of course automatically occur upon re-entry in the atmosphere.

Disperse a large number of fine metal wires (*chaff*) over a large portion of the sky. If the wires are of a length close to the radar wavelength and are sufficiently dense, they would provide a large radar reflection within which the re-entry vehicle could be concealed.

Equip the offensive missiles themselves or shoot alongside with them appropriate vehicles, carrying *jamming devices* designed to confuse and impede the ABM radar.

Produce a large region of complete *radar blackout* by exploding a nuclear weapon above the atmosphere, and subsequently shoot the offensive missiles through it. The radar blackout may be induced by a sufficient level of ionization, which may be produced either by the fireball or by beta particles emitted from the fission products.

Design into the offensive missile some capability to *modify its trajectory* in mid-course or near its terminal part.

Provide each offensive missile with several (say, ten) *independently targetable* warheads, which may be released at different stages of the missile trajectory. This is the MIRV development. The task of the ABM system requires that each of these warheads be separately tracked and intercepted.

Generally these penetration aids are mostly (and, in some cases, only) effective against the long-range interceptor necessary for the area defence, which requires an extra-atmospheric hit. But this does not mean that point defence is easier, for it has its own obvious disadvantages: short warning times (very few seconds), no possibility to correct mistakes, danger for the population below from the explosion of the ABM warheads themselves, and easiness to be by-passed by appropriate attacker's strategies.

Attacker's Strategies to By-Pass ABMs

A number of strategic options are open to the military planner, who is confronted with the task of maintaining the capability to wage a successful missile attack against a territory defended by an ABM system. These options include:

Increase the number of available offensive missiles. A variation on this theme—and presumably a much cheaper one—is the introduction of MIRVs.

Saturate the ABM system by shooting many missiles at the same point, so that the ABMs defending other parts of the territory remain idle and useless (but they had to be paid for to begin with).

Kill first the "eyes" of the ABM system, concentrating a saturation attack on the relatively few key radar facilities: then follow up with the actual missile attack on the strategic targets.

Target the missiles at undefended areas upwind of densely populated regions (cities) and design the warheads to maximize fallout; this strategy is appropriate for a countervalue strike against a territory with urban areas protected by point defences.

Detonate very large weapons (say, 100 Megatons) very high above the ground (say, 30 km), and rely on the incendiary effects of such an explosion, which would affect a very large territory; this strategy is also appropriate for a countervalue strike against a territory protected by point defences, which would have no chance to operate at such a distance from the ground.

Use FOBS; the FOBS is a ballistic missile, which travels around the earth on a very low orbital trajectory of almost constant altitude above ground (say, 150 km); the normal trajectory of an intercontinental ballistic missile (ICBM) has much more curvature and rises as high as, say, 1300 km, instead. Because of its low altitude a FOBS would not cross the radar horizon (and therefore could not be sighted by an ordinary radar on the ground) until it was about 1400 km (three minutes) away; the corresponding distances and warning times for a missile on an ordinary trajectory are 4000 km and ten minutes. Moreover a missile on a fractional-orbit trajectory could approach the radar "from the back" (modern phased-array radars operate only in a fixed direction over an angle approximately 90° wide).

Finally, both the possibility of "unconventional" uses of nuclear weapons (weapon hidden in cargo ship in harbour, weapon detonated in a submarine off the coast to produce tidal waves, etc.) and of the shift to other weapons of mass destruction (biological, chemical, etc.) provide drastic alternatives which, if feasible, would make the ABM system completely useless.

A Remark

The above analysis indicates that the technical situation provides a strong inducement to try and kill the enemy's missiles as soon as possible—ideally, in the boost phase—because in such a way most of the countermeasures and counterstrategies of the attacker would be automatically precluded. At the moment this does not appear feasible; however a proposal by the US Navy to install ABMs on ships or on the seabed close to the enemy's shores is a move in this direction. It is obvious, but worth emphasizing, that the more an ABM system is designed to kill the enemy's offensive missiles in the early stages of their trajectories, the less it is distinguishable from an offensive counterforce system contributing to the eventual achievement of a first-strike capability.

Cost Effectiveness

To try and extract some quantitative conclusions from the technical considerations outlined above, the concept of *cost-exchange ratio* is introduced: it is the ratio of the cost of an ABM system to the cost of the increase in the enemy offensive forces such as to offset the ABM defence.

Because this concept has been used, at least in the United States, in the debate over the decision whether to deploy an ABM system, I felt compelled to mention it in this chapter. But I would like to stress immediately that the quantitative element which is apparently attached to the introduction of this number in the analysis is in fact largely illusory, because it depends on several estimates whose reliability is doubtful and which are inherently untestable. For instance, it is almost a tradition in the United States that the advance estimate of the cost of any new weapon system is grossly underestimated; on the other hand the actual performance of an ABM system in the real situation is essentially untestable and very difficult to guess (a proof: different analysts, all generally regarded as reliable, if pressed, will venture predictions which cover the whole spectrum, from almost complete success to complete failure; an explanation: the ABM system set up by a Superpower is going to be the largest technological enterprise ever engineered by mankind, yet it is required to operate properly on its first, and presumably unique, full-scale test under real conditions).

Moreover the determination of the cost-exchange ratio depends on the damage level whose maintenance is deemed essential by the attacker to preserve deterrence; thus, a value judgement is injected in the picture, moreover one referring to options which, in spite of their apparent quantitative precision, relate in fact to situations which the human mind is certainly unable to envisage (e.g. the environment, way of life, values, after an attack resulting in the prompt death, within one country, of several tens of millions people).

After the above qualifications, we are ready to quote some figures for the cost-exchange ratio that have been given by Robert McNamara (see Table 2.1)[2]:

The decrease in the cost-exchange ratio as the damage level increases is easily understood: if the damage level to be maintained is very high, it is easier for the defence to have some positive effect. But it should be noted that the cost-exchange ratio reaches unity only at a level of damage, which is close to that of the undefended case (no ABM system deployed at all). Moreover, in view of the arguments given above, it is likely that these figures are underestimated.

To provide some understanding for the figures here, especially those referring to low damage levels, it is worthwhile to remember the drastic difference, in terms of effectiveness, between an air defence system (such as those operative during World War II, or now over North Vietnam) and an ABM system: the former is already considered very satisfactory if it provides 10 per cent attrition (namely, if it destroys 10 per cent of the

Table 2.1 Cost-exchange ratios

Level of American prompt fatalities (by blast and fallout) which the Soviets believe will deter an American first strike (millions)	*Cost-exchange ratio (American ABM cost divided by Soviet offsetting offensive forces cost)*
40	4:1
60	2:1
90	1:1
100	Undefended case

attacking bombers); the latter is quite ineffective even if it provides 90 per cent attrition, namely if it lets through one tenth of the incoming missiles.

Uncertainty over Performance and Arms Race Effect

As emphasized above, a large uncertainty is essentially associated to any estimate of the actual performance of an ABM system. Such uncertainty is likely to produce a permanent asymmetry in the estimates of the reliability of a given ABM system, between the party who deploys it and the party who is confronted with the task of offsetting it. In fact the conservative approach which is characteristic of any estimate in such a critical area, involving life-or-death questions, almost certainly results in the defender taking the most pessimistic view of the effectiveness of his ABM system, while instead the attacker plans the increase of his offensive forces, on the assumption that the ABM system will work very efficiently. There results a drive pointing towards a divergent arms race; in the short run, a decrease in the "security" of the defending side (measured, for instance, in terms of the eventual damage incurred in an all-out attack) is produced due to the overreaction by the other side, in increasing its offensive capability. The United States' reaction to the rather limited ABM deployment by the Soviet Union—increase in number and especially performance of its offensive missiles, including a vigorous programme to develop penetration aids and the decision to deploy Poseidon and Minuteman III and to equip them with MIRVs—is a case in point.

The Present Situation

In this section the present situation concerning ABM deployment by the two Superpowers is outlined.

Historical Outline

ABM deployment by the Soviet Union began probably some years ago; the first rumours began to circulate in the Western press in 1964. At the end of 1966 Robert McNamara (then American Secretary of Defense) publicly announced that the Russians were deploying an ABM system around Moscow.

In the United States, research and development on ABMs has been going on for several years; the decision whether or not to deploy has been much debated in the last two or three years, even though it never became a central political issue. The decision to deploy was finally announced by McNamara in September 1967[3]; the speech containing the announcement is remarkable in that it presents many arguments against deployment and it describes the system to be deployed as a "thin" ABM system aimed only at the limited ICBM capability which the Chinese are expected to achieve in the early or mid-1970s. In the debate preceding the September speech McNamara had been identified as a consistent opponent to deployment; his subsequent retirement from the post of Defense Secretary has probably been influenced by this decision.

The Russian ABM System

The Russian system is apparently deployed only around Moscow and possibly Leningrad. It is presumably based on the short-range (40–50 km) Griffon missile and the long-range solid-fuelled Galosh missile. In addition the Russians have installed a defensive system along the northwestern border of the Soviet Union, but it appears that these installations, known as the Tallinn line, have no significant ABM capability.

The American ABM System: Sentinel

Much more information is of course available on the American system, named Sentinel. For the purpose of the present sketch, it may be broken down into five components: two radar systems, perimeter acquisition radar (PAR) and missile site radar (MSR), two types of missiles, Spartan and Sprint, and a huge computer network.

The PARs are low-frequency long-range radars, whose task is to sight an incoming missile: if the incoming missile is on a minimum-energy ballistic trajectory, the sighting would occur at a distance of about 4000 km, providing a warning time of about 10 minutes. Sentinel would have six PARs positioned along the Northern boundary of the United States, sometimes referred to as the "eyes" of the system.

The MSR higher-frequency radars are designed primarily to track the incoming missiles sighted by PAR and to guide the Spartan and Sprint interceptor missiles unto them.

Spartan is the longer-range interceptor, to provide area defence. They would he located in about 10–20 "farms" around the United States: assuming a 600 km range for these interceptors, the whole territory of the United States would be covered by the ranges of these sites. Spartan carries a nuclear warhead in the megaton range and would explode it at very high altitude (up to 400 km), relying on the ZAPP effect.

Sprint is the shorter-range interceptor for point defence. It is an 8-metre missile, which is ejected from its cell by a piston activated by a gas generator. Immediately after clearing its cell, the first stage of the two-stage solid-fuelled Sprint ignites, sending the missile, with an acceleration well over 100 g, to an altitude of 20 km within 4–5 seconds. The Sprint is intended to intercept those missiles, which have escaped Spartan; the intercept should occur at altitudes below 35 km. Thus Sprint must be deployed close to the point to be defended. Its nuclear warhead is presumably in the 10–100 kiloton range. Contrary to what was stated in the September 1967 speech by McNamara, it now appears that Sprints will be deployed only near radar and Spartan sites, that is to provide point defence for the key elements of the ABM system itself. The September speech envisaged deployment of some Sprints also near Minuteman sites, to consolidate their second-strike capability: a role which appeared quite inconsistent with the declared anti-Chinese (and not anti-Soviet) purpose of the Sentinel system. Deployment of Sprints to defend cities does not appear to fit within the Sentinel budget.

The computer is the brain of the system, onto which all radars and missiles are hooked. Its data handling and decision making capability would be much beyond those of any apparatus previously constructed by mankind.

The Sentinel system should be ready in 1972, operational in 1974. Its estimated cost is $5 billions; $900 millions have been approved and appropriated for this fiscal year,[4] after an attempt by those who are against deployment to prevent the appropriation from being voted by Congress, in which both strategic-political and economic arguments were used (in the Senate the majority vote for the appropriation was not overwhelming, in spite of the strong stand in favour of Sentinel by the Administration, who stated that it was essential for the protection of the American people). Of the $900 millions, $300 millions are for research and development, $400 millions for production and deployment, $200 millions to purchase real estate for the Spartan and Sprint sites; additional $300 millions have been appropriated for research and development in the ABM field but not on the Sentinel system. The cost of each missile (Spartan or Sprint) is of the order of one million dollars.

The "thin" "anti-Chinese" Sentinel system might in time become thicker and Soviet oriented. In this connection it should be recollected that in the United States the basic developmental approach has been of a flexible or "building block" nature. In this regard, another basic type radar system has been designed. This is the multifunction array radar (MAR), a very powerful radar which can perform all the functions required to cope with a large-scale ICBM attack, including target acquisition, discrimination between decoys and warheads, and precise tracking of warheads and of interceptor missiles for guidance purposes. The tactical MAR (TACMAR) is a scaled-down version of MAR, which can perform most of the same functions with a lower target-handling capacity. All the radars mentioned are based on phased ray technology, whereby many beams could be generated and instantaneously moved across the sky. The only significant disadvantage of a phased-array radar compared to a mechanically scanned radar is that the latter can operate through a full 360° of azimuth while the former is limited to a scanning angle of approximately 90°. This disadvantage can be compensated for, by deploying more than one transmitting and receiving array for a given site.

Before the decision to build Sentinel was taken, the possibility to deploy larger systems was debated. Two such systems were described by the then Secretary of Defense McNamara: a $9 billion programme (Posture A) and a $20 billion programme (Posture B).

Posture A represents a light US defence against a Soviet missile attack on American cities. It consists of an area defence of the entire continental United States, providing redundant (overlapping) coverage of key target areas; and, in addition, a relatively low-density Sprint defence of a number of the largest cities to provide some protection against those warheads which get through the area defence.

Posture B is a heavier defence against Soviet attack. With the same coverage it provides a higher Sprint defence for twice the number of cities. Posture A would provide a Sprint defence for the 25 largest cities and Posture B for the 50 largest cities. The costs given by McNamara, who warned that the estimates may be understated by 50–100 per cent, are given in Table 2.2.[5]

This breakdown suggests that the total number of interceptor missiles envisaged is about 2000 to 4000 for Posture A and Posture B respectively.

Table 2.2 Costs of United States ABM systems

	Posture A Investment Cost (billion dollars)	*Posture B Investment Cost (billion dollars)*
Radars (MAR, TACMAR, PAR, MSR)	6.5	12.6
Missiles		
Spartan and Sprint	2.4	4.8
Department of Defense	8.9	17.4
Atomic energy commission (AEC)	1.0	2.0
Annual operating cost	0.38	0.72

Presumably, except for scaling down, the breakdown of costs is similar for the Sentinel system. The large investments required by the radar network should be noted.

It should be emphasized that the $5 billion Sentinel system is not expected nor meant to be effective against a Soviet attack. In view of the general advantage that the attacker has, as described in the previous section, some analysts doubt that it would be of any use even against the kind of attack China might be able to launch in the mid-1970s.[6] The question is therefore asked: Why deploy it at all? This is discussed in the following section.

Arguments Pro and Con ABM Deployment

Arguments Pro

Defence is good. To be sure, this is an irrational feeling rather than a compelling argument; but it does carry a lot of weight. A more subtle way to express this point of view is by saying that in any case, if things go wrong, the existence of ABM defences cannot but save some lives.

The other party has it, so we cannot lag behind. Again this is not a rational argument; the logical response to ABM deployment by the "enemy" is an increase—or simply an adjustment—in one's own offensive forces, to offset whatever advantage the enemy might derive from the protection afforded by the ABM system. But also this argument has a tremendous psychological appeal, and therefore it carries a lot of political weight.

The transition from the present equilibrium based on deterrence to a situation characterized by the dominance of defence would result in a less dangerous world and might more readily lead to disarmament. Such a transition might occur if the two Superpowers, presumably by mutual agreement, concentrate on the development and deployment of defensive systems, such as ABMs, restraining at the same time the development and deployment of offensive weapons. This course of action has been particularly advocated in the United States by Donald Brennan[7] and Alvin Weinberg.[8] Of course the difficulty lies in achieving the transition from one type of regime to the other, because this requires going through a phase characterized by great strategic instability. The amount of mutual understanding and trust which would be required to do this is probably more than it would be necessary to embark immediately on the disarmament journey. In any case the immediate reaction of the United States to the ABM deployment by the Soviet Union has been in the direction of increasing offensive capabilities, as we indicated above. The danger that the deployment of large-scale defensive measures such as ABMs, rather than initiating a transition from deterrence to defence, trigger off a new round of the arms race is a very real one.

The possibility to reach a "winning" posture may again be contemplated. No doubt the stability enforced through mutual deterrence has a frustrating connotation which disturbs the most militant groups, especially within the military establishments, in the United States and probably also in the Soviet Union. Of course, to the extent that this point of view has an influence, it will press in the direction of large-scale ABM deployment (and also large-scale deployment of offensive weapon systems).

Technology makes it available (not necessarily effective); then it is done. This one also is not a sound argument; but I believe it is an important element nonetheless. In fact, the intrinsic momentum associated with the opening up of new technological options appears more and more as a major influence in shaping the future of mankind.

ABM is an insurance against possible Chinese "irrationality." This is of course the main official American argument to justify the decision to buy Sentinel. China is expected to possess a few operational ICBMs with nuclear warheads in the mid (or early) seventies. Up to now, the foreign policy stand of China has been most belligerent vocally, but quite cautious in deeds.

It provides some insurance against the (unlikely but possible) accidental launch of one, or a few, ICBMs by any one of the nuclear powers. To achieve such capability, however, the ABM system should be on the alert continuously, rather than only in crisis situation and, given the shortness of the available warning times, it should be able to intervene almost automatically, that is without any human control. This suggests the possibility that the chance of accidents associated with the very existence of such a huge and ready-to-go system overcompensates the protection that it affords against extraneous accidental firings.

It discourages the proliferation of nuclear weapons by adding credibility to the guarantees offered by a Superpower to allies and friends, and by increasing the gap between the Superpowers and any prospective new nuclear-weapon power.

It may be used to strengthen second-strike capabilities, therefore contributing to stability. As already mentioned, such a role was, at least in part, envisaged for the Sentinel system in McNamara's speech that announced the decision to deploy; but, at least for the moment, it does not appear that such a capability will be included in the Sentinel package. Such a role is of course quite inconsistent with the declared anti-Chinese, and not anti-Soviet, purpose of Sentinel.

Arguments Con

ABM defence is unfeasible; the ABM system will never be adequate to the offense existing at the time when it becomes operational. A more sophisticated argument—although perhaps only apparently more so—refers to the cost-exchange ratio favouring the offense rather than the defence (unless one is prepared to accept very high levels of damage).

It will result in a reinvigoration of the arms race, with all its side effects, including the psychological ones within each political system and the difficulty to maintain an atmosphere of detente between the Superpowers. We have already discussed the effect on the arms race of the uncertainty inevitably associated with any estimate of the reliability of an ABM system; and the likelihood that, even in the very short run, the overreaction by the other side to ABM deployment by one side (or even only to its beginning) results in a decrease in security for the latter. Note that, as the American reaction to the very limited Soviet deployment exemplifies, the increase in number and capability of offensive weapons decided in order to counter the deployment of ABMs is likely to be based not so much on what is actually being deployed, but on a hypothetical projected maximum effort.

There is danger of an unstable situation arising, in which one side might believe he could "win." Such a danger is of course implicit in any new take-off of the arms race. In this respect even more dangerous than ABMs is the advent of MIRVs. To illustrate this point consider the following hypothetical situation, in which two parties confront each other, each being equipped with 500 land-based missiles, each of which carries 6 MIRVs. Assume moreover that each MIRV has a 2/3 probability to destroy an enemy missile (thereby eliminating six MIRVs in a single shot). Then simple arithmetic shows that either side has a 60 per cent probability of destroying all the missiles of the enemy, by launching a surprise all-out attack; and 90 per cent probability that no more than one enemy missile survives such a first-strike attack.

This is of course an imaginary situation, introduced only to illustrate the point; moreover, it ignores the existence of submarine-based missiles, which add an element of stability to the picture, because they provide a retaliatory force, which is essentially invulnerable. On the other hand if ABMs are also present, they could take care of the (relatively fewer) submarine-based missiles, and also of the few land-based missiles, which would survive the attack (which might of course be more than in the previous case, if both sides possess ABMs). This is not the place to go into a detailed system analysis of such war games; this example, however, should give some idea of the possible dangers implied by the trend which would be initiated by ABM deployment. Note that, even though in practice the attainment of a "win" posture by either one of the two Superpowers is probably unrealizable, the fact that it might be perceived by the more militant military as a feasible goal may be enough to set as an important priority for them to strive for it.

The decision to deploy an ABM defence is an open-ended beginning; it will be very difficult to maintain the ABM system limited, especially in the United States. This will be the case especially if point defence of cities is undertaken, because more and more cities will then put political pressure on the decision-makers to be included within the protected élite. The resulting large-scale ABM system intended to protect the whole population against a massive sophisticated attack will also require a huge fallout shelter programme, with the associated economic and psychological costs. Up to now, at least in the United States, the investment for civil defence has been minimal; one would guess that the same is true of the Soviet Union, from the fact that a visitor in Moscow or Leningrad does not notice any preparation. It was the expected difficulty to keep ABM deployment limited, which led McNamara, in his September 1967 speech, to warn against the "mad momentum intrinsic to the development of all new nuclear weapons."[9]

ABM systems affect the prospect of disarmament; it appears that the simultaneous existence of MIRVs and ABMs makes stage-by-stage disarmament practically impossible. The reason for this may be glimpsed from the example discussed above, displaying the instability associated with the simultaneous presence of MIRVs and ABMs; the uncertainty about the performance of these weapons and the difficulty to monitor them (especially MIRVs) add to the difficulty of conciliating their presence with progressive disarmament. Yet stage-by-stage disarmament still appears as the only feasible method to proceed. More specifically, the agreement in principle between the Superpowers on the minimum deterrent (or *nuclear umbrella*) concept had been rightly hailed as an important achievement, suggesting a transition from declaratory propaganda to serious business; but the possibility to agree on a minimum deterrent posture, as an intermediate step to general and complete disarmament and on its implementation and verification, is made much more difficult, probably impossible, by the presence of MIRVs and ABMs. To some people the danger that the decision to deploy ABMs precludes any progress to disarmament is the strongest argument against this decision, but unfortunately, not one to which decision-makers are at present very sensitive.

Economic Arguments Pro and Con ABMs Deployment

Arguments Pro

The deployment of an ABM system requires large investments in the fields of electronics and aerospace; it is only natural that those who have vested interests in these fields tend to favour it. This applies prevalently to the United States, but probably to a certain extent also to the Soviet Union. Moreover the prospect to make money, or gain economic and political importance, developing defensive systems—whose value is assessed in terms of human lives saved rather than fatalities produced—has a tremendous psychological appeal to people who have up to now concentrated their efforts in the development of offensive weapon systems.

Arguments Con

The cost of any ABM system is very large; and it may be multiplied many times over by the reinvigoration of the arms race that might be triggered by the ABM deployment. On the other hand both Superpowers are at present confronted with serious economic difficulties, arising from very pressing needs at home whose satisfaction is necessary to avert major internal political unrest. This economic background should provide a strong incentive to restraint in this field. Of course a termination of the hostilities in Vietnam would make quite a difference from this point of view, especially in the United States where a considerable fraction of the whole military budget is devoted to them.

Why They Did It

After an analysis of the reasons in favour and against the decision to deploy ABMs it is worthwhile to try and understand why in fact the decision to deploy was taken, first in the Soviet Union and subsequently in the United States.

In the Soviet Union

To be sure, we do not know. Probably the traditional defence mindedness of Soviet strategic thinking and the pressure of public opinion have played an important role. Presumably decision-makers felt that resisting the pressure of those who advocated deployment would have exposed them to the accusation of letting the country undefended.

In any case it should be remembered that ABM deployment by the Soviet Union is not very extended; it is possible that it was initiated but not really pursued, perhaps as a consequence of a change of policy. It is known that at least some scientific advisors have argued against deployment, using the same arguments, which were reported above.

In the United States

Much more is known about the American decision, since the debate preceding it has been widely publicized; it is, however, not easy to assess the actual relevance of the various arguments and considerations in influencing the final decision to deploy a "thin" "anti-Chinese" system. In my opinion the determining element has been the Soviet deployment of an ABM, however limited; not because it implied any strategic need to build an ABM system in response, but because it rendered politically impossible to resist the advocates of deployment.

Thus, in the last analysis also in the United States the background pressure of public opinion at large has probably been the determining influence; not that there were demonstrations in the street asking for ABMs (in fact, the public essentially ignored the whole issue), but decision-makers were afraid that resistance against deployment would have provided the advocates of it with an explosive political issue, which might be raised whenever the convenient time came. In fact James Reston (the *New York Times*' leading political commentator) joked that the ABM system was actually deployed neither against the Soviets nor against the Chinese, but against the Republicans.

Another reason which has certainly been relevant is China, both because of a genuine preoccupation with her expected future behaviour (also in the face of the factional struggle going on within her) and because of the psychological, and therefore political, relevance of the Chinese threat for the American public.

Finally, pressure for deployment coming from economic and military pressure groups has certainly been influential; presumably these groups perceive the Sentinel decision, with its limited and "anti-Chinese" character, as just the first step towards more extensive Soviet-oriented deployment. This policy of "sticking the foot in the door" has been constantly followed by the military-industrial complex, whenever the administration was not prepared to go all the way with their requests for the development or deployment of new weapon systems.

It should be emphasized in conclusion that the decision to deploy ABMs might still be reversed, or at least greatly limited, by the Administration and/or the Congress, although it now appears that the chances of this happening are slim, unless this policy change results from an agreement with the Soviets.

A Final Remark

The threat of complete destruction associated with an all-out nuclear attack is so great that any system likely to effectively protect a country from it would be bought, irrespective of its economic, or, for that matter, political, cost. This fact is quite clear both to Soviet and American leaders; it was clearly stated in the very Sentinel speech by McNamara:

> ...an impenetrable shield over the United States. Were such a shield possible, we would certainly want it—and we would certainly build it ... It has been alleged that we are opposed to deploying a large-scale ABM system because it would carry the heavy price-tag of $40,000,000,000. Let me make very clear that the $40,000,000,000 is not the issue. If we could build and deploy a genuinely impenetrable shield over the United States, we would be willing to spend not $40,000,000,000, but any reasonable multiple of that amount that was necessary.[10]

Thus, the basic objection against the ABM system is its technological and strategic ineffectiveness. Again McNamara:

> The money in itself is not the problem—the penetrability of the proposed shield is the problem. There is clearly no point, however, in spending $40,000,000,000 if it is not going to buy us a significant improvement in security … It is futile for each of us to spend $4,000,000,000, $40,000,000,000 or $400,000,000,000—and at the end of all the spending, and all the deployment, and all the effort, to be relatively at the same point of balance on the security scale that we are now.[11]

Effect of ABM Deployment by the Superpowers on Third Countries

In view of the crucial nature of the issues involved, the decisions determining the extent of ABM deployment by the Superpowers will be mainly made on the basis of assessments of their reciprocal military postures and of the internal political pressures arising therefrom. However the secondary consequences throughout the world of such decisions also have some relevance.

It is perhaps appropriate in this School, also in view of its venue and participants, to discuss this problem in somewhat more detail than it has been done up to now. One might thus hope to contribute, in a small way, to bring to the attention of decision-makers some (secondary, but not negligible) elements which should be included in the complex evaluation of the consequences implied by the decision to deploy ABMs.

Consequences on Arms Race and Détente

These have been sufficiently illustrated above, to warrant further elaboration here; except to emphasize once more the importance of these manyfolded, if unquantifiable, effects.

Non-Proliferation Treaty

The fact that the Superpowers embark on the deployment of a new major weapon system, likely to trigger off a novel round of the arms race, is certainly against the spirit of the Non-Proliferation Treaty (NPT); some would even say that it is against its sense. This development is therefore going to provide good arguments for those who oppose the NPT; it will therefore work against that widespread acceptance of the NPT, which is essential for its success.

The deployment of ABMs provides a strong pressure for the continuation of nuclear weapons testing, at least underground. It makes, therefore, the achievement of an agreement on a Complete Test Ban Treaty quite unlikely. Yet such an agreement (or at least its prospect) had appeared as a possible quid pro quo that the nuclear-weapon countries, and in particular the two Superpowers, might offer to convince some non-nuclear-weapon countries to accede to the NPT.

The decision by the Superpowers to deploy ABMs, especially if undertaken on a massive scale, and the fact that an ABM system employs nuclear warheads, is an incentive for other countries to retain the option to acquire nuclear weapons in view of their eventual use for ABMs. To be sure, an ABM system is such a large-scale undertaking to be beyond the means of any country but the Superpowers; but this will not prevent those who are against the NPT from using the argument just quoted, also in view of the fact that the NPT has no time limit. Arguments of this kind are likely to be more important for the "increased gap" consideration as an argument whereby ABM deployment by the Superpowers should discourage nuclear proliferation.

On the basis of the example set by the Superpowers, pressure may develop in favour of a European ABM. Besides being in itself a dangerous development (see below), this would put into question the NPT, by raising again a very delicate point, namely the question of nuclear sharing within an alliance.

Credibility of Guarantees, in Europe and Asia

It has been argued that the protection afforded to a Superpower by its ABM system would enhance the credibility of any guarantee offered by it to third countries; in particular this argument is supposed to apply to guarantees against an attack (possibly nuclear) by China. The weakness of this argument lies of course in the low level of reliability of any ABM system conceived so far, as discussed above.

On the other hand, the feeling that the others are left undefended, while the Superpowers are protected by ABMs, is likely to provoke nationalistic feelings of frustration and resentment in allied and non-aligned countries, feelings whose intensity need not be correlated to the actual effectiveness of the ABM system.

It has also been noted that, by stating that the ABM system is deployed against China, the United States may have in fact enhanced the status and importance of China, both in Asia and worldwide.

Effects on Neutral Countries "in Between"

This problem has been typically raised in Sweden. What should a neutral country do to prepare itself for the possibility of missile intercepts (involving the explosion of large nuclear weapons) occurring overhead (say, over Sweden in the event of a missile "fight" between the Soviet Union and the United States)? Should it develop countermeasures (for instance, of electronic nature) to interfere with the offensive and/or defending missiles and prevent their explosion over its territory? In fact, from its point of view it might be less damaging if the ABMs fail, so that the offensive missiles hit their targets, which are located far away. Would such countermeasures—to be developed and deployed immediately—be consistent with its neutrality status?

Superpowers' Involvement in Local Conflicts

The development of a really effective ABM system would make each Superpower relatively invulnerable so that its involvement, even at the nuclear level, in local conflicts would be less risky and therefore more probable. Even though such a development is at present unfeasible, the mere indication that the Superpowers are striving towards such a posture could raise a new sense of insecurity in the other countries; the Superpowers are protected by such a system and have the above mentioned opportunity to intervene while the other powers are denied this protection, and their fear of a military intervention by the Superpowers is enhanced. This sense of insecurity will inevitably result in increased armaments of all kinds in most States.

Even the ABM systems actually deployed or planned might modify relative strategic posture sufficiently to affect the method by which the Superpowers handle crisis in the direction of less caution.

West European ABM

The fact that such a system is, for rather obvious reasons, quite unfeasible, is not going to prevent pressure in its favour from rising in Western

Europe. For one thing, the political groups who approved the idea of an American-European nuclear collaboration such as it was envisaged in the Multilateral Force (MLF) scheme would generally tend to favour a new possibility which fits into the same pattern; and for obvious, if different, reasons, in favour would also be those who dream of the necessity to acquire nuclear weapons for Western Europe (once it achieves political unity), and for the time being for the West European States. Another group likely to press for a European ABM will be found in military circles, especially in certain European countries (for instance, Great Britain) where the military have recently suffered a drastic reduction in the scope of their activities.

The unfeasibility of a European ABM need not be illustrated in detail, in view of what has been said previously and the fact that the closeness of the prospective attacker enhances enormously his advantage, both technically and strategically. But it should again be emphasized that the unfeasibility of a reliable ABM defence for Western Europe need not be a sufficient argument to exclude striving for it (and the Soviet and American examples support this contention).

Of course, the decision to develop an European ABM system (which, incidentally, is not easily distinguished from an offensive system, in view of the closeness of the opponent) would have profound political consequences; it would presumably signify the end of any hope for a peaceful if slow development through détente towards an European security system based on opening and integration rather than close blocks and confrontation.

Outlook

Clearly the outlook is not bright. One recent development is, however, of the highest importance, and it provides some ground for hope. This is the decision by the Superpowers to engage in talks with the stated purpose to reach an agreement on the limitation both of offensive and defensive missile capabilities. While some people doubt the seriousness of the Superpowers and perceive the announcement of the talks just as a trick to convince certain reluctant countries to accede to the NPT, I would suggest that both Superpowers have much to gain from such an agreement; and maybe they begin to realize that, without it, a futile—if not really dangerous—round of the arms race is round the corner, leading—at best—to a new equilibrium with more weapons and less security. Much of what has been discussed in this chapter seems to me to justify this hope. Or maybe it is only wishful thinking.

If the Superpowers are serious in attempting to reach an agreement to harness the strategic arms race, then I would suggest that as soon as possible, before the formal talks initiate or concurrently to them, measures be taken by both sides—possibly by tacit agreement—to stop or even reverse the deployment of ABM systems and to stop the development of new offensive missile capabilities—in particular, refrain from flight testing MIRVs and FOBS.

Much of the difficulty inherent in negotiations of this kind is lack of confidence; measures such as these (which are easily monitored without any inspection) would help to establish such confidence, both in the decision-makers who control the negotiations and in the public opinion, who backs them. It would instead be very bad if, in view of the coming talks, development and deployment of new weapon systems are made to proceed at maximum speed, to afford a position of strength in the negotiations.

Decision-makers should be aware that, if they yield to suggestions of this kind—which certainly are being put forward within each country by military and other pressure groups—then most likely they will not succeed in getting any agreement, much as they—and their counterparts—may wish for it; the action-reaction dynamics of the arms race, which breeds on suspicions as well as facts, will again prevail.

Post Scriptum

After this lecture was delivered [August 1968], two important events have occurred: the announcement by the United States of a successful flight test of MIRVs; the military invasion of Czechoslovakia by the Soviet Union, Poland, Hungary, the German Democratic Republic and Bulgaria. Both these developments are for the worst.

Notes

1. In preparing this chapter, I have profited from notes taken during the Third Pugwash Symposium on *Implications of the Deployment of ABM Systems* (Krogerup, Denmark, 14–20 July 1968).
2. Robert S. McNamara, *The Essence of Security: Reflections in Office* (New York: Harper and Row, 1968).

3. Robert S. McNamara, "The Dynamics of Nuclear Strategy," *Department of State Bulletin* 57 (9 October 1967): 443–451.
4. Lyndon B. Johnson, "Annual Budget Message to the Congress, Fiscal Year 1968, January 24, 1967," in *Public Papers of the Presidents of the United States, 1967* (Washington, DC: US Government Printing Office, 1968) Book 1.
5. McNamara, *The Essence of Security.*
6. See, for instance, Richard L. Garwin and Hans A. Bethe, "Anti-Ballistic-Missile System," *Scientific American* 218 (March 1968): 21–31.
7. Donald G. Brennan, ed., *Arms Control, Disarmament, and National Security* (New York: G. Braziller, 1961).
8. Alvin M. Weinberg, *Reflections on Big Science* (Cambridge MA: The MIT Press, 1967).
9. McNamara, "The Dynamics of Nuclear Strategy," 450.
10. McNamara, "The Dynamics of Nuclear Strategy," 448.
11. Ibid.

Francesco Calogero is Former Professor of Theoretical Physics at the University of Rome, "La Sapienza". He served as Secretary-General of Pugwash Conferences on Science and World Affairs from 1989 to 1997. He accepted on behalf of Pugwash the 1995 Nobel Peace Prize, jointly awarded to Pugwash and Sir Joseph Rotblat.

CHAPTER 3

International Relations and Game Theory

Anatol Rapoport

Game theory is a branch of decision theory. One can speak of decision theory in two different senses: in the sense of a normative (or prescriptive) theory and in the sense of an empirical (or descriptive) theory. A descriptive theory of decision seeks to discover patterns, regularities or principles in the way people actually make decisions in given situations. Clearly the nature of the decision maker, his goals or values, his state of knowledge or his thinking habits, his predilections or prejudices are relevant in a descriptive theory. But the question of whether the decisions are good or bad is not relevant: a descriptive theory is always concerned with what is, not with what ought to be. A normative theory, on the other hand, seeks to discover rules for making decisions, which are in some sense "best." This sort of theory must be anchored in a specific value system, which is assumed as given. To put it in another way, a normative decision theory asks how a "rational" decision-maker would act.

Originally published in Frank Barnaby and Carlo Schaerf, eds., *Disarmament and Arms Control: Proceedings of the Third Course Given by the International School on Disarmament and Arms Control* (New York-London-Paris: Gordon and Breach, 1972): 141–151.

A. Rapoport (✉)
University of Toronto, Toronto, ON, Canada

P. Foradori et al. (eds.), *Arms Control and Disarmament*,
https://doi.org/10.1007/978-3-319-62259-0_3

Another important distinction is between theories of decision where only one decision-maker is involved and those where two or more are involved. When there is only one decision-maker, the outcome may depend only on his decision, or it may depend on other circumstances. In the former case, the normative decision problem involves only the knowledge of a preference order applied to the outcomes and knowledge of what decisions lead to what outcomes. In that case, the decision-maker needs only to select the decision, which leads to the most preferred outcome.

The situation is different if the outcome depends on circumstances not controlled by the decision-maker. For instance, in deciding whether to take his umbrella in the morning, the commuter must reckon not with two outcomes (having the umbrella with him or not) but with at least four, namely having the umbrella in fair weather, having the umbrella when it rains, not having the umbrella in fair weather and not having it when it rains. Assuming that having the umbrella is the preferred outcome when it rains and not having it is preferred in fair weather, the commuter will usually attempt to estimate the likelihood of rain. His problem can be put in quantitative terms if numerical probabilities are assigned to rainy and sunny weather, provided also numerical "worths" or "costs" can be assigned to all four outcomes. In that case, normative decision theory can prescribe a choice, namely that which results in the greatest expected utility. This expected utility is calculated as the weighted sum of the utilities associated with the corresponding outcomes. The weights are the probabilities assigned to the events not controlled by the decision-maker.

So far we have assumed that the circumstances beyond the decision-maker's control are governed by chance or, at any rate, by an agent who is indifferent to the decision-maker's preferences. Thus, contrary to some people's impressions, whoever or whatever governs the weather is not deliberately turning on rain just on the days when the commuter leaves his umbrella at home. In prescribing the utility-maximizing policy to the commuter, we have assumed that Nature has no discernible interest in the affairs of men.

There are, however, important situations where the outcomes of decision are controlled not by one but by two or more decision-makers, each of whom has his own preferences (i.e. interests), different from the interests of others. If the range of decisions open to each decision-maker can be specified, if utilities can be assigned to each of the outcomes by each decision-maker and if these utilities are known to all the decision-makers who jointly determine the outcome by the totality of their separate decisions, then the decision situation is called a game, and the decision-makers are called players.

Game theory, as it is presently developed, purports to be a normative theory of such decisions situations. I should like to emphasize that game theory is not now a descriptive theory. It does not inquire into how decisions are actually made but rather how rational players would make decisions in a situation, which has the necessary attributes of a game.

There is no reason, of course, why game theory could not be extended to include an inquiry into how decisions are actually made by ordinary mortals in game-like situations, provided the situations are well enough defined to yield to rigorous analysis. Such a theory could be called a descriptive theory of games. Also the definition of the game situation itself could be relaxed, for instance, by dropping the requirement that each player knows the utilities assigned by all the other players to the several outcomes and so on.

The question we wish to examine is the following. Could game theory be useful for the study or the conduct of international relations?

Note that an application to the study of international relations would imply a descriptive theory while an application to the conduct of international relations would imply a normative theory. We shall first examine the normative potential of game theory.

The idea that the conduct of international relations can be guided by scientific principles is an intriguing one. The idea has a counterpart in the notion of military strategy, which purports to be a normative theory related to the conduct of war. There is a widespread notion, first explicitly formulated by Carl von Clausewitz, that there is continuity between international politics and war. The notion has been challenged in recent times when war began to appear as an affliction of the human race instead of a normal phase in the relations among nations, as it appeared to most princes, statesmen and chiefs of staff of the pre-1914 Europe. Institutions like the League of Nations, the World Court and the United Nations attest to attempts to find a basis of international relations other than the interplay of military capacities. However, the Clausewitzian doctrine still persists among experts on international relations and even carries an aura of "realism." Put bluntly, the doctrine states that power is the currency of politics in the same way as money is the currency of economics.

People who subscribe to these views might be expected to be favourably inclined to the idea that normative game theory (which purports to be a theory of rational conflict) may well serve as a foundation of a science of international politics, much in the same way as probability theory and statistics serve as the foundation of actuarial business (such as insurance) or as operational analysis can provide a scientific basis for industrial engineering.

As a matter of fact, however, the interest so far shown in game theory by those responsible for decisions on the international level and their advisers has, for the most part, been spotty, not to be compared, for example, with the alertness of the military profession to new developments in the physical sciences.

There are several reasons for this comparative lack of interest on the part of diplo-military strategists in a discipline, which, on the face of it, might be taken as a sound theoretical foundation of their profession. The people recruited into political life and its appendages are, for the most part, ignorant of mathematics. They are not in a position to acquire an entirely new conceptual framework. (Some of them cannot undertake to learn Russian, let alone mathematics, which is a much more difficult language.)

Moreover, the politicians and their advisers tend in our age to be practical men. Their interest in a scientific discipline is usually aroused only by demonstrable practical applications. So far, explicit applications of game theory to conflict have been confined to only very special problems of military tactics and to a limited analysis of power distribution in legislative bodies. There have been no practical applications to large strategic problems and none to practical diplomacy.

Third, decision-makers (or more likely their advisers) who tend to have a broader outlook or some intellectual curiosity may have actually consulted specialists in game theory with a view to discerning a potential for future applications. If so, they must have been told by honest game theoreticians (and I assume, in the absence of evidence to the contrary, that all of them are honest) that an actual application of game-theoretical principles in international relations is beset with well-nigh insuperable difficulties.

The conditions, which must be satisfied if a situation is to yield to game-theoretical analysis, are extremely stringent.

First, the range of choices open to every relevant decision-maker must be exactly specifiable if the problem is to be at all defined. Moreover, the range of choices must not be too large if the problem is to be tractable.

Second, all the possible outcomes resulting from any combination of strategy choices by all of the relevant players (including Chance, which is often included as a player) must be known in advance.

Third, each of these outcomes must be evaluated on a numerical utility scale (except in special cases where an ordinal scale suffices). To take an example, imagine Austria as a player in the game known as *Der Drang nach Osten*. At a certain point in the game, Austria is faced with the decision of whether to send an ultimatum to Serbia. Among the possible outcomes might be (a) the ultimatum is sent and accepted; (b) the ultimatum is rejected, and Russia does nothing; (c) the ultimatum is rejected, and Russia mobilizes; or (d) the ultimatum is not sent. Here it would not be enough to know that, as Austria saw the situation, the best outcome was (b) (ultimatum rejected, Russia does nothing); that the next best was (a) (ultimatum sent and accepted), which is preferred to (c) (ultimatum rejected, Russia mobilizes), which is better than (d) (the ultimatum is not sent). Austria would also have to specify how much value (or cost) is associated with each outcome. When the outcomes are expressed in money or battle casualties or votes, such a demand can sometimes be met by using the corresponding numbers as utilities (although this, too, is usually a gross simplification). Where intangibles are involved (i.e. "national honour"), the problem of assigning utilities to outcomes becomes extremely vague. It is, of course, possible to assign such utilities arbitrarily or intuitively; but then the solution of the decision problem (assuming a solution can be deduced) reflects no greater precision than the assignment of utilities.

Fourth, to apply game-theoretical analysis, as it is now developed, not only must the decision-maker know his own mind to the extent of assigning numerical values to outcomes, but he must also know the minds of all his opponents in the same sense.

Besides these technical difficulties, there are conceptual ones. Frequently, the solution of a game (i.e. the "optimal choice") turns out to be a so-called mixed strategy. The use of a mixed strategy can be best illustrated in poker. A poker player, playing against skilled opponents, cannot afford to make bets in proportion to the strength of the hand he holds. If he adopted this strategy, the other players could eventually deduce the strength of his hand from the size of his bet and would use this knowledge to their own advantage. Therefore a poker player must occasionally bluff, that is, make high bets when he holds a weak hand or low bets when he holds a strong one. The exact proportion in which he must mix his strategies, that is, make bets of various sizes, given various hands, can, in principle, be calculated by game-theoretical methods. (Actually this problem is so enormously complex that no one has yet undertaken to solve it.)

Now, in poker the optimal strategy mixtures (if they were ever discovered) could be effected by choosing the strategies with frequencies corresponding to their assigned probabilities in repeated plays of the game. In international relations, however, a mixed strategy solution (assuming one could be deduced) could hardly be effected in this way. History may repeat itself, but situations which are exact replicas of each other do not. The only way one could resort to a mixed strategy is by delegating decisions to some device governed by chance.

If the "game" were repeated many times, these chance decisions would cumulate and would occur with appropriate frequencies. However, if the "game" could be played only once, the outcome would be determined by chance. If it turned out to be the "wrong" outcome, the decision-maker could only console himself by the thought that the expected utility of the outcome had been the largest possible under the circumstances.

Nevertheless, proposals involving leaving the decision to a chance device have actually been made by some advisers on military policy in the United States. For example, in the game called *Nuclear Deterrence* (alias the *Balance of Terror*), one strategist, possibly inspired by the concept of mixed strategy, proposed to modify the then dominant policy of massive retaliation by the use of a probabilistically determined threat. The point is that a threat of a nuclear war in response to a minor aggression lacks credibility especially since counter-retaliation has become a distinct possibility. To remove this shortcoming, a proposal was made to design a range of threats. The destruction threatened in each case was to be of the same devastating magnitude. (In those days the exquisite gradation of nuclear warfare had not yet been described by Mr Herman Kahn[1]). The variation of severity of the threat (to make the punishment fit the crime) was to be achieved by attaching to each threat a varying degree of probability of its being carried out. Thus, if the Russians (who at that time were assumed to be the arch-enemies of the United States) were preparing, say, to subvert the government of Nicaragua, a nuclear war would be unleashed only if a thrown pair of dice showed a double ace (i.e. with probability 1/36). Should the Russians make a more serious encroachment on the security of the United States, say, by installing missiles in Cuba, a nuclear war would be initiated if a thrown coin showed tails (i.e. with probability one-half).

Practical men are not likely to be impressed with decision rules of this sort. Contrary to some popular beliefs, diplo-military decisions are not yet dictated by electronic computer printouts. Statesmen and political scientists still insist that political expertise is more a matter of judgement than technique, that there is no substitute for deeply internalized experience and an intuitive grasp of the essentials of each unique situation.

Nevertheless, although bizarre proposals derived from mathematical formalism are likely to be rejected, the pressure of scientism on the orientation of decision-makers is heavy. By scientism I mean the attitude, which places great weight on arguments backed up by the trappings of scientific discourse (facts and figures, charts and graphs and so on) at the expense of arguments which appeal to perceptions, interpretations, ultimate goals and values. The danger of scientism is that it is seductive. The power of science actually does derive from a self-imposed discipline of objectivity, the primacy of "What is" over "What ought to be." Hence the displays of facts and figures rather than searching inquiries into the sources of values, established doctrines and commitments command the attention of people who respect the scientific attitude and think of themselves as "realists." Moreover, the commonly appreciated rewards of science reside almost totally in the power conferred by science over the environment. Therefore, the striving to extend the method of science to all areas of endeavour in order to extend the domain controlled is understandable in men of affairs.

Now, science deals with data, and the sort of data which are amenable to "hard analysis" are quantities. Dollars, kilowatts, shipping tonnage, fire-power and so on, are naturally expressed as quantities. These are items of interest to organizers of economic and military operations. The logistic calculations involved in the allocations of resources, economic and military, for the purpose of extending, maintaining and exercising power are often intricate and beset with challenging problems. In this way, the decision-making apparatus of a superstate attracts men of considerable intellect, especially those whose talents are directed towards organization of large-scale projects and whose emotional commitments are bound up with the exercise of power.

At the same time, the political goals of a superstate tend to become simplified, especially if its power is so great that it seems to the leaders to be invincible. The complex intrigues of pre-1914 Europe with its shifting alliances and subtle diplomatic gambits no longer represent the international political reality of our day. Power has become polarized in two superstates and, except for the recent emergence of a third candidate for this status and for some political manoeuvring in the ex-colonial world, international relations, so-called, were overwhelmingly dominated during the first two decades after World War II by preparations for a final showdown. The situation has not been conducive to the development or the exercise of political skills in the international arena. Perhaps this was why the intellectual efforts associated with decision-making on the international level have been so heavily concentrated on military and associated technical problems instead of on problems directly related to human needs, such as relief from hunger, want, fear and disease.

To the extent, then, that "scientific methods" are applied at all to decisions on the international level, they tend to be applied on the lowest level of decision theory, for example, operational research, which is concerned only with finding efficient ways of carrying out a given task, or the type of strategic analysis which involves two parties with diametrically opposed interests. The latter situation is represented in game theory by the paradigm of the two-person zero-sum game—a conflict in which the gains of one player are synonymous with the losses of the other.

Since I believe that a conflict so conceived and conducted between the chief contending power blocs can lead only to disaster for the populations of both, I cannot see what benefit is to be derived from conducting this conflict in a "rational" manner, which, in this context, can only mean in the most efficient, hence the most ruthless manner.

It is asserted, however, by those who are conducting the Cold War (and its hot episodes) that they are conducting the conflict "rationally," that is, with restraint, keeping the door open for negotiations and so on. That is, they are giving evidence of understanding that the game is not a zero-sum game and that some outcomes of this "game" can mean an unacceptable loss for both sides. Indeed, the question of how force can be used "rationally" in the conduct of foreign policy (in the modern showdown-avoiding sense) occupies a central position in contemporary strategic thinking. Robert Osgood begins his book *Limited War: The Challenge to American Strategy* with the question: "How can the United States utilize its military power as a rational and effective instrument of national policy?"[2] Herman Kahn, in his book *On Escalation: Metaphors and Scenarios*,[3] discusses the game of *Chicken* (*Brinkmanship*) in connexion with the techniques of international blackmail. Thomas Schelling, in his book *The Strategy of Conflict*,[4] pleads for a development of the theory of non-zero-sum games. Characteristically, however, Schelling sees non-zero-sum game theory as simply an enlargement of the strategist's repertoire of concept. The aim of strategic analysis remains the same: to design safer and more effective ways to pursue "national interest."

In my opinion, a thorough study of "higher game theory," that is, the theory dealing with only partially conflicting interests and with conflicts involving more than two players, would indeed be instructive, but not in the sense in which the utility of a theory is generally appreciated by decision makers. Higher game theory leads to insights into the nature of complex conflict situations, not answers about how to use the knowledge in the pursuit of self-interest.

This aspect of game theory is not generally understood. It can be summarized so: the game-theoretical method of strategic analysis, if pursued far enough, lays bare its own limitations. These limitations have to do not so much with the complexities of real life situations as with the paradoxes, which reside in the very notion of a "rational decision." These paradoxes are not apparent in situations where there is only one decision-maker, for in those instances a rational decision in the sense of maximizing one's utility (or expected utility) is at least conceivable, given sufficient knowledge about the facts of the case. Likewise, in situations where there are exactly two decision-makers, whose interests are diametrically opposed, a rational decision can still be defined as that which extracts the most benefit at the expense of the opponent. However, once these situations are transcended, for example, in cases where the interests of the decision-makers are partly opposed and partly coincident, and also in situations where more than two decision-makers are involved, the very concept of rational decision becomes riddled with contradictions.

A simple situation of the first kind is seen in the *Balance of Terror* between two nuclear powers. Suppose it is in the interest of both powers to dismantle their nuclear establishments. A decision to do so cannot be rationalized on the grounds of self-interest. For, should the opponent disarm, it is more advantageous to remain armed (because of the intimidating power of nuclear monopoly). Should the opponent remain armed, one must remain armed to avoid being intimidated. Thus, it is in the interests of each power to remain armed regardless of the state of the other. This conclusion contradicts our original assumption that it is to the advantage of both powers to disarm. The contradiction arises from the circumstances that the interests of the two "players" are in this case only partially, not diametrically, opposed.

The simplest paradox involving three parties can be illustrated by a game in which the three parties, A, B and C, are to divide a shilling among them by majority vote. Clearly A and B, being a majority, can take the whole shilling. If bargaining and shifting coalitions are allowed, C can offer A 7d of the shilling to lure him away from B. Clearly C gains in the transaction (getting 5d instead of nothing) and so does A (getting 7d instead of 6d). But if A, in the pursuit of self-interest, accepts C's offer, he is in danger of losing all. For then B and C, according to the new arrangement, are to get 0d and 5d, respectively, and so it is in their joint interest to outvote A; for then they can split the shilling evenly, and both will gain thereby. B and C can realize this outcome, for they too are a majority. At this point A, the loser, can approach B or C with a new offer, which will make both him and his new partner richer at the expense of the third; and they are back where they started. Under these conditions, the merry-go-round pursuit of self-interest leads absolutely nowhere.

We see, therefore, that game theory fails as a normative theory in certain cases: it cannot prescribe an unambiguously rational decision to any of the players. However, game theory points up rather dramatically, I think, the nature of the dilemma, which lurks in conflict situations more complex than a clash of two diametrically opposed interests. Once one has recognized the impasse, one can seek ways out. One way is to abandon the notion of constructing a normative theory to fit all conflict situations and to concentrate instead on a descriptive one, which does not depend on a rigorous definition of rational decision. This course is now pursued by those who see game theory as a point of departure for the empirical study of conflicts. Laboratory experiments provide an excellent vehicle for such studies. Whether the results of such experiments can be used as a point of departure for studying real life conflicts, particularly international relations, remains to be seen. One cannot tell without trying.

Another way out is to redefine rational decision in such a way as to bring out clearly the sharp differences between individual rationality and collective rationality. For instance, in the divide-the-shilling game it seems to be in the collective interest of A, B and C to take 4d each, simply because it is impossible to question the fairness of this solution. But in order to agree to the even division, each pair of the players must forgo the opportunities (which they clearly have by the rules of the game) to take the whole shilling. A similar argument can prescribe the disarmament of the two nuclear rivals, although it is in the interest of each of them (separately) to remain armed.

Suggestions of this sort are not likely to be received enthusiastically by the designers of current diplo-military politics of the major contending powers. Being preoccupied with the central problem of finding optimal decisions each from the point of view of the power they serve, the policy makers and their advisers may well look askance at the prospect of subjecting to intense scrutiny the goals they are pursuing and the values they are guided by; for that is what any attempt to escape from the impasse involves. In particular, the decision-making process itself must be subjected to scrutiny, which interferes with the pursuit of pre-set goals.

In order to study how decisions are actually made, it is not enough simply to label the actors (as is done in normative game theory); one must also identify and examine them. It is not enough to accept the stated goals and values (utilities); one must inquire into the actual values of the actors. If one does this and examines the decisions in the light of what is discovered, one may well find that it is difficult to apply the term "rational" to the actors of the international drama even in the traditional self-interest–seeking sense, let alone in the collective sense.

In fact, if one persists in examining the referents, one may find that many of the key terms in which discussions of international relations are couched appear to be devoid of meaning. Once upon a time the king of France was France; an affront to the foreign minister of Austria was an affront to "Austria." It was natural to speak of decisions made by "states," because such decisions were made by individuals or small groups of individuals in direct contact with each other. In our political life, we have come a long way from identifying a country with its prince or his privy council. Decisions on the international level are resultants of innumerable pressures, not the least of which is the pressure exerted by the burgeoning military technology itself, one's own and the opponents'. A thorough analysis may reveal that "decisions" on the international level may have given way to quasi-mechanical interactions over which no individual "player" has control. Still, international relations continue to be pictured by expert as the same old game played by the same old players for the same old stakes. Predominant thinking about international relations continues in channels laid down by Clausewitz and Bismarck, except where it degenerates into the nightmare scenarios of "nuclear strategy," which Clausewitz and Bismarck would have dismissed as ravings of madmen.

What is needed is not a theory of how to get ahead in the game, or how to be the last survivor in a holocaust, but how to stop playing the stupid and deadly game or, at least, how to tame the "players." I think it is a mistake to personify the players of the current international game as rational beings. They can be more accurately conceived as representatives of a new species on the face of the earth. The species might be called the war-waging state (*status bellagerens*). We need a natural history of this beast. We need to know the structure and function of his nervous system, that is, the chains of communication and command in the war-waging state, which enable it to mobilize its fury and to wreak unspeakable destruction regardless of the inclinations of its "cells," that is, the human beings which constitute the chains of communication and command. A beginning of such a study appears, for example, in Karl Deutsch's book *The Nerves of Government.*[5] We need to know more about the sources of manic fixation of *status bellagerens*, called "National Interest." Above all, we need to know how it came to pass that we, human beings, so readily identify our goals and aspirations with those of the beast even though he appears to be unable to think, to feel, to love, to dream or to possess any human quality which we as human beings ordinarily admire.

Inquiries of this sort would take us far afield from decision theory as it was originally conceived. Nevertheless, decision theory can play an important part in the coming conceptual revolution, namely, in revealing a profound truth of our age: at the present stage of man's development, pursuit of self-interest on the international level is incompatible with genuine rationality.

Notes

1. Herman Kahn, *On Thermonuclear War* (Princeton, NJ: Princeton University Press, 1960).
2. Robert E. Osgood, *Limited War: The Challenge to American Strategy* (Chicago: University of Chicago Press, 1957).
3. Herman Kahn, *On Escalation: Metaphors and Scenarios* (New York: Praeger, 1965).
4. Thomas C. Schelling, *The Strategy of Conflict* (Cambridge, MA: Harvard University, 1960).
5. Karl W. Deutsch, *The Nerves of Government* (New York: The Free Press of Glencoe, 1963).

Anatol Rapoport (1911–2007) was Professor of Mathematics and Psychology and Professor of Peace Studies at the University of Toronto. Previously, he was Professor of Mathematical Biology at the University of Michigan at Ann Arbor. He served as Director of the Institute of Advanced Studies (Vienna). He contributed to general systems theory, to mathematical biology and to the mathematical modelling of social interactions.

CHAPTER 4

The Origins of MIRV

Herbert F. York

MIRV Description

One of the most important military inventions in recent years is MIRV. This acronym stands for Multiple Independently Targetable Re-entry Vehicles. A single ballistic missile with MIRV capability can deliver one or more warheads with a very high accuracy to each of several different targets. Multiplicities as high as 14 have been reported. This is accomplished by means of a "bus," a device sometimes more formally called the Post Boost Control System (PBCS) and sometimes the Sequential Payload Delivery System (SPD).

The MIRV system works as follows: initially, the main rocket booster puts the bus on a course that would cause it to impact somewhat near target number one. The bus contains several re-entry vehicles (RVs), a guidance and control system, and some small rocket propulsion units. The guidance system instructs these small rockets to modify the velocity of the bus so that it is aimed as precisely as possible along the orbit leading to target number one. When this has been accomplished, the bus

Originally published in David Carlton and Carlo Schaerf, eds., *The Dynamics of the Arms Race* (London: Croom Helm, 1975): 23–35.

H.F. York (✉)
University of California, San Diego, San Diego, CA, USA

P. Foradori et al. (eds.), *Arms Control and Disarmament*,
https://doi.org/10.1007/978-3-319-62259-0_4

very gently ejects one of the RVs. While this first RV vehicle continues inexorably on its course to target one, the bus guidance system instructs the propulsion units to modify its course so as to put it on a course orbit leading to target two, and repeats the process until each RV is on route to its prescribed target.

What follows is an attempt to show just how it was that MIRV came to be. We shall see that several independent military requirements led to several different lines of technological developments. As time went on, ideas and personnel were interchanged among the various programmes, resulting in a very complex web of technological developments and inventions. This web could have been cut in a large number of places, and the ultimate result would have remained about the same: MIRVs on intercontinental ballistic missiles (ICBMs) at the beginning of the 1970s.

The Polaris A-3 MIRV

A convenient date to pick up the oldest strand in the web leading to MIRV is 1957. In the Autumn of that year the first Sputnik was launched. Shortly after, and at least partially in response, the Department of Defense (DoD) set up a so-called Re-entry Body Identification Group, a committee to study the question of whether the designers of offensive missiles (ICBMs and Polarises) should take seriously the possibility of defences against missiles (ABMs), and, if so, what they should do about it. The committee was fully informed about the American ABM programme (then the Nike-Zeus) and was aware that the Soviet Union was probably working on something similar. In early 1958, the committee concluded that the possibility of missile defence should be taken seriously, but it described a number of counter-measures for the offence. These included decoys, chaff, reduced radar cross-sections for the RV, blackout, tank fragments and, most important for our purposes here, multiple warheads. All except the last are designed to confuse the defences; multiple warheads, on the other hand, penetrate defences simply by saturating or exhausting them. Collectively, these counter-measures are called penetration aids.

The committee's report was unusually influential, coming as it did at a rather critical juncture, and eventually its conclusions were widely accepted. The Air Force incorporated several of the deception devices such as decoys in its early penetration aids packages, and the Navy developed a multiple warhead system for the A-3 version of Polaris.

The Polaris A-3 warhead system consists of a cluster of three separate RVs each of which contains a nuclear warhead. This type of system is called an MRV, the acronym standing for Multiple Re-entry Vehicle. The cluster of three is launched as a unit and is aimed as accurately as possible at some particular target. After the boosters burn out, the three RVs are mechanically separated from each other and given an additional velocity of some tens of feet per second relative to the velocity of a cluster as a whole. As a result, the three warheads impact in a triangular pattern having dimensions of the order of a mile, and centred more or less on the target. Thus, against soft targets the damage radius of one RV is, very roughly speaking, about the same as their separation, and the destruction caused by such an MRV is about the same as that which would be caused by a single warhead.

The first MRVs were deployed on Polaris A-3s in 1964. It has been estimated that the yield of an individual RV is 200 kiloton (kt).

In the first few years after the decision to deploy these MRVs, considerable progress was made in the design of American anti-ballistic missile (ABM) systems, and American perceptions of what the Soviets were doing also changed. As a result, it was realized in about 1962–1963 that the separation of the RVs in the A-3 MRV was too small to cope with any except a first-generation ABM, and that by the late 1960s it might become possible in principle to intercept all three RVs with a single ABM. This problem could not be solved by simply increasing the spread of the impact points; doing so would mean making the separation distance bigger than the dimensions of most targets. We will return to this matter later.

Early Multiple Satellites

Following the launch of the first Soviet ICBM in August 1957 and of the first Sputnik in October 1957, there was an outburst of new ideas in the United States about how to make and how to use missiles and satellites. Some of these involved launching multiple satellites with a single booster.

One early multiple satellite idea had missile defence as its goal. The basic design objective was to intercept ballistic missiles during the first few minutes after launch while their booster engines were still operating. Missiles are especially vulnerable during that period because simply puncturing their propellant tanks with shrapnel will cause them to fall thousands of miles short of their targets. In this scheme, the detection and destruction of the missiles was to be accomplished by defensive satellites overhead at the time of launch. Satellite systems designed to achieve these objectives were collectively known as BAMBI, for Ballistic Anti-Missile Boost Interceptor.

One version of the missile defence multiple satellite idea involved a "mother ship" which housed a number of sub-satellites. Typically, the mother ship was conceived of as having on board some kind of sensors and computer intelligence which could detect and track enemy ICBMs as they rose from the atmosphere. Using this intelligence, it was supposed to orient itself appropriately and to determine when, at what rate and in what direction to launch its sub-satellites. The sub-satellites were to contain their own propulsion systems, and enough of a guidance and control system so that they could accomplish the final steps of the intercept.

No BAMBI satellite system was ever built, but many versions were studied in depth by the Advanced Research Project Agency (ARPA) of the DoD. During those early years after its founding in February 1958, there was an especially rapid interchange of key technical personnel between ARPA and industry, and among the industrial groups most heavily involved in missile and space technology. As a result, many of the persons involved in working out these paper concepts later turned up in the groups that designed and developed the real hardware to be discussed below.

The first multiple satellite launch, involving real hardware, took place on 22 June 1960. The satellites were Transit IIA and a Solar Radiation Satellite of the Naval Research Laboratory. The first stage booster was a Thor and the second stage was an Able-Star. The booster plus upper stage were first used to place the upper stage with the two satellites still coupled to it on a trajectory with an apogee of 500 miles. The velocity necessary was achieved after the second stage engine had burned about four minutes, and well before apogee was reached.

Then they continued on an altitude-controlled coast for about 18 minutes, on up to apogee. There the Able-Star engines were restarted and operated for 14 seconds more, placing the combination on what was intended to be a circular orbit at 500 miles above the earth. The pair of satellites was then decoupled from the Able-Star stage and separated from each other by a compressed spring, which gave the smaller one an additional velocity of 1.5 feet per second.

This Able-Star second stage engine was designed and built by Aerojet-General, and used hypergolic propellants, that is propellants which ignite simply on contact. The Able-Star also included a number of sub-systems designed and integrated into it by Space Technology Laboratories, including the restart and the guidance and control units, and a programmer and accelerometer. All of these capabilities, techniques and sub-systems are essential for the MIRV bus, and the Able-Stars thus represented a major step in the development of MIRV technology.

Another interesting case, involving an additional small step, occurred on 16 October 1963, when the Atlas-Agena combination was used to launch a pair of Vela satellites. In this case the requirement was more complicated than in the case of the Transit launches: the two satellites were to be placed in two very different positions.

The Agena is the oldest American space vehicle with its own propulsion and guidance and control system. It was developed under a contract awarded by the US Air Force (USAF) to Lockheed in 1956, well before Sputnik went into orbit. The Vela satellites were designed to detect nuclear explosions in space and in the upper atmosphere. Their purpose was to monitor compliance with the Partial Test Ban Treaty of 1963, which prohibited nuclear tests anywhere except underground.

In the October 1963 flight, the Agena with the two Velas on board first placed itself on a very elongated equatorial orbit having a perigee near the earth but an apogee at an altitude of 64,000 miles. On first reaching apogee, the Agena oriented and released one of the Vela satellites. The Vela had a solid propellant rocket motor, which it fired to give it enough additional velocity to circularize its orbit. The Agena and the second Vela then made a round trip down to perigee and back to apogee. By that time the first Vela was roughly halfway around the earth. The second Vela was then oriented, released and accelerated like the first. The result was two satellites, launched together from the earth, but now about 180° apart on orbits varying from 62,000 to 72,000 miles above the earth. Lockheed Missiles and Space Division, which was responsible for the Agena stage, later designed the Poseidon MIRV, and the Space Technology Laboratory, which was prime contractor for the Vela system, did the systems engineering and technical direction for the Minuteman programme.

Titan III and Transtage

The Transtage is a highly flexible PBCS or "bus," as it is usually called. The origins of Transtage, and the Titan III launch vehicles, which are used to boost it into orbit, are to be found in 1961. The first successful launch of a Titan III with Transtage took place on 10 December 1964, and the first fully successful delivery of multiple sub-satellites into multiple orbits took place in 1966. In 1968, when John Foster, the Director of Defense Research and Engineering was being questioned about why he was so confident that the Minuteman III and Poseidon MIRVs would work, he cited the successful operation of Transtage as proof that all the essential engineering problems had been solved.

The technical requirements that led to the development of Titan III and Transtage arose in the first months of Robert McNamara's term as American Secretary of Defense. Shortly after taking office, the new Secretary issued to his various chief subordinates a very long list of questions. One of the questions asked why it was that in the previous three-and-a-half years the Soviets had been so much more successful than the Americans in launching satellites, and what might be done to change that situation. Even while the study was in progress, Yuri Gagarin became the first man to orbit the earth; the importance of the study was re-emphasized, and its due date was moved forward.

The study concluded that part of the answer lay in the fact that the Soviet Union had used the same rocket booster for all their space launches while the Americans had used a wide variety of rockets, often developing a new launcher for each new satellite. One of the unfortunate results of the Americans' approach was that they seldom got the "bugs" fully worked out of one system before they moved on and began to use a newer one. The solution proposed was that the DoD and National Aeronautics and Space Administration (NASA) should jointly agree on the development of a very limited family of standardized boosters of different sizes, each with enough flexibility to launch a variety of satellites to different orbits.

It was further recommended that the largest member of this family of "standardized workhorses" be based on the Titan II, the largest and ruggedest of the missile boosters. The first two stages were to be the two stages of the Titan II. The upper stage was to be a new unit, known as Transtage equipped with the Titan II guidance system and having a propulsion system capable of coasting and restarting as in Able-Star and Agena-B. Further flexibility was to be achieved by making it possible to strap two large solid rocket boosters onto opposite sides of the Titan II first stage, thus giving a much larger initial thrust, and making it possible to launch still heavier payloads.

The Transtage was to be designed so that it could reach the "stationary" or "24-hour" circular orbit at 19,000 miles above the earth. There was no specific military requirement for such a booster, but it was expected one would develop, so McNamara and James Webb, the Administrator of NASA, concurred in the recommendation. Hardware development was initiated through a contract issued to Martin Marietta on 20 August 1962.

Development of the system proceeded smoothly and rapidly, but the anticipated need for such a vehicle failed to appear. Finally, it was decided to use the Titan IIIC/Transtage to launch the Initial Defense Communication Satellite Program (IDCSP) satellite.

The question of what kind of satellites should be used in a defence communication system, how many there should be, what altitudes they should be at and who should build and control them were all matters of very heated controversy, largely beyond the scope of the present discussion. However, two generalizations are pertinent here. Usually it is held that a military communication satellite system should involve many satellites at very high altitudes. They should be many so that broad coverage of the whole world can be achieved and so that the sudden failure of one or two, accidentally or deliberately, will not seriously degrade the system. They should be very high, again for coverage, but also so that extreme rapid slewing of ground receiving antennas is not necessary. The question of exactly how high was one of the most complicated and heated of the arguments.

The coming into being of the Titan IIIC/Transtage, without a specific mission but with plenty of capacity and flexibility, helped to resolve some of these questions. Finally, on 16 June 1966 this new launch vehicle was used to place eight communications satellites in eight different predetermined equatorial orbits, all at an altitude of approximately 21,000 nautical miles.

As has already been stated, the Transtage is a true PBCS or "bus." Using its coast and restart capacity, the Transtage first achieved a very nearly circular orbit varying from 20,913 to 21,051 miles and having a period of 1334.2 minutes. It then gently nudged off one of the subsatellites. Then, using its 450-pound thrust vernier motors for controlling pitch and yaw, it added the very small increment of velocity needed, and dropped off the second satellite at essentially the same altitude, but with a period of 1334.7 minutes. It repeated this manoeuvre on through number eight which was dropped off three minutes later with a period of 1347.6 minutes. More of these 100-pound IDCSP satellites were added in multiple launches on 18 January 1967, 1 July 1967 and 13 June 1968.

The systems engineering on the Titan III and the Transtage programmes was performed for the space and missile office of the USAF by the Aerospace Corporation. The same organizations were involved in the Minuteman III MIRV programme, which was initiated after this Transtage programme was conceived but before it achieved its first flight.

The Poseidon MIRV

In the early 1960s, a number of events, developments and situations brought about the requirements for and the development of the Poseidon MIRV. The most important of these were: the completion of the Polaris A-3 development programme; changing ideas about ABM design and possibilities, particularly changes in American perceptions about Soviet ABM; and further developments in strategic thinking, as exemplified by Secretary McNamara's "Counter-force" speech.[1]

As the A-3 development programme was nearing its end, the Special Projects Office, then under the direction of Rear Adm. Levering Smith, turned its attention to the next step, then designated the B-3. From the beginning it was clear that the B-3 would be a bigger and more accurate rocket than the A-3. The improved version could be used to deliver a bigger bomb more accurately or, at the other extreme, could be used for making a higher multiplicity MRV, thus further reinforcing its defence penetration ability and enhancing its deterrent or "counter-value" role.

As mentioned earlier, ABM had moved along steadily since the earlier days when the A-3 MRV was first proposed as a solution to ABM penetration. Similarly, accumulating intelligence information about the Soviet programme, plus Khrushchev's famous boast that "You can say our rocket hits a fly in outer space," plus what the Americans knew about the Soviet nuclear tests at high altitude in 1961 and 1962, led the Americans to ascribe to the Soviet ABM the capabilities which they knew that they could achieve, in principle, on their own. In particular, it was deduced that the explosion of a single large ABM warhead could simultaneously destroy all three of the A-3 MRVs. Thus, it was concluded that the separation of approximately one mile between warheads as in the A-3 MRV was too small, and that separation of tens or even a hundred miles might be necessary. Clearly, simply expanding the size of the triangular pattern of the A-3 MRV would not do. Doing that would simply mean that, even in the case of a large soft target like a city, at the most only one of the RVs would be aimed at the city, and that one RV in theory is the one any defence system would concentrate on. Thus, the simple straightforward "shotgun" approach to multiple RVs was seen as obsolescent, and so, from an early date, a different approach was sought for the B-3.

The third important factor that influenced the Poseidon/B-3 warhead decision was the counter-force strategic philosophy as enunciated in Secretary McNamara's speech at Ann Arbor in 1962. McNamara did not originate the idea of a counter-force strategy; it had been a natural part of strategic thinking before the advent of nuclear weapons, and had never been entirely forgotten. However, the notion of "deterrence" by threatening cities and the industrial base ("counter-value" targets) had gained the upper hand everywhere except in Air Force circles in the years before the Ann Arbor speech. Furthermore, McNamara was surprised by the reaction to his speech, especially by the Air Force interpretation of it as justifying their most extreme missile deployment plans. As a result, he soon afterward backed away from the views expressed in his Counter-force speech, much to the disappointment of those who advocated expanded deployments.

But despite their transitory nature, the ideas in that speech put a "pulse" through the technological community which stimulated the kind of thinking that promoted MIRV-like ideas as a means of expanding the number of points that could be targeted.

From the beginning of the nuclear age there has always been an intense rivalry between the Navy and Air Force over roles and missions, especially in the realm of strategic warfare. The Polaris and the Minuteman were both part of and involved in that rivalry. When the principal mission of each missile was thought of as deterrence of nuclear war by threatening to destroy cities and industry (i.e. so-called counter-value targets), Polaris and Minuteman could play roughly equivalent roles. Then, the arguments about which was "better" revolved around such issues as cost-effectiveness and survivability in the face of a pre-emptive attack. But when the mission became an enemy's well-protected missiles and other military forces (i.e. so-called counter-force targets), then the Minuteman with its accurately delivered single bomb was seen as possibly gaining a decisive edge over the Polaris "shotgun" MRV in the roles and missions argument.

The transformation of MRV to MIRV neatly solved this last problem. By using a "bus," or PBCS, as described earlier, it was in principle possible to deliver multiple warheads to one single target, but along trajectories having different apogees. Thus, they could all be aimed at a single (hard) target as accurately as guidance technology would allow, but they would all arrive at different times; being spaced as much as a hundred miles apart. Therefore, they would still have the same potential for exhausting a newer, more powerful ABM as the original MRV had against a smaller, cruder, ABM.

In principle, a bus system providing multiple shots with smaller bombs is more effective against hard targets than a single shot with a larger bomb, provided not too much total yield is sacrificed in dividing the explosive up into smaller pieces and in providing the weight needed for the bus.

This conceptual development deserves to be emphasized because it clearly shows that the PBCS's "independently targetable" feature was essential for solving the ABM penetration problem alone, independent of whatever other military uses it may also have.

The development of Poseidon MIRV was approved in the Autumn of 1964, its deployment was approved in 1966 and the first boatload was deployed at sea on the James Madison SSBN on 1 April 1971. The Poseidon MIRV is usually described as having from 10 to 14 individual warheads of 50 kt each.

Minuteman MIRV

The Air Force at first concentrated its missile warhead development programme on single, large, accurately launched warheads protected against interception by the various types of confusion devices described in the Re-entry Body Identification Group's report, such as decoys and low radar cross-sections. But in 1962 and 1963, stimulated, like the Navy, by progress in ABM and the Soviet high altitude nuclear test series, the USAF began to give serious consideration to installing multiple warheads on Air Force missiles as a means of improving their ability to penetrate defences. At about the same time, a steady growth in the perceived number of vital individual military targets in the Soviet Union stimulated interest in the use of multiple warheads as a means of improving force effectiveness. The Air Force was as usual aided in these considerations by its extensive advisory apparatus (RAND, The Aerospace Corporation and various ad hoc committees of outside experts). In 1968, John Foster, US Director of Defense Research and Engineering, made it clear in Senate testimony that the Air Force had this dual motivation for turning to multiple warheads in the early 1960s and that this was still the case in 1968.

But despite there being two complementary reasons for turning to MIRV, the argument about whether multiple small weapons were preferable to a single large one persisted within the defence establishment for some time. The controversy reached a point where a contract which was about to be let for the development of a large single weapon RV was held up because of the dispute over the question. Since information about commercial contracts is normally in the public domain, news of this struggle thus reached the missile press. *Aviation Week* reported in its issue of 29 July 1963 that

> Selection of a contractor to develop Mark 12 [the RV for Minuteman III] had been stymied over the past six months by the Department of Defense Research and Engineering's [DDR&E] criticism of USAF's original development plan ... DDR&E intervened during the Air Force's first competition to pick a contractor ... last Fall, charging that the USAF development plan was insufficiently advanced and lacking in multiple warhead capability. The Air Force's concept of Mark 12 was referred to as just another "rock."[2]

While this argument was still going on, two entirely different versions of MIRV were being considered. In one concept, a single missile would launch a cluster of small, one-stage missiles each with its own self-contained propulsion and guidance systems. After the cluster as a whole was placed on an approximately correct trajectory to the target area, the cluster would break up, and the individual sub-missiles would then adjust their velocities so as to get on to precise trajectories to their predetermined targets. The extra weight (and even cost) of the individual guidance and control system made this a rather poor option compared to the "bus," and its development was never authorized.

The other version given consideration in the period 1962–1964 was the "bus" or PBCS, virtually identical to the Transtage already being developed at that time under the direction of the same Air Force office.

This really was the better idea, and since the necessary technology was already under development, it was finally selected as the appropriate technological solution in 1964. Full-scale development of the bus-type MIRV followed promptly. The decision actually to deploy MIRV was made in 1966, and the "first flight" of ten Minuteman III missiles with MIRV was turned over to Strategic Air Command on 19 June 1970.

In addition to the specific developments discussed above, bits and pieces of MIRV technology were invented independently or re-invented in some other Air Force programmes.

One such programme was the development of an SPD system whose purpose was to deliver a sequence of RVs from California to Kwajalein Atoll where they were used as test targets in the ABM development programme underway at the latter location. The main stimulus for this SPD development was simply economics; delivering several targets with a single launch was cheaper than providing a separate launcher for each one.

This programme was approved in 1964 and the first successful application was in 1966. It was done by the same organization which was responsible for the Transtage.

Of course, practically all of the technological techniques that make MIRV possible were also eventually developed in the civilian space programme, particularly in the lunar exploration programme. However, emphasis has been placed here only on those that had their origin in some military requirement.

The Decision to Deploy

I think that for all practical purposes the decision to deploy the two MIRVs was made inevitable by the decisions to develop them. Even so, the matter continued to be argued after the development decisions were made, and the matter was not really finally and formally resolved until the deployments actually took place in 1970 and 1971.

The argument had several facets. The basic reasons for supporting MIRV deployment were those already discussed: ABM penetration and increasing the number of points that could be targeted.

The ABM penetration feature of MIRV was most strongly put forward by those in the United States who believed the historic Russian penchant for defensive measures would stimulate the Soviets to press forward rapidly with the development and widespread deployment of ABM systems. The Soviets had indeed followed just such a pattern in the case of defence against bombers. The most extreme form of this argument held that the Soviets might be secretly preparing to modify their very numerous and ubiquitous SAM-2 bomber defence system so as to upgrade them into an ABM system.

The expansion of the number of targets that could be hit was primarily of interest to those who took the counter-force mission most seriously. Counter-force strategies are closely connected to and encourage notions about actually fighting and winning a nuclear war. Such notions in turn lead to requirements for a substantially larger force than is needed for simple deterrence. In their extreme form, these notions lead to essentially open-ended requirements for ever more offensive weapons.

These two ideas obviously reinforce each other and were often combined into a single argument favouring MIRV. Thus we find John Foster in 1968 defending the need for MIRV by saying that we need enough warheads so that we can "be sure of exhausting their defence capability, and then being able to deliver enough to provide assured destruction."[3]

Secretary McNamara had another interest in MIRV. His Counter-force speech unintentionally gave support to calculations of numbers of offensive weapons needed that were very much bigger than the force he had in mind. Some argued for very substantial increases in the total number of Minuteman and Polarises. In this context, the MIRV development programme became a tool in McNamara's arguments against force expansion. In the continuation of the Foster testimony just cited, he said:

> Now, we could choose, for example, to put in a number of Polaris boats. We could increase the Polaris force by a factor of five or ten. ... But we can get the same equivalent military capability against the Soviet Union by taking the existing ... boats and changing ... to the Poseidon missiles.[4]

Hence, from McNamara's point of view, MIRV was a device with an effective positive arms control feature.

Those directly responsible for American policy in arms control did not see it that way. Officials in the US Arms Control and Disarmament Agency (ACDA), especially Herbert Scoville and George Rathjens, as early as 1964, foresaw ways in which MIRV deployment could upset the "balance of terror" and destabilize the arms race.[5] They predicted that the deployment of MIRV by one side would be seen by the other as part of a possible preparation for making a first strike. And indeed, an anticipated deployment of MIRV by the Soviets on their large SS-9 ICBMs was seen exactly in that light by Secretary Melvin Laird and his associates in 1969. The Arms Control Agency people therefore opposed MIRV deployment from the start of the development programme.

These important controversies were not known to the public, nor even to any but a selected few in the Congress, until the 1968 Presidential Campaign. Then, Eugene McCarthy echoed the ACDA point of view and said:

> The introduction of sophisticated anti-ballistic missile systems and new missiles equipped with multiple warheads threaten to make the situation unstable. With the deployment of such weapons systems, each side will become concerned as to whether in the event of a pre-emptive attack it will be able to inflict sufficient damage in retaliation—if not its deterrent will not be credible. The arms race will thus be impelled to a new intensity. In crises, there could be an incentive to launch a first strike.[6]

Later, Senator Brooke introduced a resolution calling for a suspension of MIRV testing, and some other Senators and a number of former officials supported him, but to no avail; the first deployment of MIRV took place while the issue was still being debated.

In summary, the MIRV programme had many roots and branches. Important decisions were made by many persons only loosely connected with each other, and over a period of more than a decade. Of all the stimuli that gave rise to MIRV, the most important was the perceived need to penetrate with assurance ABM systems whose theoretical capability was slowly improving with time. However, all the technologies needed for MIRV had other reasons underlying their development, and so MIRV would very likely have emerged at about the same time even if the need for ABM penetration had not been perceived until much later, and possibly even if it had not arisen at all.

Almost all the important decisions were technologically determined. Economics entered mainly as an added stimulus to the development of the capability for making multiple satellite launches. Strategic analysis entered only fairly late (and indecisively) in a relatively narrow argument over the relationship of MIRV to the arms race. More general strategic thinking and political considerations did not enter into the process until it was too late for them to have any effect.

Notes

1. Robert S. McNamara, "Defense Arrangements of the North Atlantic Community," *Department of State Bulletin* 47 (9 July 1962): 64–70.
2. "Washington Roundup," *Aviation Week,* 29 July 1963, 24–25.
3. Senate Armed Services Appropriation Committee, *Hearings on Department of Defense Appropriations, Fiscal Year 1969* (Washington, DC: US Government Printing Office, 1968): part 4.
4. Ibid.
5. US Arms Control and Disarmament Agency, *Documents on Disarmament 1964* (Washington, DC: US Government Printing Office, 1964).
6. Eugene McCarthy, *The Year of the People* (Garden City, NY: Doubleday and Company, 1969).

Herbert F. York (1921–2002) was Professor of Physics at the University of California, Berkeley, and founding Chancellor of the University of California, San Diego. During World War Two he worked on the Manhattan Project. He was the first director of Lawrence Livermore National Laboratory and held numerous positions in both government and academia. He served as US Ambassador to the Comprehensive Test Ban negotiations in Geneva.

CHAPTER 5

The Importance of Agreements

Thomas Schelling

Most of this chapter has to do with identifying the motivational structures that might underlie arms agreements or arms understanding. Some of it has to do with figuring out what the purpose of an agreement is, what we would want to call an agreement, what kinds of agreements there are and how to evaluate them.

We can identify a number of interest structures. There is a variety of ways that two participants can view the outcomes of a possible arms negotiation, and maybe at least classify and somewhat clarify the different kinds of situations in which it may be said that both parties want an agreement or one party wants an agreement or neither party wants an agreement. Let us take a question like deciding whether or not to have anti-ballistic missile (ABM) system with the over-simplified assumption that no ABM will be pretty close to zero and some ABM will mean

Originally published in David Carlton and Carlo Schaerf, eds., *The Dynamics of the Arms Race* (London: Croom Helm, 1975): 65–77.
This was an impromptu presentation and not a formal paper. The original editors, with the kind cooperation of Kosta Tsipis, have prepared it for publication from tapes of Thomas Schelling's talk.

T. Schelling (✉)
School of Public Policy, University of Maryland, College Park,
College Park, MD, USA

P. Foradori et al. (eds.), *Arms Control and Disarmament*,
https://doi.org/10.1007/978-3-319-62259-0_5

unrestricted ABM. We want ABM if the Russians have it, we want ABM if the Russians do not have it; they want ABM if we have it, they want ABM if we do not have it. But we both prefer that neither have it to both having it.

In game theory this has the structure that is familiarly known as *Prisoner's Dilemma*. In other words, if our behaviour had no influence on their behaviour and if there were no bargains, we would choose to have ABM irrespective of what they did, which is what they also would choose, and yet our desire to have it ourselves being less than our desire that they do not have it, and similarly on their side, means that we can both prefer that both not have it and reach an agreement, the agreement being basically that I will not if you do not. Notice that this agreement requires some kind of enforcement because even if he does not, we would still independently prefer to have ABM, unless our having it will induce him to break his side of the agreement, or unless there is some penalty on violation. I think this is what usually people have in mind when they talk about genuine necessary agreements, agreements that have to be monitored and enforced. It is getting both parties to do what they otherwise would not do, but doing it conditionally with both sides benefiting.

Then there is a second, very different structure, which arises if we want ABM if they have it but not if they do not, and equally they may want it if we have it but not if we do not. If we decide first, we determine their choice. If we have it, they will; if we do not, they will not. If we know that their interest corresponds to ours, we know that if we do not, they will not; that if we do, they will. Since we prefer not to have it if they do not and to have it if they will, we decide whether we would rather both have it or both not have it, and if the hypothesis is that arms control in this circumstance makes sense, then we decide not to have it. That is sufficient to determine that they do not have it. At least it is if they know our decision, if they can observe our action and if they can respond to it. In this case it does not matter who goes first. Whoever commits himself not to have it—if he can do it visibly and reliably—takes care of the other person's decision. Notice that in this case there is no incentive to violate or cheat; as long as he does not, we do not want it.

There may be a lot of instances of this, and it is possible that during some periods of the 1960s, to the extent that governments can be said to have had preferences or interest structures rather than simply a multitude of agency and individual positions, this was the situation with respect to ABM. At least one can find endless testimony before American

Congressional Committees to the effect that ABM might not be worth having if the Soviets did not have it, but if they have it for a variety of reasons, usually vague psychological reasons, we cannot not have it. I would guess that there are any number of exotic weapons systems that have the characteristic that if they have it we cannot afford not to also for very ill-specified reasons, such as that somebody will think we do not know how. If they have a fractional orbital ballistic system (FOBS), we have to have a FOBS. If they have a nuclear-powered aeroplane, we have to have a nuclear-powered aeroplane. If they build an aeroplane that flies sideways, we have to have an aeroplane that flies sideways, just in case somebody will think that we are lying down on the job or do not know how to build one or will think that they know something we do not.

We can combine those two interest structures by supposing that we want it whether or not they do and they want it only if we do. In other words, if our interest structure is of the first kind I mentioned, and if their interest structure is of the second kind, so that in effect they would follow our lead but that if we had no influence on them we would go ahead and have it, we can determine the situation by voluntarily not having it; they cannot. If they decide not to unconditionally, we would go ahead and have it. We want it even if they do not; they want it only if we do. If we do not perceive that they will respond to us, we will have it and they will have it. On the other hand, if we can abstain unilaterally, we take advantage of the fact that they do not want it unless we do. So we do not, and they do not.

Again in this case if they are not afraid that we will surprise them by changing our mind, if they are not afraid that we can do it clandestinely and they will not know it, if it is all open and above-board, there is no need for any enforcement. The abstention is enforced on us by the knowledge that if we buy it, they will. Compared with both having it we prefer that both do not. As long as we understand that their reaction will be not to have it if we do not have it, we can unilaterally have the agreement, have the mutually observed restraint. Here again one can say this is only getting what we both want, but of course what they want depends on whether we have it. What we want is the opposite of what we got in terms of wanting it independently for our side, but the joint outcome, both not having it, we prefer. So in a sense viewing the two decisions of the two sides jointly, we get the best we can get. But we would still rather have it. That is, we are still abstaining from something that we would proceed with except for the notion that if we did they would too.

There are a lot more cases, and one can pair the motives on both sides. Then there are the interesting cases in which one does not quite know what the other side's interest is. For example, suppose that under no circumstances would we build ABM. We have decided that it is not worth the money, that it will scare the population, accelerate the arms race, give aid and comfort to the militarists and we are against it. Suppose that on the other side they like ABM, they would like it if we do not have it, but they might be willing to do without it if that were the only way to get us to do without it. If they know our motives, they know that no matter what they do we will not waste money on the contraption. They may freely go ahead and have their ABM because they want it, and it looks as though there is nothing we can do about it. On the other hand, if we pretend successfully that we would want it if they had it, then to keep us from having it they may have to abstain. This is one version of the bargaining chip notion.

There is the particularly interesting case, and this probably does sometimes arise—it could, for example, have arisen with ABM—in which each side believes the other wants it but neither side does. Each hopes to negotiate a limitation. Each feels that it has nothing to negotiate with unless it is believed ready to go ahead with it. Both sides simulate strong interest, both sides negotiate arduously. They finally reach an agreement. Each feels satisfaction in having successfully forestalled the other's pursuit of, let us say, ABM. But if they had only known to begin with that both sides were posturing with respect to their interest in ABM, they could have relaxed; they would not have needed an agreement.

Another important interest structure deserves mention. Suppose it is the case, as it may be, that with respect to something like ABM the other side prefers both that we do have it and if we do not, have it. In this case it looks as though there is no possible basis on which to appeal to them for an agreement. Suppose we prefer not to have it or at least not to have it if they do not. If the only item on the agenda is ABM, there is nothing we can offer. If we threaten to build it if they do, they would still do it. If we offer not to build it if they do not, they would still do it. On the other hand, if we can find another item, a new bomber, let us say, which has the same interest structure in reverse, namely we want it whether or not they have it, they do not want it, but we prefer both to have it to neither having it because we want it so badly. There again there is no basis for agreement on the bomber. But if one couples the two together, it may be that while we want the bomber badly enough to accept whatever they do about bombers, we more strongly prefer that they not have ABM; and it is possible that though they would prefer ABM independently of what we did about ABM, they even more strongly prefer that we should not have bombers. If the two are put on the same agenda together, one can then find a basis for both agreeing on neither having ABM nor bombers, which is a state preferred by both parties to what they would get if the two were kept on separate agendas and in each case no agreement was reached.

There is yet another possibility. Let us suppose that we both prefer no ABM and that if it was an independent item on the agenda it would be easy to agree. On the other hand, we want that bomber very badly, and in the discussion restricted to bombers there is no way that the other side can appeal to us to have a restriction. We do not care that much whether or not they have a bomber. They can still say no ABM, no bomber. No bomber treaty, no ABM treaty. In other words, they can couple together not two things that are asymmetrical—we want one, they want the other—but two things, one of which is symmetrical—we both want it—the other asymmetrical—only one wants it. And if we want badly enough the thing that we both want, maybe they can hold us up and obtain a bomber treaty by saying they will not go along with the ABM treaty.

Then we say that if you have a perfectly good ABM treaty, why jeopardize it by tying it together with, let us say, an offensive missile agreement or with anything else. I think the answer there is that one calculates one's interests, and the other party's interests, one thinks about bargaining tactics, one runs the risk and then one says that this is the price of that. One can even say that this ought to have been the American policy over the ABM treaty, namely that we should have held out for zero, which is precisely the same as saying that we should have held out for a better offensive weapon agreement.

That is to say, in the abstract it is holding up an agreement acceptable on its own merits, mutually desirable, perfectly satisfactory, as leverage on the other side to go along with another agreement that may be harder to get, either because he does not want it or because it is simply too hard to negotiate. And indeed if we were more careful about our language, we might say that we got a zero agreement, a perfect zero agreement, on ABM population defence. But the real question is whether we ought to have held it up in order to get zero agreement on ABM command centre defence and whether we would then have wished to hold that one up in order to get zero agreement on ABM strategic forces defence. It may be a very correct judgement to answer in the affirmative but I think it is important to realize that in principle one is simply saying we have an agreement, and by a certain definition it is a zero ABM defence, or zero defence of particular well-specified important kinds of targets, and whether one holds it up to get an exchange of ballerinas, a wheat sale, a vote in the United Nations or zero ABM defence of Moscow and Washington, the fact is one is using it as what one might call a bargaining chip.

Indeed, the term "bargaining chip," which we use so often, ought also to be written down very carefully in order to see precisely what is meant by it. One kind of bargaining chip, and I attribute this to Henry Kissinger because he is quoted as having used it with respect to ABM and, I believe even more recently, with respect perhaps to Trident systems, is either to pretend one wants or commit oneself to get a weapon one does not really want in order to trade it away. This incidentally is a tactic that has a long tradition, not entirely successful, in tariff bargaining. It has always been customary in most governments, at least in most Western governments, that before going off to a big tariff negotiation one makes sure that all possible tariffs are on the books and will go into effect unless a general agreement on tariffs and trade is reached. One does not go empty-handed into a tariff negotiation with no tariff on chickens, tobacco or automobiles. One at least goes through the motions of getting them.

This is what is supposed to be the bargaining chip approach to at least the nominal financing of ABM, and it takes two forms. One is pretence and the other is commitment. If it was pretence in the United States, it probably fooled nobody. One cannot tell a hundred Senators, who will blab it to the newspapers, that we would not touch ABM with a ten-foot pole but we would like to make the Russians think we are thirsting for it so that they will come to the bargaining table and trade it away, because the Russians can read the same newspapers I read, and if I get the impression that ABM is the last thing Henry Kissinger wants but he is pretending to go ahead to fool the Russians, the Russians can also get the impression that he is pretending only to fool them, and they are not fooled.

The other form, though, is more deadly serious, and it is to commit oneself to go ahead even though one would rather not. One says, I am not going to kid you that I want this; I do not. But since I will have no leverage on you unless I get it, I will pay for the leverage by running the risk that if the negotiation fails I am stuck with it. I will spend the money on something I do not want. And then if the other side calls our bluff and insists that if the negotiation fails we will not spend the money, the answer may be yes we will, we have ways of making ourselves spend the money.

There are legal legislative ways, and there is the more dangerous way of generating such elaborate domestic expectations that we would do it, so much confusion as to whether this was a bargaining tactic or a genuine decision to go ahead if the other side did, that we end by having cultivated a desire for it by the bargaining tactic itself. Cultivating a desire is a metaphor; what I particularly have in mind is that we probably increase within the government the bargaining power of those who want it by going through the motions of pretending we do so. In short the likelihood that we feel obliged, the likelihood that we will reach a decision to go ahead, is increased merely by going through the otherwise transparent bargaining chip technique.

The second kind of bargaining chip is one referred to earlier, namely the acceptable agreement that is made part of the bargain. We can either commit ourselves to get ABM as a bargaining chip to reach an agreement or we can commit ourselves to refuse an ABM agreement in order to get an offensive weapons agreement. I mention this because it seems to me that a very common kind of bargaining chip is the one in which one takes an otherwise acceptable agreement and puts it at risk in order to couple it to something else, which is not altogether different from the bargaining chip which takes the form of committing oneself to buy a weapon. And indeed if we think of the same weapon involved in both cases, one is the bargaining chip that says we will get ABM unless something, and the other says we will not have an ABM agreement and therefore get ABM unless something else happens.

The reason why I have presented the foregoing analysis in such detail is that I often find it convenient when I try to think about what we are doing with ABM, submarines, bombers or whatever it may be, to see if I can identify not just for the two parties but for different elements perhaps within the American Government how they rank their different outcomes. In the ABM case there are four outcomes: we both have it, neither has it, they have it and we do not, we have it and they do not. Just to try to rank these outcomes in order to see what kind of bargaining situation we have, what the motivational structure is and then to superimpose on top of that what the other side probably perceives our motivational structure to be. One can go a little further: what they think we perceive theirs to be.

At some stage one has to stop the gyrations of listing what each misperceives the other's misperceptions to be, but I think in the case of, say, a test ban of multiple independently targetable re-entry vehicle (MIRV), a lot of confusion in the American Government in 1969 may have been due to the fact that nobody was quite clear on what they meant by we want or do not want an MIRV test ban, they want or do not want an MIRV test ban. I even suspect that there was even a brief fleeting moment when both the Soviet Union and the American Governments each thought the other would insist on an MIRV ban of some sort, and I think possibly they both thought wrong. There is an interesting case where one could go quite astray even in characterizing the nature of the bargaining situation unless one had straight whether they really want an MIRV ban or they want it only because they think we want it, or they are pretending to want it, and we are pretending to want it.

I mentioned several different cases, and an interesting question arises with each of them: does any limitation that may be arrived at in any of these interest structures depend on some kind of enforceable agreement? Here a lot hinges on what we mean by enforcement. Most agreements are enforced by reciprocity. I agree not to burn rubbish in my backyard if you will not burn rubbish in your backyard. I am free to burn rubbish in my backyard anytime I want to, but I expect that you will begin to burn rubbish in your backyard, and we will end up each creating a nuisance for the other that we agreed not to do.

Here I think we should distinguish two important things. One is means of observation. With some interest structures there is indeed a motive to cheat. That is to say, you like the agreement, a no ABM agreement, but if you could secretly have it, you would like that even better. Once the Soviet Union and the United States really became interested in a test ban, if they ever did, I think this was probably the result of the excruciating and frustrating dilemma they were in, namely, both might wish it were impossible to test secretly and lament that it is possible and be unable to reach an agreement merely because it is possible to hide tests. Yet if it were impossible to hide tests, impossible not only to hide them but to make them ambiguous; if testing were a well-defined notion as it comes pretty close to being with nuclear explosives, and if testing could not be hidden, then probably a simple moratorium is as durable as a treaty signed in blood by the heads of state.

The interesting question thus arises as to why then they bother with treaties. One answer is that it does not occur to people to do without the treaties: the treaties look good, and they are what diplomats consider to be the money in the game. Enforcement in most of these cases, in important cases at least, is almost solely by reciprocity. Usually there are three broad types of enforcement with agreements. One is based on the analogy of criminal law in which a violation or an unauthorized withdrawal, a renunciation of the treaty, is a bad act to be punished. The second is based on the analogy of civil law in which it is not a bad act but it does open one to a suit for damages; that is to say, one must suitably reimburse the other, and to the extent that he can prove harm, take care of him. The third is that one owes him nothing, and is susceptible to no pain or punishment, but he is released from his obligation to one. Not all agreements involve reciprocity of release, but most international agreements do. That is, it is typically construed that if one ratifies a treaty swearing one will never do something, elsewhere in the treaty this is made conditional on the other party's living up to the treaty. Hence an act of violation is in effect a denunciation of the treaty or the agreement by the other party, one's own part of the agreement ceases to exist, and the enforcement on him then is merely the expectation that one will feel free to reciprocate.

If non-compliance can be identified, why should a treaty have anything more than instantaneous duration? That is, why should not the treaty say we hereby inform you that until further notice we will not build ABM, and then set up a teletype system so that it never takes more than five minutes to change one's mind. One could even say that if we change our mind we will let you know within a week rather than beforehand. In terms of the construction time of ABM, a week before, or a week after is not important. Why should one say 5 years, 25 years, perpetuity?

Aside from tradition, a very important reason is to get the subject off the agenda, and not to have to discuss it tomorrow and the day after and the next year. Very specifically, this means that nobody has to buy the same agreement twice. If you rent my house, you may want to sign a two-year lease, one reason being that if I ask a certain price for the house wondering whether you will pay it and you accept it, I am not supposed then to raise the price. In other words, once I know that you will accept a treaty, then I may believe that you would accept a treaty slightly less favourable to you, so I may want to back out tomorrow and try again for a higher gain. I think one of the purposes of the treaty is to say no, we settled that. If we have an agreement, this is the agreement. You can break the agreement any time you want to, but the agreement is not automatically open for renegotiation; it is settled. You can break the lease and leave my house, but you cannot argue about lower rent once you are in. Maybe I can throw you out, but as long as you stay I cannot ask for more rent. One of the reasons for a treaty is simply to say that settles it, we do not bargain about it anymore; we might wish we had charged a higher price for something we gave away, but there is no chance. In other words, it puts the negotiation beyond easy reach so that what the duration of the treaty really determines is not how long the limit lasts but how long the limit on negotiation lasts, the limit on renegotiation. The duration of the treaty specifies for how long it is in poor taste to suggest renegotiation.

A possible other reason, somewhat related, is that some agreements are better if they are written down. Once one has worked out an agreement in detail it is a shame not to record the details one thinks one has agreed on. One may want to do this merely by an exchange of understandings, but if one does they tend to get printed on both sides and treated virtually as official interpretations of the agreement, so a second purpose of a treaty may simply be to produce a document, and somehow diplomats do not like ephemeral documents, that is, documents with a date which means that when tomorrow comes they cease to be enforced. They like to make them last. Again, if negotiating details is arduous, difficult, even risky in terms of success, it may be nice to put beyond reach the amendment and renegotiation of the terms themselves.

Still another purpose of a treaty, which has little to do with the agreement itself, is to generate expectations about either the subject of the treaty or other subjects. If the ABM Treaty lasted 365 days, more people would be plotting next year's campaign to get the Senate not to ratify the new treaty. There is something about the formality of a treaty, including making people stand up and vote, twisting their arms if necessary, something even about making it costly to have a treaty, costly in legislative time, that makes a lot of people who otherwise would work to undermine the treaty give up the attempt. This works on one's own populace, it works on agencies of one's own government, and it probably works on allies. Again the treaty is a way of tying your own hands visibly so that others won't importune you to change your mind or attempt to extort from you a change in position through blackmail of some sort, and sometimes it ties the hands of others, or at least seems to. Even if internationally the rules of the game are that when matters of vital interest are at stake the treaty can be forgotten, domestically the existence of a treaty can have powerful effects and inhibit the kind of political action that might otherwise either undermine the treaty or cause various other kinds of nuisance.

An example concerns the partial Test Ban Treaty (TBT). It has been claimed that the TBT is an anti-pollution agreement and as such is working well. Somewhat facetiously I would say no, it is an anti-noise agreement: the purpose was to stop all of that endless chatter about test bans which for six years had kept arms control negotiations from getting anywhere. I also think the test ban negotiations were enormously mischievous in getting at least the United States and probably the Soviet Union officially on record with hypocritical statements of all kinds. It was an inflammatory popular issue, and what the TBT did was to stop test ban negotiations in the best way. I was not one who thought this was a first step; I thought this was a very final terminal step and now we clear the boards and start all over again. I did not think the test ban led directly to anything more: in a literal sense it puts nuclear weapons underground, off the front page. One rarely, since 1963, in the United States has seen in a newspaper a photograph of a mushroom cloud, whereas prior to the TBT a week hardly went by that one did not see one of those traditional photographs of nuclear explosions.

What was the test ban about? In one literal sense the concern was all about fall-out, partly because fall-out became a very simple, human, domestic, homely issue, and a typical housewife or husband, who had no great thoughts about world strategy, could nevertheless be alarmed about the milk he drank or fed their children, that is to say, a sufficient argument to many people against this awful spectre of nuclear warfare was that the testing was bad for children. My personal experience with people who were not in the strategy business was that they almost all thought that somehow nuclear testing was magically associated with the arms race, that it was the only issue of arms control up for decision, and that therefore the best way to put the ogre back in the bottle was to put testing underground or out of existence.

Something else even more important happened with the test ban. Notice that we did not have a ban on the testing of incendiary weapons; we did not have a ban on the testing of means of delivery. The ban was on nuclear weapons, and primarily it said that nuclear weapons are ugly, inhumane and not like other weapons—they must go underground. The implication was not only do not test them but do not use them or even more do not have them. It was like sending lepers off to a colony; we do not do it with tuberculosis or smallpox but because traditionally leprosy has been considered cursed in spite of the fact that the epidemiology of it makes it comparatively innocuous.

The successful effort in the test ban was to continue to put the curse on nuclear weapons, and in that respect it was very ceremonial, very symbolic, and if somebody had discovered a way to eliminate all fall-out from nuclear tests, my guess is that most people would still have wanted a test ban as an almost magical substitute for banning weapons, and that as an expressive act it had a very powerful effect. The willingness of any head of government to use nuclear weapons was reduced by the test ban; perhaps even the willingness of certain governments to possess nuclear weapons was at least temporarily reduced. A slight reaction may set in; once the testing goes underground and the weapons become invisible, the milk gets clean, the Geiger counters stop clicking, people may stop worrying, and it may become possible to talk rationally and coolly about nuclear weapons again, just as when the danger of nuclear war seems to recede people lose interest in arms control all over the world. It may be therefore that we should arrange a few dramatic limited nuclear test accidents just to remind people that, as with smallpox, maybe we had better go on vaccinating because the risk of an epidemic is just around the corner. After all, who would ever bother to vaccinate if no one has ever heard of a case of smallpox in the last 50 years?

I have raised a question, why have an agreement? I think there is a lot to be said for unwritten agreements, for understandings arrived at but not written down, for letting the other side know what would and what would not concern one and what one would be prepared to do on certain conditions, but to leave it slightly vague. A reason occasionally for leaving it vague is that if each party wants the agreement badly enough, and if the terms are not written down in fine print, each may bend over backwards to avoid seeming to take advantage of the other or seeming to cheat or violate. One finds this, let us say, in relations between neighbours. If one has an understanding about who mows the lawn up to what line or who shovels snow up to what point or who parks his car where, if one writes it out in fine print one's neighbourly relations become like commercial relations and one does everything right up to the line and one haggles if one thinks he has violated the agreement even slightly. If one leaves it all on a matter of gentlemanly good behaviour, everybody bends over backwards not to seem to be taking advantage of the other or to avoid any possible clash, and very often if there is a vague borderline where genuine altercation may occur, both parties may stay behind the vague borderline to avoid altercation. This often will not work if the subject is so complicated that one has to negotiate out details, but I think one of the reasons why a zero agreement is often so much more viable than any other kind of agreement is that it is typically the one kind that does not have to be written down, and often if one cannot write it down one gets zero; if one can write it down, one gets zero with 99 footnotes of small exceptions.

There is a great advantage in the Strategic Arms Limitation Talks (SALT) in two respects. First, particularly between the United States and the Soviet Union serious non-posturing conversation between high government officials concerned with arms control was exceedingly difficult until very recently, is still very difficult and will go on being at least difficult for a long time. Just learning how to converse about these subjects, which is partly learning vocabulary, partly learning concepts, is important. It is equally important to get some appreciation of how the other side values the things you do. It sometimes turns out that the things that concern him most of all are things that one may not think matter terribly much, in which case one can do him lots of favours cheaply. But it may be awfully hard to find that out, and possibly SALT talks, if they go on long enough, will even develop techniques for better discovering the values that each side places on these things. I have to confess that I am on record as not minding if no agreement ever came out of the SALT talks, partly in the belief that the talks themselves were so important that an effort to get written agreements would be disruptive. I now think that was wrong; I think the effort to get an agreement was not altogether disruptive, although there are people who think that the accelerated effort to combine it with a summit visit may have somewhat impeded better agreements than we might have got from SALT I. But most of the interest structures that one can identify will lead to fairly self-enforcing understandings without treaties if there can be a sufficient exploration of where the common interest lies. One might get a lot less arms control as posturing if what one has is private understandings, non-committal understandings, arrived at between two sides, non-binding understandings with merely better appreciation on each side of what the other's behaviour would be.

I think that since about 1967 it has become clearer that either side probably had it in its power to influence the other to take it easy on ABM by simply standing pat himself. There was a kind of implicit moratorium, somewhat disturbed by the bargaining chip notion, which may have been, or could have been, exceedingly mischievous if the agreement had not been arrived at. But the treaty probably was not required for the no ABM; all that might have been required would have been for Gerard Smith and his counterparts to look each other straight in the eye and say, I cannot bind my government, but for the time being we really would not have any interest unless we thought you were going ahead. In that respect talks on matters that can be monitored and observed may often be as good as and frequently superior to treaties.

On the other hand, I have to confess error in that I believe now that the enormous advantage of the ABM Treaty, perhaps like the enormous significance of Nixon's visit to China, is not that it binds the Russians but that it binds the Americans, that it settles an issue internally in a way that it could not have been settled by informal understandings we might have reached with the Russians, and since ABM to Americans became a symbolic issue of the arms race, of militarism and all of that, I think it is spectacularly good to have the President of the United States indicating that trading with the enemy is not wrong, but that it is the most important kind of trading a country can do. In that respect we have done far more than to settle for the time being the ABM issue in the United States; we have established in a way the legitimacy and the propriety of arms control in a way that it had not been established before.

That is to say, SALT is the first genuine recognized effort to try to use an agreement to save money, to slow down the arms race and to make things a little safer. General complete disarmament did not involve any of that; nor was the test ban very much perceived to involve that. An enormous amount of arms negotiation as posturing has discredited arms control all over. But here was a case in which it looked as though a rather stingy Secretary of Defense, Robert McNamara, in effect said I do not want ABM, but if they have it we are bound to have it; let us try not to; and in five years with a treaty succeeded. This probably establishes, particularly with the politically conservative people in the United States, what had already been established with a great many military officers in the United States over the past ten years, namely a belief that really one of the best ways to disarm your enemy is to negotiate.

Thomas Schelling (1921–2016) was an economist and former Professor of Foreign Policy, National Security, Nuclear Strategy and Arms Control at the School of Public Policy at the University of Maryland, College Park. Schelling previously taught as Professor of Political Economy at Harvard Kennedy School. He was awarded the 2005 Nobel Memorial Prize in Economic Sciences for "having enhanced our understanding of conflict and cooperation through game-theory analysis." He served with the Marshall Plan in Europe, the White House and the Executive Office of the US President.

CHAPTER 6

The Fallacy of Thinking Conventionally about Nuclear Weapons

Hans J. Morgenthau

It is unsound to think in conventional terms about nuclear problems and, more particularly, about nuclear disarmament. But what is obvious to people reflecting theoretically about certain issues of the contemporary world is not necessarily obvious to the policymakers. In other words, there exists a profound and wide gap between, on the one hand, our traditional modes of thought and action, and, on the other hand, the objective conditions under which we live.

The availability of nuclear power, more particularly in the form of nuclear weapons, has ushered in a new period of history, which is at least as different from all of recorded history until 1945 than are, say, the Middle Ages from the ancient world or modern times from the Middle Ages. The very conceptions of nuclear "weapon" and of nuclear "war" are misnomers. For when we speak of weapons, we have in mind a rational relationship between a means, an instrument and an end. That is to say, we can use a gun to kill a man, we can use a cannon to breach a wall, and if we have set our mind upon killing a man or breaching a wall, then the use

Originally published in David Carlton and Carlo Schaerf, eds., *Arms Control and Technological Innovation* (London: Croom Helm, 1977): 256–264.

H.J. Morgenthau (✉)
City University of New York, New York, NY, USA

P. Foradori et al. (eds.), *Arms Control and Disarmament*,
https://doi.org/10.1007/978-3-319-62259-0_6

of a gun or of a cannon is a perfectly rational means to a rational end. The same is true of conventional violence in the collective sense, that is war. War, in the conventional sense, is a perfectly legitimate instrument of national policy in a society, which is composed of sovereign nations, that is to say, of nations which have no secular superior above them, which cannot be forced to do something, which cannot be compelled to engage in certain behaviour, by legitimate superior authority.

Violence, for better or for worse, its threats or actual application, is the inevitable result of the anarchic character of a society composed of sovereign nations. Thus it was from the beginning of history to 1945 perfectly legitimate, perfectly rational to use the threat or the actuality of war for the purpose of defending or promoting the interests of individual nations.

All this has been radically changed through the impact of the availability of nuclear weapons. For a nuclear weapon is not a weapon in the conventional semantic sense. It is not a rational means to a rational end. It is an instrument of unlimited, universal destruction; hence the threat or the actuality of a nuclear war is not a rational instrument of national policy because it is an instrument of suicide and genocide. It is exactly for this reason that for more than a quarter of a century the two major nuclear powers have been extremely careful not to come too close to the brink of nuclear war, both being fully aware, at least in a general philosophic sense, that nuclear war is a self-defeating absurdity.

However, from the beginning of history to 1945, when mankind thought naturally in pre-nuclear terms, it developed certain conceptions about weapons and war, which have not yielded in the minds of certain theoreticians, or even in the minds of practitioners, when they have time to think in theoretical terms, to the impact of an entirely novel phenomenon, the availability of nuclear weapons and of what we call euphemistically a nuclear war. So we have a disjunction between the conventional ways we think and act about nuclear weapons and the objective conditions, under which the availability of nuclear weapons forces us to live.

Let me give a simple example of this disjunction from recent history. One of my former students, who has reached a kind of eminence, was for a considerable period of time one of the leading members of the Central Intelligence Agency (CIA). One of his tasks was to brief the Joint Chiefs of Staff about the basic issues which arose in the foreign and military policy of the United States. This official said to me that when he talked to General X about the difference between conventional and nuclear weapons, the latter said, of course, obviously, but when the former read this general's position papers, there was no trace of that recognition in them.

There is a psychological and sociological problem here, which is of the most crucial importance for the future of humanity, that is our seeming inability thus far to adjust our conventional modes of thought and action to the objective conditions which the nuclear age imposes upon us. This is particularly true when consideration is given to the approaches to a nuclear strategy, which have followed each other in the last 25 years in the United States. From the clean H-bomb through graduated deterrence, to the counter-force strategy of ex-Secretary of Defense, James Schlesinger, there is one impulse tying those different strategies together: to find a way by which a nuclear war can be fought in a conventional way, that is, to conventionalise nuclear war in order to be able to come out of it alive. In other words, there is what I would regard as an absurd attempt, not to adapt our modes of thought and action to the new objective conditions of the nuclear age but to transform those objective conditions in the light of the pre-nuclear modes of thought and action.

We have tried, then, instead of adapting our modes of thought and action to the objective conditions of the nuclear age, to conventionalise nuclear war in order to be able to fight and win it and to come out of it alive. And this is true not only with regard to military strategy but also with regard to disarmament and the attempts to develop a defence against nuclear war. For it is one of the characteristics of the nature of nuclear weapons that their destructiveness is so enormous that it has simply destroyed, disintegrated like an atomic bomb, the very conceptions from the beginning of history to the beginning of the nuclear age.

Let us very briefly consider the different attempts at a new strategy, which would allow us to use nuclear weapons without the universal, uncontrolled destructive effects, which, in theory, we correctly associate with nuclear weapons. Take the so-called clean H-bomb, which made its appearance at the beginning of the 1950s, that is an H-bomb that would not have the devastating indiscriminate effects which even the kiloton bombs dropped on Hiroshima and Nagasaki had. It was a bomb which would have very little if any fall-out and whose effects would be those of a gigantic conventional bomb. In the immortal words of a former Chief of Staff of the American Air Force, Gen. Curtis LeMay, the nuclear bomb is just another bomb. The American Atomic Energy Commission in a book entitled *The Effects of Nuclear Weapons*, published in 1957,[1] made short shrift of this idea when it said that there is no such things as a clean

H-bomb, and that all H-bombs are more or less unclean even though the distribution of blast, fire and radiation effects may be different in different designs of the bomb. But the idea that it is possible to devise an H-bomb, which is not essentially different from a conventional bomb, is utterly mistaken.

Take the conception of graduated deterrence, that is a method of waging nuclear war, which does not escalate almost immediately into all-out war, but in which in a rational, almost predetermined way, similar to a chess game, one side makes a move by, say, taking out one city, and the other side makes another move taking out a city of its opponent. Thus each side in a perfectly detached, rational way inflicts a certain degree of damage upon the other. It is this idea, which has gained wide acceptance in certain think-tanks where this kind of playing games with survival issues is highly developed.

There is something to be said in favour of them in terms of the hypothetical possibilities that exist. However, in practical terms, it is inconceivable that living human beings, with the ideological conceptions and values which the policymakers in the Soviet Union, the United States and China possess, would look at, say, the destruction of Chicago by the Soviet Union or at the destruction of Minsk by the United States with the same detachment with which chess players would look at the exchange of pawns. They are bound to arrive very quickly at a point at which, aside from the aroused emotions, one side or the other or both sides will feel that, in this rational simulation of a chess game, one or the other side will take advantage of the other, that is the equivalence, which is theoretically assumed between Minsk and Chicago, will not be self-evident to the players of the game. The Soviet Union will inevitably find that Minsk is more important than Chicago and the United States will find that Chicago is more important than Minsk. Thus they will find that this type of graduated deterrence is really not deterrence at all, because it leads inevitably by its own dynamism to escalation and to an all-out strategic war, which it was the first purpose of the enterprise to avoid. For once the United States has arrived at the conclusion that Chicago is more important than Minsk, it will take out two Soviet cities, which are regarded as the equivalent of another American city, whereupon the Soviet Union will take out two American cities, which are regarded as the equivalent of one Soviet city and before we know it, we shall be in the middle of the all-out nuclear war which we wanted to avoid in the first place.

Furthermore, we have to reckon with the emotions, the passions of the people at large and of the policymakers, for in such an undertaking started in the rational way I have indicated, enormous powers of passion are of course involved on both sides. The population of the United States will not look with equanimity at the successive, however rational, elimination or partial destruction of American cities and their inhabitants, nor will the people and the government of the Soviet Union, and hence again we have a force which almost inevitably will lead to escalation and to the various effects which the graduated deterrence was intended to avoid.

Incidentally, those conceptions of limiting nuclear war, of making it possible to wage it without destroying oneself and one's enemy, are peculiarly American and are the result of a humanitarian impulse within the framework of an utterly inhuman enterprise. Since we are confronted with the possibility of nuclear war, we want to make nuclear war as painless as possible, as limited as possible, one might even say, if so grotesque a juxtaposition is allowed, as humane as possible. On the other hand, the official military doctrine of the Soviet Union has never accepted those distinctions. That doctrine assumes that a war, especially a European war, which starts as a conventional or limited nuclear war and whose stakes the belligerents regard as being of prime importance, is bound to escalate into all-out nuclear war. Thus the idea of the firebreak or the pause between either conventional war or limited nuclear war, on the one hand, and all-out nuclear war, on the other, is alien to the military doctrine of the Soviet Union.

We shall next examine the counter-force strategy, whose philosophy the then Secretary of Defense, Robert S. McNamara, explained to an audience at the University of Michigan in 1962.[2] It was an attempt to limit nuclear war, to make it acceptable as an instrument of national policy. The counter-force strategy is very simple; in fact, it assumes that a nuclear war can be waged and ought to be waged not against population and industrial centres, but against strictly military objectives.

The revival of this doctrine in very recent times starts with the same assumption, fortified by the increase in the sophistication of nuclear weapons during the last 12 years. For in 1962 one could well make the case that it was impossible, in view of the character of nuclear weapons, to distinguish in practice strictly between military and civilian objectives, that the indiscriminate and widespread destructiveness of nuclear weapons was so enormous that a nuclear weapon aimed at a military objective was bound to destroy, by virtue of the mere proximity of civilian objectives, the latter as well. That argument is still valid, in my opinion. If the Soviet Union tries to take out the missile sites near Phoenix and Cheyenne, to give only two examples, the Soviet missiles are not likely to be so accurate as to be capable of destroying the missile sites without having any negative effect upon the adjacent population and industrial centres. But recent increases in accuracy may have improved the situation somewhat in this respect, and therefore there is a grain more merit in the recent revival of the counter-force strategy than in its original formulation.

There is, however, a more profound argument against the counter-force strategy, namely its ultimate military purpose. In the case of the American version, which is the main version of the counter-force strategy, the United States will not initiate a nuclear war by a first strike. It will wait, in other words, until the other side has initiated a nuclear war by a first strike and then will attack the military targets, which the other side presents not merely in the form of missile sites, but of the missile themselves. But the first strike has already emptied most or many of the missile sites. So it is necessary, then, to make a distinction—I have been assured that it is possible, even though I cannot see how it can be made—between missile sites, which still contain their missiles and the other missile sites from which the missiles have already departed. Now let us suppose that this distinction can be made, and that this exchange operates as intended. This of course depends on the enemy who started the war with the first strike not having destroyed all of one's missile sites, and on one having a sufficient number of missiles left with which one can destroy the enemy missile sites, which still contain missiles.

Accept all of this and assume that the two belligerents knock out their land-based missiles reciprocally, what have they gained? They are in the same position that they were originally, except that now they have to rely exclusively upon the sea-borne deterrent and perhaps upon the airborne deterrent. So one would have the same distribution of destructive power with the same deterrent effect one had at the beginning of the war with only the difference that the mechanics of deterrent would have changed from land-based to sea-based missiles. Now there are people who say that land-based missiles are obsolete anyhow, and that they should be phased out through the arms control negotiations between the United States and the Soviet Union. If this position is correct, then in the projected counter-force encounter we would simply have engaged in a mutual disarmament enterprise by knocking each other's land-based missiles out and be in the same position as we were before, that is to say, there would be no victor and no vanquished.

This brings us to another point, which is the basic distinction between victory and defeat in war, which is again a distinction that is deeply ingrained in our consciousness, because it has been imposed upon us through millennia of historical experience. Thus the military, in particular, have found it unacceptable both in Korea and Vietnam that a conventional war should not end in the clear-cut victory of that side whose cause is regarded to be just, which is of course one's own side. The same reluctance to give up the distinction between victory and defeat can be noticed in our thinking on nuclear war. The idea that a nuclear war should necessarily end in a stalemate or in the mutual destruction of the belligerents is simply unacceptable to people who have made it their business to prepare for victorious wars. They are in the position of a banker or a business in general, whose purpose in life is to make a profit for his company, and all of a sudden he is faced with the contingency that the best he can hope for is to break even. He will never make a profit and pay a dividend on the stock of his company, which goes against his grain, against his nature. As far as nuclear war is concerned, this is the objective situation we face, a situation which again is utterly different from any situation which any nation has faced in the past: war itself becomes a completely senseless, irrational enterprise in that if it can be limited in terms of counter-force strategy, it will simply end in the same kind of equilibrium with which it started, only that the composition of the forces through which the equilibrium is presented will be different.

Take the concept of defence. It has been axiomatic throughout history that any new weapon will call forth sooner or later a counter-weapon, a defence against it. Let us assume that this axiom is borne out by historical experience. But it is still true that the destructiveness of nuclear weapons is so enormous, is so staggering to the imagination, that it is inconceivable in view of present technology to devise a defence against nuclear weapons. Thus, the abolition, for all practical purposes, of anti-ballistic missiles (ABMs) by the two Strategic Arms Limitation Talks (SALT) negotiations simply recognises an objective fact of nuclear life. But it should again be kept in mind how insistent the attempts were on both sides to find a defence against nuclear weapons, for once we have a defence against nuclear weapons; we have removed the main deterrent against nuclear war. If we can expect to come out of a nuclear war alive, then to wage or not to wage nuclear war becomes simply like the approach to conventional war, a matter of pragmatic, expediential calculation.

Let us now consider tactical nuclear war, which is another attempt to wage a nuclear war, which will not lead to the destruction of both sides, and which can lead to the victory or defeat of one or the other side. The conception of tactical nuclear war, that is to say, of the battlefield use of nuclear weapons, first of all is up against the impossibility of drawing an objective, generally recognised and recognisable line between tactics and strategy in general. The military schools in all countries have debated this question without ever arriving at a conclusion. For the distinction is not so much in the objective situation on the battlefield as in the minds of the military planner or director of military operations. Incidentally, the fact that the Hiroshima and Nagasaki bombs today are classified as tactical nuclear weapons shows how far tactics can be stretched to cover what generally would be regarded as strategy.

What might be intended by one side as a tactical manoeuvre may thus be interpreted by the other side as a strategic move, and the reply of the other side may either be interpreted in tactical or strategic terms by the first side. Since in such situations both sides are inclined to use a worst-case approach to the problems, that is to assume the worst in terms of the intentions of the enemy, the distinction is bound to break down very quickly. This is true not only of nuclear but of conventional war as well. But the problem is aggravated by the nature of nuclear war.

Assume for a moment as a hypothetical case that a conventional war breaks out in Central Europe, in which the United States and the Soviet Union are involved. The United States, which has about 7000 so-called tactical nuclear warheads in Europe, uses some of them against the military objectives presented by the Soviet Union, such as bridges, military concentrations, ammunition dumps and logistic installations. But that may lead to the destruction of, say, certain cities in White Russia. The Soviet Union replies in the same tactical spirit by attacking the Channel ports, Brest, Cherbourg, Le Havre and so forth, in order to inflict upon the Western armies the same tactical disadvantages the Americans have tried to inflict upon the Soviet Union. But since the tactical targets are different in nature, asymmetric in the extreme on both sides, the Americans will ask themselves when they see Le Havre, Brest and Cherbourg going up in flames, what are the Soviets after? Is this tactical or is it strategic? Applying the worst-case interpretation, the Americans take out some of their cities as a reply to their move. The Soviets then reply in kind: if the Americans take out some of our cities, we take out some of theirs. And one morning we wake up, if we wake up at all, and we find that both sides are engaged in an all-out strategic war, not because either side wanted it, but because the objective dynamism of the initial act leaves neither side a choice.

Another rationalistic fallacy is causally connected with the attempt to make a distinction between all-out nuclear war and a civilised nice little nuclear war out of which both sides will come alive, namely the idea that the nature of war, as it appears to the historian in retrospect, is a result of conscious designs of the war-makers. This may sometimes have been the case, but it is by no means typically the case. It is much more likely that you take one step, to quote Goethe's Faust, which you are free to take, yet from the second on you are a slave of the first. That is to say, the consequent action is predetermined by the first step one has taken and one is not able to escape the inner logic, the inner dynamism of the first step.

So it is a naïve, rationalistic illusion to think that the war-makers remain in control of the war. They are in control before they take the first step. Once they have taken the first step, the dynamism of that first step pushes them in the direction, which that first step indicates. Abraham Lincoln said at the end of the American Civil War: "It is sure that I have not controlled events, events have controlled me."[3] So here is a philosophic fallacy, which attaches to the conception of the nature of man as a war-making animal, and that is another factor in the confusion and delusions to which the attempts at developing a rational strategy of nuclear war and nuclear disarmament are exposed.

Take, finally, the problem of nuclear disarmament, which is, because of the nature of nuclear weapons and of nuclear war, fundamentally different from the problem of conventional disarmament. The conventional arms race is indeed an inescapable function of the balance of power. To simplify the situation by only speaking of two nations, we may say that both nations want to maintain an equilibrium between them. But they can never be sure whether they have calculated correctly their own military strength or that of the other side. So they need a certain insurance against miscalculation in their disfavour. In consequence, if the quantity of x would establish and maintain a balance between themselves and the prospective enemy, they must add to x a y, say 10 per cent more, in order to be sure that even if they have made a mistake in their disfavour, the balance is still maintained. The other side, seeing this addition of 10 per cent or y, must add z to its military power in order to make sure that it is not disadvantaged by the increase in the military power of the other side. The other side notes again that the enemy has added z to its power, so they add more and vice versa. So there is a cumulative inevitable increase in military power, which is another term for the conventional arms race.

To stop this arms race by disarmament agreements has proved to be possible only if the underlying political conflicts, which have given rise to the arms race in the first place, have been mitigated or eliminated. When one reviews the history of attempts to secure conventional disarmament from the end of the Napoleonic Wars to the present—there have been scores of them—one realises that there have been only two successes, one temporary, the other permanent: the Washington Treaty for the limitation of naval armaments of 1922 and the disarmament of the American-Canadian frontier, both having been the result of the permanent or temporary elimination of political conflicts. So there is an element of hopelessness in the numerous attempts at conventional disarmament by means of an isolated technical approach.

The situation with regard to nuclear disarmament is, however, utterly different. For the dynamism which characterises the conventional military balance of power policies of nations does not apply to nuclear weapons. When it comes to machine-guns, one can never have enough of them, because there are always many more possible targets available than there are weapons to eliminate the targets. When it comes to nuclear weapons, there exists an optimum, which does not exist with regard to conventional weapons, beyond which to go is utterly irrational. If we are capable of destroying our enemy 10 times over, under the worst of conditions, it becomes utterly irrational to compete with him for the sake of being able to destroy him 15 times over, and our enemy, who is only capable of destroying us 6 times over, is by the same token not inferior at all. This simple and obvious syllogism has not escaped policymakers in theory but it has escaped them in practice, because the impulse to get more and better nuclear weapons has proved to be irresistible.

So the modes of thought and action, which are perfectly appropriate for conventional weapons and for the conventional arms race, have been transferred to the nuclear field, where they are bound to prove, and have already proved to be, to a certain extent, catastrophic. They have proved to be catastrophic in the economic sense, and they are bound to prove to be catastrophic in the very vital sense of the survival of Western civilisation, if not humanity, if the nuclear arms race is not stopped. In theory the Americans have recognised this, as have the Soviets. For this reason we have had the SALT talks. But when politicians get down to business and when they need the approval of their military establishments, they find themselves handicapped in transforming their theoretical insight into practical measures of nuclear arms control and later on disarmament.

We may thus conclude as we began: the issue of nuclear disarmament or at least of arms control is a literally vital issue, not only for the superpowers, not only for their allies, but for humanity. For with proliferation now having started in earnest, there is no doubt in my mind, and I think in the minds of most experts, that a nuclear arms race not limited to two superpowers having responsible governments mortally afraid of each other, but spreading over the whole globe, is bound, sooner or later, to lead to an unspeakable catastrophe. For history shows, if history shows anything, that all nations have been governed at times by fools and knaves, and even a combination of both. That was bad enough before nuclear weapons existed. But imagine a fool or knave or a combination of both in the possession of nuclear weapons, and nuclear war will be unavoidable. So it is the almost inevitable danger of actual nuclear war, inherent in the dynamism of a generalised unlimited nuclear arms race, which makes nuclear arms control and in the end nuclear disarmament a question of life or death for all of us.

Notes

1. Samuel Glasstone, *The Effects of Nuclear Weapons* (Washington, DC: US Atomic Energy Commission, 1957).
2. Robert S. McNamara, "Defense Arrangements of the North Atlantic Community," *Department of State Bulletin* 47 (9 July 1962): 64–70.
3. Abraham Lincoln, "Letter to Albert G. Hodges, 4 April 1864," in *The Collected Works of Abraham Lincoln*, eds. Roy P. Basler, Marion Dolores Pratt and Lloyd A. Dunlap (New Brunswick, NJ: Rutgers University Press, 1953): 88–93.

Hans J. Morgenthau (1904–1980) was one of the major twentieth-century figures in the study of international politics. He taught at the University of Chicago and took a professorial chair at the City University of New York. Morgenthau was a consultant to the US Department of State during the Kennedy and Johnson administrations until he was dismissed, when he began to publicly criticise American policy in Vietnam.

CHAPTER 7

Strategic Arms Limitation and Military-Strategic Concepts

Michail A. Milstein

It is well known that the improvement in Soviet-American relations is of decisive importance for the entire process of improving the international situation. The general atmosphere in the world and the stabilisation of the entire international situation, of course, depend on the evolution of the relations between the two mightiest states, economically and militarily, and on their attitude to the solution of the most important problems of our time and, above all, to the problem of arms limitation.

The problem of strategic arms limitation is, as it was before, the focal problem in Soviet-American relations. First, strategic arms constitute the foundation of a country's military might. Second, these armaments are the main means for waging nuclear warfare and are, as it were, the main means of containing war. Third, any accord between the Soviet Union and the United States on the limitation of and even more so on the reduction of strategic arms increases trust between the two countries, strengthens the stability of relations between them and in so doing favourably influences

Originally published in David Carlton and Carlo Schaerf, eds., *Arms Control and Technological Innovation* (London: Croom Helm, 1977): 198–209.

M.A. Milstein (✉)
US and Canadian Studies Institute, Soviet Union Academy of Sciences, Moscow, Russia

P. Foradori et al. (eds.), *Arms Control and Disarmament*,
https://doi.org/10.1007/978-3-319-62259-0_7

the entire international situation by reducing the threat of war and strengthening international peace. Fourth, accords on strategic arms limitation will facilitate the solution to the problem of limitation of the arms race in other fields and, in particular, can help to solve the problem of limitation of armed forces and armaments in Central Europe. Conversely, the expansion of strategic arms arsenals stimulates the build-up of armaments as a whole. Fifth, the development and stockpiling of new, even more destructive, systems of strategic arms call for considerable military expenditure. Therefore the banning of the development of new, even more destructive strategic arms systems might lead to a reduction in military budgets.

On the whole we may say that it is impossible to think about the removal of the threat of nuclear war and at the same time not to take practical and effective measures on the limitation and, eventually, the reduction of strategic offensive arms. The interests not only of the peoples of the Soviet Union and the United States but also of all the world's nations demand that the Soviet Union and the United States which possess colossal might in nuclear arms spare no effort in finding a solution for this complex problem.

The strategic arms limitation problem is not simply an important but also a very complex and diversified problem, which could not and cannot be solved all at once by any single agreement and which demands, as before, to a certain extent a stage-by-stage solution. That is why the talks on the limitation of strategic arms are conducted in definite phases, which in their turn may be conditionally divided into a number of stages, the completion of which is the signing of corresponding agreements. At present the first phase of these talks has been completed and the sides are conducting the second phase. The first phase was completed by the ratification in October 1972 of the Treaty on the Limitation of Anti-Ballistic Missile (ABM) systems and the approval of the Interim Agreement on Certain Measures with Respect to the Limitation of Strategic Offensive Arms. Later both sides made new and important agreements: the Agreement on the Prevention of Nuclear War of 22 June 1973; the Treaty on the Limitation of Underground Nuclear Weapon Tests; and the Protocol to the Treaty on the Limitation of Anti-Ballistic Missile systems of 3 June 1974.

The whole history of Soviet-American relations has lasted over four decades, but the history of military détente, in the field of strategic arms limitation, is comparatively short—only a few years. The agreements that have been signed are unprecedented in nature. Henry Kissinger, the

American Secretary of State, has written that never before have two important powers divided by ideology, historical traditions and contradictory interests established officially agreed-upon limitations on the development of their principal armaments.

At the same time, if we speak about strategic offensive arms then the Interim Agreement, for all its positive significance, was of a limited nature. It was limited in time—for five years from 1972 to 1977—and, which is the main point, it dealt only with the first two of the three important components of strategic offensive arms, namely inter-continental ballistic missiles (ICBM) and submarine-launched ballistic missiles (SLBM), but not with strategic bombers. Moreover, the Interim Agreement also hardly imposed any limitations on the qualitative development of these arms. Nobody of course ever argued or could argue that the goal of the very first agreement would be to impose limitations on the entire complex of strategic arms and on all of their aspects. Both sides clearly realised that this would constitute the subject of subsequent talks.

The Vladivostok Accord, reached at the summit meeting in November 1974, was thus of great significance for the further successful solution of the strategic arms limitation problem. First, as distinct from the Interim Agreement, it imposed limitations not only on ICBMs and SLBMs but on strategic bombers as well. Second, it established the total number of strategic arms delivery means for both sides and in so doing imposed firm limitations upon a further build-up in their numbers. Third, an important and fundamentally new aspect of the accord was that a limit was established on the numbers of ICBMs and SLBMs equipped with multiple independently targetable re-entry vehicles (MIRV). Limitations have thus been imposed upon one of the most important elements in the qualitative strategic offensive arms race—the development of MIRV missiles. Fourth, the new agreement will be long-term, from October 1977 to 31 December 1985. It will also incorporate the relevant clause from the Interim Agreement of 26 May 1972, which will remain in force until October 1977. The stabilisation imposed upon strategic offensive arms for so long a period will have a favourable effect upon the development of Soviet-American relations and on the entire international situation. Moreover, these limitations will stop the race in the development of strategic offensive arms. For, as experience indicates, development, production and deployment require approximately ten years. Fifth, the signatories also declared their intention to make further progress not only in imposing limitations on strategic offensive weapons but in their reduction as well.

The new agreement, to be drafted on the basis of the principle of equality and equal security, will of course contribute greatly to the improvement of Soviet-American relations, to reducing the threat of war and to strengthening international peace. Naturally many details have to be agreed upon and a number of technical and not only technical problems have to be solved in order to compile the ultimate text of the treaty. And that is essentially what the Soviet and American delegations, having concrete instructions from their governments, are now doing at the talks in Geneva. Any objective analysis of the treaty shows that its realisation would impose serious limitations upon the strategic arms race and would develop good preconditions for reduction in the future.

The Soviet Union proposed not to stop only at the limitation of the already existing means, but to reach accord on the banning of the development of even more destructive systems, such as the Trident submarines and the B-1 strategic bombers in the United States and similar systems in the Soviet Union. The American side, however, rejected this proposal.

On the whole, we can say that the agreements already signed constitute the first steps along the road to strategic arms limitation and prove that the Soviet Union and the United States can reach accord on such complex questions, which deal with their vital interests.

Many specialists and non-specialists—both those who are sincerely interested in ending the arms race and those who are for a build-up in armaments—continue to discuss the significance of these and future agreements, and especially their possible influence upon the strategic arms race and upon the stability of the world situation. What would have happened if the two sides had failed to make these first steps? Can the agreements reached be considered adequate in all respects? Who got more out of these agreements—the Soviet Union or the United States? The need for correct answers to these questions is obvious, the more so if we take into account that the existing agreements and treaties are being attacked now in the United States, mainly by spokesmen of the militarist circles and by those who have vested interests in the arms race, in building up of military efforts and in returning to the former "position of strength" policy.

Certain military, political and research circles in the United States are even of the opinion that the Soviet-American agreements on strategic arms limitation, signed and ratified in 1972, had a very small influence upon the very process of the armaments build-up. Spokesmen of these circles declare that each of the sides continued, in fact, to do whatever it wanted to do in respect to the build-up of strategic arms and did not do anything it would not have been doing if there had been no agreements. We cannot subscribe to this opinion. First, the agreements stopped the race in the most costly and destabilising field of strategic arms—in the field of the ABM systems—and thus strengthened the stability of the situation as a whole. Second, they have halted the build-up in the most modern strategic offensive arms—ICBMs and SLBMs. Third, concrete obligations were undertaken on preventing a nuclear war. Therefore, these were the first concrete and important stops along the road of containing the arms race and reducing the greatest threat which has loomed over mankind in the last decades—the threat of a world thermo-nuclear war. They narrowed down the material base for the nuclear missile arms race. It is obvious that if this accord had not been reached the world would have seen a new escalation of the arms race—which is senseless politically and unimaginably costly economically.

It has been estimated in the United States that if a new agreement on strategic arms limitation had not been signed then the American military budget would have had to be increased by another $11,000,000,000. The Pentagon is planning a whole number of measures, if such should be the case. In such a strategic arms build-up, the following would be included: an increase in the Minuteman-3 missiles from 550 to 800; the development of a new generation of ballistic rockets; and the development of cruise missiles. All this would produce a new spurt in the arms race.

A strategic arms race, as had been already stressed, would have triggered off a chain reaction in the building up of conventional armaments. Moreover, and most important, the threat of a nuclear war would increase and the world would go back from the peaceful coexistence policy to the state of cold war between countries with different social systems and to the senseless arms race.

Are the existing agreements adequate? Of course, the complete liquidation of strategic offensive arms, the banning of production and use of nuclear weapons in general, and universal and complete disarmament would be the ideal state of affairs. Unfortunately, the time has not as yet come for this. But those who are really interested in strengthening security cordially greet any progress towards it. Therefore, the existing agreements and treaties signify a definite headway made for the ending of the arms race and for strengthening peace. The Soviet Union has proposed, in order to abate the war threat even more and in order to create favourable conditions for making progress towards disarmament, a world treaty on the non-use of force in international relations and a treaty on banning the development and production of new types of mass destruction weapons and new systems of such weapons. It will be readily understood that agreements on these questions, together with accords in the field of strategic arms limitation, would be of great importance for strengthening universal peace.

As for the question as to who has got more out of these agreements, the answer to this is very easy. The fact is that when they make these agreements both sides must be strictly guided by the main principle on which they are based—the principle of equality and equal security. Therefore, neither of the sides receives, nor can receive, any advantages as compared with the other side. The agreements are made for mutual benefit, for the benefit of the peoples of both countries. Hence it is the cause of disarmament and strengthening of peace that gets the greatest benefit. And those who declare that the existing agreements provide advantages for any one side are deliberately distorting the truth in order to justify the slowing down of talks or a build-up in armaments and an increase in military spending.

It is impossible to overestimate the significance of the efforts made by the Soviet Union and the United States to limit strategic arms in the name of universal peace and universal and complete disarmament. The time will come when the question will arise whether other nuclear powers should also join the process of strategic arms limitation.

Leonid I. Brezhnev, General Secretary of the Central Committee of the Communist Party of the Soviet Union (CPSU), said at the 25th CPSU Congress when referring to the current Soviet-American negotiations on further strategic arms limitation:

> We are conducting them in an effort to carry out the 1974 Vladivostok Accord and to prevent the opening of a channel for the arms race, which will nullify everything achieved so far. An agreement on this issue would obviously be of very great benefit both for the further development of Soviet-US relations and for building greater mutual confidence, and for the consolidation of world peace.[1]

Such is the Soviet Union's position.

From time to time, one hears responsible quarters in the United States urging accelerated arming, which they justify with reference to the dragging out of the talks with the Soviet Union—a protraction, to put it bluntly, that has occurred through no fault of the Soviet Union whatsoever. Meantime certain influential circles in the United States are trying their best not only to get approval from the Congress for developing new weapon programmes but at the same time to complicate the ongoing talks on limitation of strategic weapons. Their chief method is the same as they used ten and twenty years ago, namely, to frighten the public, the politicians and particularly the Congress with the imaginary "Soviet menace." Thus the so-called Soviet threat is the main motive for the arms race given by its advocates in the United States. In fact, of course, there is no Soviet threat. But fictions concerning the "Soviet menace" have become part and parcel of American politics. And the threats grow, in particular, when the American military representatives are on their way to the Appropriations Committee. This created obstacles even in the early stages of Soviet-American talks. And it creates obstacles to the implementation of the agreement reached at Vladivostok. If things continue in this way all efforts to stop or even to limit the arms race will be endangered. Therefore the task of peace-loving forces is to assist in every possible way the conclusion at the earliest possible date of a successful agreement on the problem of limitation, and thereafter also on the reduction of strategic arms. Those people who are slowing down, under different pretexts, the process of strategic arms limitation are, in fact, against making agreements and are thus opening new channels for the arms race. They should be opposed resolutely and with sound arguments. The struggle to end the arms race and for disarmament must be intensified. Political détente needs to be fortified by military détente.

We should bear in mind that measures aimed at strategic arms limitation, at improving relations between states with different social systems and at strengthening the détente process by measures on military détente will not produce durable results unless they are accompanied also by measures aimed at containment and limitation in the field of military-strategic concepts and military doctrine as a whole. We know that military doctrine and military-strategic concepts are important components in the entire mechanism of military policy and among the more important factors having a substantial influence upon the arms race, both quantitatively and qualitatively. The present military-strategic concepts of the Pentagon provide particular evidence of this.

The present military policy and strategy of the United States and especially the process of its formulation and realisation reflect the complex interaction of the two main trends in the world situation—the dominating tendency towards the development and expansion of détente and the opposite tendency leading to material preparations for war and interference in the affairs of others. It is not by accident that the United States has been reconsidering her military-strategic concepts in the last few years.

The most acute discussion is carried on around the main problem of military strategy and military doctrine—the problem of using nuclear weapons. There is much discussion as to what a nuclear war may be like, what strategic forces the United States must have to wage it and what demands should be placed on these forces, which targets should be hit first and whether theatre nuclear forces could interact with strategic nuclear forces. Essentially the idea is how to be able to utilise their tremendous nuclear weapon arsenal as a means of dealing with international problems and at the same time avoid the threat that the United States could suffer "unacceptable" losses. In other words the discussion centres around the far from new problem—how to make nuclear war "acceptable."

James Schlesinger, the then American Secretary of Defense, declared at the beginning of 1974 that some principles of American military strategy had been changed and that a new strategic concept had been adopted.[2] He said that at present both sides have and would have in the future indestructible means of delivering a retaliatory blow. It was inevitable, therefore, that a strike made by one side upon the cities of the other in the course of a total attack would almost immediately bring a disastrous blow at its own cities. As a result of this, the range of circumstances under which the question of a total strike against the enemy's cities might arise had been considerably narrowed down. The need arose to find "alternative versions" for the use of strategic forces. This apparently signified a "limited strategic nuclear war," in which "it is impossible to make strikes against a large range of targets," but in which strategic forces should be used in such a way so as "to limit the damage to both sides" taking part in the nuclear conflict. Schlesinger also declared that the American strategic nuclear forces were able to strike at the enemy's military targets with adequate yield and precision in order to destroy the chosen targets, but avoiding "accompanying destruction" while so doing, and that this had supposedly provided the Pentagon with the opportunity "to force the enemy, in case a nuclear war starts, not to attack the cities of the United States and its allies."

This concept has officially become an inalienable component of the American military doctrine since 1974. For the changes in personnel in the Pentagon made in autumn 1975 were not accompanied by substantial changes in the military-strategic concepts formulated before that. Judging from the report made to the Congress by Donald Rumsfeld, the New Secretary of Defense, on the American military budget for the 1976–1977 fiscal year,[3] all the main concepts (including "limited strategic war") remain in force. True, more often than not "approximate equality" is mentioned instead of "substantial equality" but this purely semantic difference does not change either the essence or the trend of the Pentagon's approach.

The present Pentagon military concept envisages a spectrum of nuclear wars including one more type of "limited strategic war," which is understood to be a "limited exchange" of strategic nuclear strikes on a small number of military targets. It is assumed that neither side would be interested in expanding the nuclear conflict and would take measures to localise and even maybe to liquidate it. This kind of war is considered to be the most probable one in the conditions of strategic nuclear equality.

The effectiveness of this concept is seen in the context of constant ability to inflict "unacceptable damage" on the other side in any conditions. Therefore, a guaranteed annihilation is not excluded. At the same time it is planned to retarget a part of strategic nuclear forces from cities and industrial projects at military projects, and above all at launching sites and at air force and naval bases. It is thus deemed necessary to improve further the American strategic nuclear forces so that they would be able to hit military targets with greater precision. For the effectiveness of the so-called American nuclear umbrella for the North Atlantic Treaty Organisation (NATO) is being questioned.

The grave threat embodied in this concept is that its advocates regard nuclear weapons as a means of waging warfare and consider a nuclear war quite acceptable because a limited strategic nuclear war is not connected with big risks and therefore may be regarded as something like a conventional war. It is therefore appropriate to mention here that Henry Kissinger himself wrote in 1960, in his book *The Necessity for Choice*,[4] that even though a theoretical model could be designed for a limited nuclear war the fact remains that no such model had yet met with general recognition ever since the beginning of the nuclear age. He said that it would be nearly impossible to get any reasonable definition of what should be understood as a "limited nuclear war" from the American armed forces. He arrived at the conclusion that since arguments about targeting were usually decided by permission being given to each branch of the armed forces to destroy whatever they deemed necessary for performing their tasks, limited nuclear war waged in such a manner could quite well become indistinguishable from a total war.

The acceptance of the Pentagon's new concept may lead to a stepping up of the strategic arms race because the advantage gained in improving the precision of missiles creates a potential for acquiring the capability for delivering a first strike. And the possibility is increased that conventional war would develop quickly into a nuclear war. In short the so-called nuclear threshold is objectively being lowered.

Moreover, it is impossible to distinguish a ballistic missile attack upon military targets from an attack on civilian targets by the same means. But even if we say that such a war is possible, there is the obvious risk that it can very easily develop into a total nuclear war. No wonder that this approach is being subjected in the United States to sharp criticism in the most diverse circles.

A critical view of the current military strategy is held by many sober-minded American politicians, scientists and public personalities, who stress its danger with a feeling of apprehension. So where is the way out? The possibility of using nuclear weapons as a "flexible" and "acceptable" means of foreign policy should be rejected once and for all. It is useless to make attempts at obtaining strategic advantages in the conditions of our time because such attempts fail to bring any political dividends and lead only to more resources being squandered on the arms race. Nuclear war cannot and must not serve as a means of settling international disputes.

The Pentagon leaders also regularly stress that the reliability of "containment" depends, along with other factors, on the existence of the three-sided potential, namely strategic forces, theatre nuclear forces and conventional forces. And the importance of each of the three elements is always pointed out as well as the impossibility of replacing one of them by any other. At present certain changes are being made in this approach—a certain re-stressing of accents, while the essence is being preserved. These changes are being made first of all in the so-called theatre nuclear forces.

It is being stressed that NATO troops may make first use of tactical nuclear weapons in the theatre of hostilities. A Pentagon report indicated that it was possible to forecast hypothetical circumstances under which non-nuclear NATO forces would not be able to stand their ground faced by an attack by non-nuclear forces. Therefore, the report said, it was not possible to exclude the possibility of the first use of tactical nuclear weapons. The report then clarified what these circumstances were. It said that examples were the possibility of the loss of a large tract of territory or a big grouping of NATO troops, in which case the political leadership of NATO would, maybe, decide to take the risk and deliver the first strike.

Thus, the possibility of delivering a first strike with tactical nuclear weapons is now openly emphasised. There are more than 7000 American tactical nuclear weapons in Western Europe. What would happen if only a small proportion of them were used? Here is one of the examples cited in *Defence and Retaliation* written in 1961 by Helmut Schmidt, later Defence Minister and Chancellor in Federal Germany.[5] NATO held a war game under the code name "Carte Blanche" in Western Europe in 1955. Three hundred and thirty-five tactical nuclear weapons were used in the three days of the game of which 265 had theoretically exploded on the territory of Federal Germany and the German Democratic Republic. As a result, between 1,500,000 and 1,700,000 people were "killed" (without taking into account deaths from radioactive fall-out) and 3,500,000 people were injured. As a comparison we may note that 305,000 people were killed and 780,000 people were wounded throughout World War II during raids by the British and American air forces on Germany. Such are the possible results of using only a "small" number of tactical nuclear weapons in Europe. All this sharply aggravates the danger of conventional conflict developing into a nuclear war.

Such are some of the new elements in the military-strategic approach of the United States. These are not of course all the changes that are characteristic of contemporary American military-strategic thinking. The examples cited here were needed only to show that in spite of the fact that the peaceful coexistence principle is meeting with ever-greater recognition in the world and much is being done to normalise relations between countries with different social systems, there still exist in the US influential forces which are opposing this process and strive to preserve the atmosphere of tension in order to justify every new stage in the arms race. These circles more often than not try to justify their ideas by declarations about the so-called Soviet threat. Leonid I. Brezhnev said at the 25th CPSU Congress:

> In fact, of course, there is no Soviet threat either to the West or the East. It is all a monstrous lie from beginning to end. The Soviet Union has not the slightest intention of attacking anyone. The Soviet Union does not need war. The Soviet Union does not increase its military budget, and, far from reducing, is steadily augmenting allocations for improving the people's well-being.[6]

This enables the conclusion to be drawn that the Soviet military doctrine is not aimed at preparing but at the repulsing of an attack. Soviet military doctrine is not aimed at preparing war but at deterring it.

Quite a lot has been done in the field of Soviet-American relations in the last few years to facilitate headway for détente. At the same time there is no doubt that the supplementing of the political détente with a military one presupposes the adoption of not only practical measures on curbing the arms race but also the imposing of limitations in relation to military-strategic concepts which slow down the détente process and, on the contrary, lead to a stepping up of the arms race.

The realities of our time imperiously demand from the Western circles, which elaborate military policy and strategy, that they stop thinking in categories of war in general and of nuclear war in particular. Any attempts made to widen the "usability" of nuclear weapons or to threaten nuclear war of any type contradict the very spirit of détente, the international agreements that have been achieved and the constructive nature of the Soviet-American relations that have been developing in the last few years.[7]

Notes

1. Jessica Smith, David Laibman and Marilyn Bechtel, eds., *Building a New Society: The 25th Congress of the Communist Party of the Soviet Union* (New York: New World Review Publications, 1977).
2. James Schlesinger, "Strategic Forces," *Department of Defense Annual Report to the President and the Congress* (Washington, DC: US Government Printing Office, 4 March 1974).
3. US Department of Defense, *Report of Secretary of Defense Donald H. Rumsfeld to the Congress on the FY1977 Budget and its Implications for the FY1978 Authorization Request and the FY1977–1981 Defense Programs. January 27, 1976.* (Washington, DC: US Government Printing Office, 27 January 1976).
4. Henry A. Kissinger, *The Necessity for Choice: Prospects of American Foreign Policy* (New York: Harper, 1961).
5. Helmut Schmidt, *Defence or Retaliation* (Edinburgh and London: Oliver and Boyd, 1962); German edition published as *Verteidigung oder Vergeltung* (Stuttgart: Degerloch, 1961).
6. Smith, Laibman and Bechtel, *Building a New Society.*
7. The author wishes to emphasise that this paper was prepared before the American presidential election of November 1976. Hence his comments about American policy apply only to the Republican Administration and do not necessarily relate to that of President Carter.

Michail A. Milstein (1910–1992) was a Lieutenant General in the Army of the Soviet Union. He was a senior lecturer at the Department of Foreign Armed Forces of the Soviet General Staff Military Academy and senior research fellow and head of the Military-Political Studies Institute of the US and Canadian Studies Institute of the Soviet Union Academy of Sciences.

PART II

The Hard Times, 1978–1989

Introduction to Part II

The second part of the book covers some of the most dramatic years of the Cold War from the end of the seventies until the fall of the Berlin Wall in 1989. It is a period of worsening relations between the United States and the Soviet Union, with growing tensions and expanding nuclear arsenals, peaking to a total of 70,500 warheads in the mid-eighties.

With the Doomsday Clock dangerously moving close to midnight, the engagement of ISODARCO in non-proliferation and disarmament education focused on the confrontation between the two nuclear-armed superpowers and on the ways the risks of a nuclear Armageddon could, if not be eliminated, at least be reduced.

Two main issues absorbed the energy and time of the community of ISODARCO senior experts and junior participants over more than a decade. The first question, continuing seamlessly from the previous period covered in the first part of the book, has to do again with how to ensure stability and security through negotiation and agreement to reduce nuclear dangers. Lawrence Freedman's chief concern is with the deployment of thousands of sub-strategic nuclear weapons in Europe, and the difficulties in negotiating an arms control agreement for this specific category of nuclear weapons. George Bunn and David Carlton, facing the impossibility of achieving the ideal goal of complete disarmament under the existing circumstances of the time, are interested in how to overcome distrust between the United States and the Soviet Union in order to achieve a successful arms control agreement and doable, gradual measures that could induce restraint upon armament policies.

The second key issue relates to the challenges facing any state attempting to incorporate nuclear weapons into its broader politico-military planning. ISODARCO's commitment to non-proliferation and to the longer-term objective of nuclear disarmament cannot be omitted from the deepest understanding of nuclear doctrines and strategies, and the conditions in which they can face the many security paradoxes, puzzles, unintended consequences, and dilemmas of a nuclear-armed world. The strategy of deterrence obviously received most of the attention and scrutiny, engaging the brightest minds in the attempt to reconcile the possession and planning of possible use of nuclear weapons with the hope that in doing so they will never actually have to use them. The serious problems of credibility of a strategy of extended deterrence in the context of the North Atlantic Treaty Organization (NATO) is discussed by Jane Sharp, while Richard Ullman examines the notion of minimum deterrence and its practical implications in terms of the smallest force possible to provide deterrence against a nuclear attack. The analysis of Cui Liru focuses on the nuclear first-use option from China's perspective and discusses possible ways to overcome the possible catastrophic consequences of this high-risk strategy.

Parallel to these two overarching themes, the interest in the technological dimension of nuclear weapons and policies which characterizes the ISODARCO approach to non-proliferation and disarmament education is present in the opening chapter by Nobel Prize winner Joseph Rotblat. His chapter discusses the radiation produced in the civilian fission fuel cycle, with the consequent health hazards; his analysis also pertains to the military production of fissile materials, being the two cycles identical, and hence reminds us that the very production of the basic material for the nuclear weapons has negative effects on the environment and public health.

The value of the following chapters of the Cold War period to our understanding of nuclear weapons and policies is not just historical; rather, many issues maintain their relevance in the present day. Important lessons can be learned from the US–Soviet Union rivalry that can be applicable to a time when US–NATO and Russia relationships have dramatically worsened and likely hit the lowest point since the end of the Cold War over the Ukraine crisis. The dilemmas and paradoxes of the nuclear strategies of the

Cold War continue to afflict international relations, which have grown, one might argue, even more complex from the simplified patterns of the bipolar system. Finally, the insights on negotiations and arms control agreements of the eighties proved not only instrumental to achieve the successful results of the bilateral reduction treaties agreed after the end of the Cold War, but can teach us a lot even today: that we have entered a new phase where further steps towards a world free of nuclear weapons can be realized.

CHAPTER 8

Radiation Hazards in Fission Fuel Cycles

Joseph Rotblat

Introduction

Just like nuclear weapons, radioactivity is intrinsically linked with nuclear energy and cannot be separated from it. The raw materials, uranium and thorium, are radioactive; the by-products, plutonium and other actinides, are radioactive; the waste products, the fission fragments, are radioactive. Moreover, any substance placed in a reactor is transformed into radioactive material. The amount of radioactivity associated with nuclear power production is truly colossal: a reactor of 1000 megawatt electric power produces in one day of operation the same amount of radioactivity as two bombs of the Nagasaki size. The radioactivity produced in nuclear reactors is indeed of the same origin as in A-bombs, and some of which all of us have in our bodies from the tests of nuclear weapons.

The radioactive nuclides have half-lives varying over an enormous range. For the fission fragments, they range from half a second for

Originally published in a slightly different form in David Carlton and Carlo Schaerf, eds. *The Hazards of the International Energy Crisis: Studies of the Coming Struggle for Energy and Strategic Raw Materials* (London: Macmillan and New York: St. Martin's Press, 1982): 107–126.

J. Rotblat (✉)
St Bartholomew's Hospital, London, UK

P. Foradori et al. (eds.), *Arms Control and Disarmament*,
https://doi.org/10.1007/978-3-319-62259-0_8

germanium-73, to 16 million years for iodine-129. Among the actinides, they range from a few minutes to millions of years. Plutonium-239 has a half-life of 24,400 years. The parent nuclides, uranium-238 and thorium-232, have half-lives of the order of 10^{10} years. The amounts of individual nuclides produced also cover a wide range. Moreover, most radioactive nuclides decaying give rise to new radioactive nuclides, and for some the equilibrium may not be reached for a very long time. For this reason, the overall variation of the radioactivity with time is very complex; it depends on how long the reactor has been in operation and on the type of reactor. The rate of decay is different for different types of radiation emitted. The decay of gamma-rays shows an initial decay by six orders of magnitude in the first 500 years, followed by a slower decay of three orders of magnitude over the next million years. For alpha-particles, the decay is much slower in the early years.

At this stage, mention must be made of the unit of activity. The new unit is the becquerel, which is the activity of a substance in which there is one disintegration per second. However, in all textbooks, the old unit, the curie, is still used, and it will also be used here. The curie is approximately the activity of one gram of radium, but by definition it corresponds to 3.7×10^{10} disintegrations per second. Thus 1 Bq = 2.7×10^{-11} Ci or 27 picocuries. To give an idea of the orders of magnitude, the natural radioactivity of potassium-40 in our bodies is 130 nanocuries (1.3×10^{-7} Ci). In a typical nuclear reactor, the total activity is 20,000 megacuries (2×10^{10} Ci).

As already stated, the radiations emitted are of different types. Some of them, like alpha-particles, are very easily absorbed: a sheet of paper will stop them completely; but if an alpha-emitting substance gets into the body, by inhalation or ingestion, it can deliver a huge dose to the layers of tissue in the immediate vicinity. Beta-rays are more penetrating, but a few millimetres of aluminium, or a centimetre of tissue, will stop them. Gamma-rays are most penetrating and may require many centimetres of lead to bring their intensity down to a low level. In reactors, we also have neutrons, which have penetrating properties similar to those of gamma-rays, but are attenuated fastest in light substances, for example water.

All these radiations carry a great deal of energy, and when they are absorbed in matter, most of the energy is converted into heat. If a large quantity of radioactivity is put in a container, the heat will soon raise the temperature to hundreds or even thousands of degrees and melt the vessel. For this reason, the fuel elements must be cooled continually during storage until they decay considerably.

Exposure to ionising radiation is harmful to living organisms. In large doses, they can kill in a matter of days or minutes. In smaller doses they may induce cancer, sometimes after a very long delay (10–30 years after the exposure) and they may also produce genetic damage which could show up as a defect in the children born to exposed persons, or in future generations. It is now generally accepted that any increase in the dose of radiation, however small, may cause such harmful effects. Therefore, one should avoid being exposed to radiation. Of course, there are many things which we have to avoid if we want to stay alive. Often they cannot be avoided completely, and we have learned to accept certain risks, either because they are associated with greater benefits, or because attempts to reduce a given risk may result in an increase of other risks. A sensible balance between benefits and risks is an important task for society. Sometimes, the balance can be arrived at easily. For example, if you are ill and have to have an x-ray taken, you run a radiation risk, but the x-ray may help to diagnose a serious disorder and save your life. In other cases, it is much more difficult to strike the balance; this is particularly so when the persons who run the risk are not those who reap the benefit, as happens with occupational risks.

Practically every occupation has some hazard attached to it, but some occupations carry much greater and recognisable risks. In such cases, special regulations are needed to reduce the hazards to acceptable levels. In the case of radiation hazards, a comparison of the risk with those in other industries has led to the establishment of dose limits for radiation workers. Similar considerations have led to criteria for the exposure of whole populations. We shall discuss later the rationale of radiation safety regulations and the validity of the dose limits.

First, it is appropriate to explain the units in which we measure the radiation dose. The basic unit is the gray; this is the dose received when the radiation energy absorbed is 1 joule per kilogram of tissue through which is passed. The old unit was the rad, which equals 1/100 of a gray. Different types of radiation, say gamma-rays and neutrons, produce different amounts of biological damage for the same dose. Thus 1 rad of neutrons may give the same biological effect as 10 rads of gamma-rays. To take this into account, we use another quantity, the dose equivalent, which is measured in sieverts; the old unit was the rem, which equals 1/100 of a sievert. One rem is the dose of the given radiation, which produces the same biological effect as one rad of gamma-rays. As mentioned, for neutrons, 1 rem may be equal to 10 rad; for alpha-particles the factor may be 20. Since the gray and the sievert have not yet come into general use, rads and rems will be the terms used in this analysis.

Once again, to convey an idea of the order of magnitude of radiation doses, reference may be made to the natural background to which we have always been exposed, and which therefore is a useful yardstick (see Table 8.1).

As is seen in Table 8.1, the natural background consists of a number of components, the main ones being cosmic radiation, natural radioactivity in the earth's crust and the radioactivity in our bodies, the biggest contributor to the latter being potassium-40. The total natural background varies somewhat for different organs, but for most it is about 80 millirad per year. This is an average value, since the natural background depends on a number of factors, such as locality, type of building and altitude. Indeed, it is the fluctuations in the background that are often used as a criterion for the hazard from additional radiation. The diagnostic use of x-rays in medicine gives, on average, 16 millirad per year, although it is considerably higher in the United States. Tests of nuclear weapons gave an average of 7 millirad per year. The limiting dose to radiation workers is 5 rad, or 5000 millirad per year.

In the nuclear power programme, radiation exposure can occur in any part of the fuel cycle, and we shall discuss each of them in turn. For each part we shall consider the occupational dose, that is, the dose of radiation

Table 8.1 Annual per capita doses from normal exposure to natural sources of radiation (in mrad)

	Gonads	*Whole lung*	*Bone-lining cells*	*Red bone marrow*
External irradiation				
Cosmic rays	28	28	28	28
Terrestrial radiation	32	32	32	32
Internal irradiation				
Potassium-40	15	17	15	27
Radon-222 (with daughters)	0.2	30	0.3	0.3
Other nuclides	2	5.5	9.1	4
Total	78	110	84	92

Source: United Nations Scientific Committee on the Effects of Atomic Radiation (UNSCEAR), *Sources and Effects of Ionising Radiation* (New York: UNSCEAR, 1977)

received by the workers in the industry, and the population dose, that is the radiation exposure of the rest of the population, arising from the release of radioactivity into the biosphere. There are several ways in which these doses could be expressed. One could, for example, discuss the doses actually received by the workers in the given plant, or by the population in the vicinity of the nuclear facility. But release of radioactivity is not confined to a given locality; a release at one point may add to the dose received by the population in another point of the globe. For this reason, it is preferable to calculate the population dose on a global basis. The same applies to the occupational dose. If we assume that the biological effect is proportional to the dose, and as long as the dose is small enough so that the probability of one person getting cancer, or a genetic change, is small, then the total effect will be the same whether a small number of people received a large dose, or a large number of people received a correspondingly smaller dose. We can then express the global dose in terms of man-rad, or man-rem. Thus, if, say, 100,000 persons each received a dose of 4 rad, the global dose would be 4×10^5 man-rad, and the effect would be the same if every one of the 4,000,000,000 persons on the earth received 0.1 millirad. Since, to a first approximation, the exposure goes up with the energy produced in nuclear reactors, it is best to express it in terms of the electricity generated, for example man-rad per megawatt-year.

The estimates of the doses from the nuclear fuel cycle made by different investigations differ somewhat. Here we shall use mainly the figures given in the latest report of the UNSCEAR.[1]

Mining

The first step in the cycle is the mining of uranium ore. Uranium itself does not present a significant radiation hazard, since its specific activity is low due to its very long half-life, but some of the daughter products, which are in equilibrium with uranium, have very high specific activities. The important element is radon-222, an inert gas, with a half-life of 3.85 days; its products have much shorter half-lives and most of them emit alpha-particles of high energy. Normally, these products are contained within the ore and never get out, but during mining, when the mineral is broken up, the radon is released. Thus there is a high concentration of radon in the air in uranium mines. The miner inhales the radon and its decay products are deposited in the lungs, becoming a cause of lung cancer.

Long before the discovery of radioactivity, lung cancer was recognised as a major occupational disease of miners in Germany and Czechoslovakia. A large increase in incidence of cancer of the lung was also observed in other mines where there was a high concentration of radon, and a correlation with the radiation dose seems to have been established. The relationship between excess incidence of lung cancer per million persons per year and absorbed dose appears to be linear, with a factor 1.6, although the errors are large.[2] The natural incidence is about 500 cancer of the lung per million persons per year, so the increase is very large for these workers, that is between 2000 and 10,000 cancer of the lung per million persons per year, due to the large doses received (between 1500 and 6000 rem).

It is a relatively simple matter to reduce the hazard, namely, have a high rate of ventilation of the air. Once the hazard was recognised, ventilation became obligatory in mines, and maximum permissible levels of the radon concentration were laid down. Those mines in which these levels could not be reached had to close down. As a result of this, present-day exposure levels of miners are considerably lower, and the average occupational dose is about 1 rem per year. Since, as has been seen, one needs about 180 tonnes of natural uranium for 1 megawatt-year of electricity from thermal reactors, which means more than 100,000 tonnes of ore per year, and considering the amount of ore mined by one worker and the number of miners, we arrive at a figure of 0.1 man-rad per megawatt-year for the occupational exposure.

The forced ventilation of mines means that the population in the vicinity is more exposed, but when spread over the world population this would constitute only a tiny addition, less than 1 per cent of the radon present in the natural background. UNSCEAR ignores the population doses altogether, but the Nuclear Regulatory Commission (NRC) in the United States puts it down as 0.5 man-rem per MWe-year, a very high figure.

Milling

The next step is the separation of the uranium from the ore in the milling plant. The uranium content of the ore is only 0.1–0.2 per cent, which means that practically the whole of the ore goes into the tailings. These still contain some residual radioactivity; in particular, they are a source of radon. The tailings are generally dumped into piles near the mills. In some places, however, they have been used as fill for foundations for buildings, including homes, schools and shops. This happened particularly in Grand

Junction, Colorado, where radon concentrations up to 1000 times greater than in the natural background were found inside certain buildings. This practice has now stopped, but the radon release from the tailing piles still constitutes a source of radiation hazard. So far this seems to have been neglected, although, according to some calculations, in the long term this population hazard may turn out to be much more important than the high level of wastes from fuel reprocessing, which has been receiving most attention.

Incidentally, this exposes a fallacy in the argument used by some proponents of nuclear energy, namely that the radioactivity is not really a problem as it all comes from the radioactivity that existed since the earth was born. What this argument conceals is that practically all the natural radioactivity is retained within the ore, and its effect is only to raise the temperature of the earth's crust, whereas in mining the ore, it is broken up and the products come out into the open. The radon emission rate from the tailings is about 1000 times greater per unit area than from the natural radioactivity in the surface of the earth.

Reactors

Normal Operation

In the normal operation of reactors, the radiation exposure is the result of unavoidable releases of radioactivity into the atmosphere or waters, due to small leaks in the cladding of the fuel elements and the cooling system. It consists of gases or volatile fission products, or radioactivity induced by neutron bombardment.

The personnel operating the reactors are subject to external irradiation with gamma-rays. About 80 per cent of the exposure is incurred during maintenance and only 20 per cent during routine operation of the reactor. All workers wear badges to monitor their doses. The average dose per worker in the period 1969–1975 was 1 rem per year. The collective dose from reactor operation is between 1 and 2 man-rad per MWe-year.

The dose to the public is mainly the external exposure to the population in the vicinity of the reactor. The biggest contribution comes from krypton-85, a fission product, although less than 1 per cent of the krypton in the reactor is allowed to escape. Other releases of significance are tritium and carbon-14. Tritium is produced by ternary fission in nuclear fuel and by neutron activation. From 0.5 to 5 per cent of the tritium is released, depending on the type of reactor. Carbon-14 is also produced in ternary fission, but mostly in *neutron-alpha* and *neutron-proton* reactions with oxygen and nitrogen. The dose rates associated with the release of carbon-14 are very low, but due to its long half-life (5730 years) it can make an important contribution to the collective dose commitment.

The total population dose from the normal operation of reactors is 0.2–0.4 man-rad per MWe-year.

Accidents

An accident involving the release of radioactivity into the atmosphere, or unplanned exposure of personnel, can occur at any stage of the fuel cycle, but accidents to reactors have received most attention, because there are many more of them than of other nuclear plants, and because for the general public, the potential consequences of reactor accidents are much greater than for other stages of the fuel cycle.

Apart from the accident in the Calder Hall reactor in the United Kingdom in 1957, no other accident with commercial power reactors, involving exposure of the public, had been reported by 1978, although there are persistent rumours of a serious accident, or accidents, in the Soviet Union around 1958–1961; it is not clear, however, whether a reactor or a reprocessing plant was involved in them. (This chapter was prepared before the accident at Harrisburg in the United States.)

Since past experience does not provide guidance for the direct estimate of the risk of a reactor accident, any analysis must be based on a probability approach. The most comprehensive study was carried out in the United States and resulted in the so-called Rasmussen Report.[3] It is based on a study of the sequence of events, or faults, which might lead to the release of radioactivity. The study used a stochastic approach in a consequence model, which took into account reliability of components, distribution of populations and atmospheric conditions. Some of the findings in the Rasmussen Report were criticised by the American Physical Society, which carried out its own study on the safety of light water reactors.[4] The human factor also appears not to have been given enough emphasis in the Rasmussen Report; the probability of an accident may therefore be considerably higher.

There could, of course, be a great variety of types of accidents, ranging from small leaks inside the containment vessel, to a complete rupture of the container with the release of a large proportion of the radioactivity. The probability of an accident decreases rapidly with its severity, since the more serious accidents require the successive failure of a number of barriers. One can thus calculate the probability per reactor-year of a release of radioactivity, which would result in a certain collective dose to the population. For example, the probability that the collective dose will be 10^4 man-rad was calculated to be about 10^{-4} per reactor-year, whereas for a dose of 10^8 man-rad, it goes down to 10^{-7} per reactor-year. By integrating along the curve of the probabilities, one obtains a value for the expected collective dose due to accidents. This value is 0.25 man-rad per MWe-year, which represents only a few per cent of the total collective dose from the nuclear fuel cycle.

It should be noted, however, that one cannot simply add this value to the total collective dose, because unlike the other radiation risks, which are due to very small increments in dose to the individual, in the case of an accident, very high doses may be received by some persons, resulting in acute effects; for such effects the proportionality between effect and dose does not hold.

Reprocessing

Next we shall discuss reprocessing plants, which are the largest contributors to the radiation dose, to both workers and the population. At the reprocessing plant, the spent fuel is broken up and dissolved in acid. A series of chemical processes follow, leading to separate streams of uranium, plutonium and waste products. Although everything is done by remote control, nevertheless the workers inevitably receive radiation doses at each stage of operation. Certain of the fission products are deliberately released into the biosphere, thus becoming a source of population exposure.

A survey of doses received by radiation workers in different parts of the nuclear fuel cycle, or in other work involving radiation, has shown that the highest doses are received by workers in reprocessing plants. A histogram of doses received in 1975 by Windscale workers shows a bimodal distribution,[5] with some 30 per cent of the workers receiving more than 1.5 rem per year; the average dose is 3.5 rem per year as compared with an average of 0.4 rem for all radiation workers. At Windscale, even if we include all personnel whose doses were monitored, the average is 1.2 rem, three times the overall average.

The population dose results from the effluents from the reprocessing plant, the release of radioactivity into the atmosphere and water. From these media they may cause radiation exposure of man through different pathways.[6] They are divided into two main methods of exposure: external, mostly gamma-rays, and internal, which would include alpha- and beta-rays. Swimming could also be unhealthy, even if the water is not drunk. The internal exposure is the more important; apart from the direct inhalation of atmospheric releases, there is ingestion directly through drinking water and indirectly by eating food, fish, meat and vegetables contaminated via atmospheric deposition or aquatic releases.

The atmospheric release consists of tritium, carbon-14, krypton-85 and iodine-129. Almost all of these elements, which are produced during reactor operation, are released from the fuel during shearing and dissolution. The biggest contribution to the dose rate comes from tritium and krypton in the first years, but carbon-14 and iodine-129 become important after 100 years.

Tritium is also released into the water, but the aquatic discharge contains other elements also; the most important are caesium, strontium, ruthenium, zirconium and niobium, as well as some plutonium and other actinides. The amount discharged depends strongly on the technology used in the plant. With more sophisticated technology, which of course means greater expense, it is possible to retain a larger proportion of the radioactivity. Conversely, sloppy procedure can give rise to large releases. Thus, for example at Windscale in Great Britain, the discharge of caesium-137 rose by a factor of ten in the course of a few years, due to the corrosion of fuel cladding from magnox reactors while in the cooling ponds. This caused a great deal of worry to the local population, particularly the fish eaters. With oxide fuel, this type of discharge is much less likely to occur. Even then, aquatic discharges contribute over 90 per cent of the dose to the public. The collective global dose from reprocessing plants is estimated to be between 1.1 and 3.3 man-rad per MWe-year.

All these doses are expected to result from normal operation of the plants. Very little study appears to have been undertaken to estimate the risk of accidents in reprocessing plants. Small accidents, involving the release of radioactivity and exposure of personnel, have happened already, and a full analysis of this problem is overdue.

Management of Wastes

We now come to the important problem of what to do with the materials after they have been processed, or with the spent fuel elements, if it is decided not to reprocess them. The question of the safe storage and/or disposal of radioactive waste has received more attention than any other part of the nuclear fuel cycle, because a time factor of thousands or millions of years is involved, and obviously there is no experience in ensuring safety over such long periods, although there is some evidence from the natural reactor in Gabon.

Next to proliferation, the problem of waste disposal is the major objection to a commitment to a large-scale nuclear energy programme. In Great Britain, a detailed investigation into the environmental hazards of nuclear power was carried out in 1976 by the Royal Commission on Environmental Pollution.[7] It came to the conclusion that there should not be such a commitment until we know for certain how to contain the radioactive wastes. On the other hand, views were expressed that this is not a real problem; it certainly appears easy when stated in the following terms: "When the wastes are prepared for disposal, the total volume produced annually by a 1000 megawatt nuclear reactor is about 2 cubic metres, an amount that would fit comfortably under a dining-room table."[8] Here our concern is with radiation risks; hence it is not appropriate to go into detail on all aspects of the waste-management problem which have in any case produced a vast literature already. A mere outline of the problem will suffice.

When the radioactive wastes come out from the reprocessing plant, they contain a number of fission products with relatively short half-lives. Allowing these nuclides to decay would make it easier to tackle the rest. For this reason, the first stage is interim storage in tanks for a period of 10–20 years. With the once-through cycle, this would apply to the fuel rods coming straight from the reactors without any reprocessing. Because of the large volumes involved, and the need to cool them, huge tanks are needed, and a large number of them. Each tank has a capacity of nearly 5000 cubic metres (1.3 million gallons) and is made of stress-relieved carbon steel, with double walls. It is enclosed in concrete, and earth backfilled around it. An array of coils inside, with water circulation, cools the liquid. Sixteen such tanks are being built at present in the United States. This may be sufficient for the immediate needs, but in the long run many more will be required. But already local environmental groups are mounting campaigns against having reprocessing plants in their vicinity. Although the storage of wastes from the weapons programme has indicated the design of leak-proof tanks, some leakage cannot be avoided, and continuous monitoring is essential. What comes after the tank storage period? Let us look at the time scale of the decay of the radioactivity. We can distinguish here three stages. In the first, the activity is due mainly to the fission products. The most important are strontium-90 and caesium-137.

Although their half-lives are only about 30 years, after 100 years their activity will be reduced to only 10 per cent, which is still a lot if one started off with a huge activity. Thus it takes about 500 years before the other substances take over. This is the period of the actinides, in which we meet half-lives up to 400,000 years. Their decay takes us to about 100,000 years. Finally we have the very long-lived fission products technetium-99, tin-126 and iodine-129, but mostly nuclides produced from the decay of the actinides. Their abundance is very low, and would not contribute significantly to the population dose, but in the long run their production adds irreversibly to the global radioactive inventory.

An important aspect of the problem is the volumes and amounts of radioactivity to be handled. If—as has been projected—by the year 2000 there is an installed nuclear capacity of 2000 GWe, then the fuel which would have to be reprocessed every year would weigh about 50,000 tonnes, in a volume of 5000 m^3. Each tonne of material reprocessed contains, after one year, 2 megacuries of radioactivity. This means that every year there would be an additional 100,000 megacuries to be disposed of. All this is just from fission products, and does not include the plutonium and other actinides.

There appear to be three approaches to the waste disposal problems: dilution-dispersion, delay and decay, concentration and containment. The first is being practised in relation to some elements such as tritium and krypton, but with the growth of installed nuclear capacity, even these gases will have to be contained. The second has been practised in relation to short-lived materials and is now being proposed for untreated fuel elements. Concentration and containment is the present-day policy. This requires the solidification of the wastes and safe storage in some geological structure. Various methods of solidification, in particular vitrification—that is, the incorporation of the materials in glasses or ceramics—have been proposed, but there is doubt whether they will retain their integrity over millennia. More recently, it has been suggested that they should be incorporated into rock-like crystals; this probably occurred naturally in the Gabon reactor, and therefore offers greater promise.

Returning to our main theme, the radiation hazard to the population from waste products, UNSCEAR was unable to make an adequate assessment of it, because of the lack of a decision about the precise method of disposal. The occupational risk associated with waste storage was considered to be very small compared with other parts of the cycle. Thus waste storage and disposal do not figure at all among the items which contribute to the radiation hazard from nuclear energy.

Summing up the Radiation Hazard

A summary of the radiation hazards, from all parts of the nuclear fuel cycle, is shown in Table 8.2. This is the UNSCEAR estimate of the collective dose commitment.[9] The individual items have already been discussed except the last, research and development. UNSCEAR considered that part of the occupational exposure incurred in research and development establishments is attributable to supporting future development of the nuclear power industry. They estimate this to be 1.4 man-rad per MWe-year. The total comes to about 7 man-rad per megawatt-year; about 60 per cent of this is occupational exposure.

At the present time, with an installed nuclear capacity of about 100 GWe, the per capita dose is about 0.2 millirad per year, which is a quarter of one per cent of the natural background. This is much smaller than the fluctuations of the natural background, and thus quite insignificant. By the year 2000, however, if the installed capacity were to be 2000 GWe, the dose would rise to 5 per cent of the natural background. From then on, with further growth of nuclear power, the radiation hazard would become more and more significant.

We may thus conclude that at the present time the radiation hazard from the normal operation of the nuclear fuel cycle—ignoring the potential hazard from accidents and the disposal of waste products—is negligible, but that in the course of time it is likely to reach a stage when it will

Table 8.2 Radiation hazards from the nuclear fuel cycle

	Collective dose commitment (man-rad/MWe-y)
Mining, milling and fuel fabrication	
(a) Occupational exposure	0.2–0.3
Reactor operation	
(a) Occupational exposure	1.0
(b) Local and regional population exposure	0.2–0.4
Reprocessing	
(a) Occupational exposure	1.2
(b) Local and regional population exposure	0.1–0.6
(c) Global population exposure	1.1–3.3
Research and development	
(a) Occupational exposure	1.4
Whole industry	5.2–8.2

become a significant factor, and will require additional measures to reduce the exposure.

Even at the present time, however, certain aspects need further attention. As we have seen, the occupational dose is the major proportion of the total dose commitment. The American Physical Society has pointed out that since not only the dose but the dose rate is also higher in occupational exposure, the relative biological effect is likely to be higher than for the rest of the population.[10] The number of workers in the nuclear industry is a very tiny fraction of the world population. It follows, therefore, that this group carries a disproportionately large burden, about 10,000 higher per capita than the population at large. This calls for a more detailed examination of the risks incurred by radiation workers.

On the basis of detailed consideration,[11] the present writer has come to the conclusion that the dose limit for workers should be reduced by at least a factor of five.[12] What would be the consequences to the nuclear industry of such a reduction of the dose limit? It would mean more elaborate protection measures, and where these are not feasible, shorter working hours and more workers employed to spread out the exposure. Both these measures would increase the cost of nuclear fuel, and therefore would be resisted. But the good image which the nuclear industry is presenting, namely that the radiation hazard is negligible, was arrived at by calculating the dose when spread over the whole world population. Is it fair to ask a small group of persons to carry such a heavy burden? The present writer believes that this is unfair, and should further study confirm that the dose limits are really too high, then they should be lowered, whatever the consequences to the nuclear industry.

Notes

1. United Nations Scientific Committee on the Effects of Atomic Radiation (UNSCEAR), *Sources and Effects of Ionising Radiation* (New York: UNSCEAR, 1977).
2. US National Academy of Sciences, *The Effects on Population of Exposure to Low Levels of Ionising Radiation, 1972* (Washington, DC: The National Academies of Sciences, 1972).
3. US Atomic Energy Commission, *Reactor Safely Study: An Assessment of Accident Risks in US Commercial Power Plants* (Washington, DC: US Nuclear Regulatory Commission, 1975).

4. Harold W. Lewis et al., "Report to the American Physical Society by the Study Group on Light-Water Reactor Safety," *Reviews of Modern Physics* 47, Supplement 1 (January 1975).
5. UNSCEAR, *Ionising Radiation*, 278.
6. Geoffrey G. Eichholz, *Environmental Aspects of Nuclear Power* (Ann Arbor, MI: Ann Arbor Science Publishers, 1976).
7. Royal Commission on Environmental Pollution, *Sixth Report: Nuclear Power and the Environment* (London: Her Majesty's Stationery Office, 1976).
8. Bernard L. Cohen, "The Disposal of Radioactive Wastes from Fission Reactors," *Scientific American* 236 (June 1977): 21–31.
9. UNSCEAR, *Ionising Radiation*, 15.
10. Charles L. Hebel et al., "Report to the American Physical Society by the Study Group on Nuclear Fuel Cycles and Waste Management," *Reviews of Modern Physics* 50, no. 1 (1978).
11. International Commission on Radiological Protection (ICPR), "Recommendations of the ICRP," ICRP Publication 26, *Annals of the ICRP* 1, no. 3 (1977).
12. Joseph Rotblat, "The Risks for Radiation Workers," *The Bulletin of the Atomic Scientists* 34 (September 1978): 41–46.

Sir Joseph Rotblat (1908–2005) was Professor of Physics at St Bartholomew's Hospital, London. He was the only physicist to leave the Manhattan Project on the grounds of conscience. A signatory of the Russell-Einstein Manifesto, he was Secretary-General of the Pugwash Conferences on Science and World Affairs from their founding until 1973. He shared, with the Pugwash Conferences, the 1995 Nobel Peace Prize for efforts towards nuclear disarmament.

CHAPTER 9

The Dilemma of European Theatre Nuclear Arms Control

Lawrence Freedman

Introduction

"It is said the appetite develops with eating." This comment is reported to have been made by President Brezhnev at the Moscow Summit with Chancellor Schmidt in July 1980. Taken as a statement on Soviet foreign policy, Brezhnev's comment might appear quite sinister, but put in context as an observation on arms control it is quite profound. Arms control has represented an attempt to codify "balances," and because these balances never seem complete, the inclination has been to include more and more weapon types in the accounting system—medium-range aircraft in this case, which prompted Brezhnev's remark. The initiative taken at the Moscow Summit in July 1980 led to a new set of negotiations, which began in preliminary form on 16 October 1980. These negotiations extend further the boundaries of arms control and exemplify all the attendant problems.

They have been stimulated by politics rather than strategy, so that they are already overloaded with expectations and responsibilities, including

Originally published in David Carlton and Carlo Schaerf, eds., *The Arms Race in the 1980s* (London: Macmillan and New York: St. Martin's Press, 1982): 235–251.

L. Freedman (✉)
King's College London, London, UK

P. Foradori et al. (eds.), *Arms Control and Disarmament*,
https://doi.org/10.1007/978-3-319-62259-0_9

sustaining what remains of détente in Europe. They have been entered into, at least on the Western side, without a clear view of their likely outcome. A principle of equality is believed to be at stake, but it is hard to see how realistically this can be turned into a credible agreement. The two sides start with completely different notions of the scope of the talks: the United States wishes to restrict matters to medium-range missiles that can hit European targets; the Soviet Union has a much broader concept encompassing all systems capable of reaching her soil.

The unpromising nature of the whole enterprise is illustrated by the difficulties found in describing the preliminary talks. To the Soviet Union discussions are "Related to Nuclear Arms in Europe," which to the West begs the question of Soviet weapons that can attack Europe from bases in Asia. The American alternative "Discussions of Questions related to the Limitation of Certain US and Soviet Forces" does not even convey a vague sense of an agenda.

Arms control has been set a critical test, which it is ill-equipped to meet. There is a real danger that, whatever the motives in reassuring domestic opinion or preserving a modicum of détente, the result will be more prolonged and acrimonious negotiations followed by disappointment and recriminations. This, in turn, could lead to the discrediting of even limited forms of East–West discussions on military issues.[1]

One objective of this chapter is to describe the background to these talks and the practical impediments to their successful conclusion. Another objective is to suggest one possible way out of the difficulties, which is to fully integrate the talks on theatre systems with those on central systems, presuming the Strategic Arms Limitation Talks (SALT) can be put on course once again. This analysis warns of the pitfalls of proceeding with theatre negotiations in the absence of SALT.

The Position of the North Atlantic Treaty Organization (NATO)

During the 1970s the position adopted by the Alliance on the advisability of negotiating on European-based systems has been reversed. At the start of SALT the threat to Europe posed by Soviet medium- and intermediate-range ballistic missiles (MR/IRBMs) was recognised but there was reluctance to raise this because it would involve conceding the Soviet point on including the forward-based systems (FBS)—the American medium-range aircraft based in and around Europe. The fear in Western

Europe was that this issue would allow the Soviet Union a means of breaking the links between the American nuclear arsenal and the defence of Europe.[2]

There were a number of reasons for moving away from this approach, six of which are:

1. The desire for arms control to provide complete, comprehensive coverage of all weapons pointed to theatre systems as a key lacuna, a "grey area" somewhere between SALT and the Mutual and Balanced Force Reductions (MBFR) talks.
2. The acknowledged strategic parity focused attention onto disparities elsewhere. The first signs of modernisation of the Soviet theatre systems in the mid-1970s (Backfire/SS-20) emphasised this particular disparity.
3. Once the Soviet Union managed to entangle cruise missiles in SALT II the fate of this option for countering the Soviet theatre advantage was bound up with the future of the negotiations. The mention of ground-launched cruise missiles (GLCMs) in the Protocol to SALT II and the promise in the Declaration of Principles for SALT III to resolve the protocol issues confirmed the entanglement.
4. In justifying the NATO programme to modernise its long-range theatre nuclear forces (LRTNF), based on the Tomahawk GLCM and the Pershing ballistic missile, great play was made with the Soviet SS-20 IRBM so that the futures of the two programmes were inevitably seen to be linked.
5. It was necessary to be sensitive to dissenting opinion in Europe, which had emerged during the neutron bomb episode of 1977–1978 opposed to the introduction of any new nuclear weapons.
6. Arms control has been seen as a medicament for détente. This came to be stressed in 1980 as the health of détente was seen to wane.

During 1979 it became accepted wisdom in NATO that the LRTNF programme could not go ahead without some "parallel" arms-control offer. On 6 October 1979 Brezhnev hinted at substantial concessions if the NATO programme was abandoned, but unpleasant consequences if it went ahead. This led to a debate on whether arms-control negotiations should precede deployment, if not development and production, of the Pershings and Tomahawks in order to allow the Soviet Union to demonstrate its good faith.[3] Although this particular approach was rejected, both

Belgium and the Netherlands have made it clear that their participation in the programme, as hosts to cruise missiles, is conditional on progress in arms control.[4]

Those drawing up the actual force plans in 1979 were working within a range of 200–600 missiles. The final figure of 572 missiles (108 Pershing and 464 Tomahawk) tended on the high side which can, in part, be seen as anticipating some future cuts in deference to arms control. As an additional sweetener it was agreed to withdraw 1000 nuclear warheads from Europe "as soon as feasible." The actual arms-control proposal involved the following principles[5]:

1. Any future limitations on American systems principally designed for theatre missions should be accompanied by appropriate limitations on Soviet theatre systems.
2. Limitations on American and Soviet long-range theatre nuclear systems should be negotiated bilaterally in the SALT III framework in a step-by-step approach.
3. The immediate objective of these negotiations should be the establishment of agreed limitations on American and Soviet land-based, long-range, theatre nuclear missile systems.
4. Any agreed limitations on these systems must be consistent with the principle of equality between the sides. Therefore, the limitations should take the form of *de jure* equality both in ceilings and in rights.
5. Any agreed limitations must be adequately verifiable.[6]

Adopting SALT as the most appropriate forum recognised that cruise missiles were already bound up with SALT, the "decoupling" implications of having completely separate talks, and the advisability of keeping the negotiations bilateral. Confusion was likely to result from attempting to involve all interested parties, and then in talking seriously within such a body. The unwillingness of the British and French to expose their small nuclear forces confirmed the bilateralism. Only American missiles from the NATO side were to be discussed. Restricting future negotiations to "land-based missiles," reflected the popular perception of the issue at hand. Excluding aircraft would also keep matters simple.

The American concept has been essentially to see SALT III as a central negotiation mainly concerned with reducing the SALT II ceilings with a series of distinct, peripheral negotiations on discrete issues, such as anti-satellite weapons and depressed-trajectory missiles as well as LRTNF.[7] There is much to be said for this concept, but there are now two obvious problems. First, notwithstanding President Reagan's expressed determination to negotiate a better SALT II, neither side seems prepared for serious negotiations on "deep cuts" for some time. This seems an unlikely sun around which all else must revolve, especially when one of the satellite negotiations is close to being joined. Secondly, as will be discussed later, LRTNF cannot easily be contained.

The Soviet Position

One reason for doubting whether a theatre negotiation can be kept short and simple is the Soviet attitude. The Soviet Union has consistently demanded that American FBS be included in SALT because they threaten the Soviet homeland. On similar grounds she has argued for the inclusion of British and French nuclear forces.

In SALT I the Soviet Union received no formal credit for either FBS or the British and French forces.[8] These same issues were raised in the early stages of SALT II. However, in the Vladivostok Agreement of November 1974 no mention was made of FBS. Henry Kissinger observed that agreement had been possible partly because the Soviet Union had dropped her insistence on including FBS in the totals.[9] However, the Vladivostok *aide-mémoire* did include "air-launched missiles," which provided the Soviet Union with an opportunity to get a handle on cruise missiles.

After signing SALT II in June 1979, she quickly placed European-based systems on the agenda for SALT III. Meeting with the press on 25 June, Soviet Foreign Minister Andrei Gromyko proposed drawing other countries and their weapons into SALT and covering the FBS issue.

A willingness to negotiate on missiles based in Europe had been indicated by Brezhnev as early as March 1979,[10] but the most direct appeal was the one made on 6 October 1979. Then he spoke of a readiness "to reduce, compared with the present level, the quantity of medium-range nuclear missiles deployed in the western parts of the Soviet Union: but, of course, only in the event that there is no additional deployment of medium-range missiles in Western Europe." In the period leading up to 12 December this offer was stressed, although its ambiguities were never clarified, particularly as to whether the SS-20 was to be included.[11]

After the NATO decision the main question was whether the Soviet Union would agree to talk at all. The initial reaction was that the "basis" for talks had been destroyed, but by July, and Schmidt's visit to Moscow, a new basis had been found. The new Soviet position was agreed by the Politburo on 4 July 1980:

> Without withdrawing the proposals put forward earlier, [the Soviet Union] could also agree to a discussion of issues relating to medium-range weapons even before ratification of SALT II. At the same time, the discussions must involve not only medium-range missiles, but also US forward-based nuclear weapons. Both these problems must be discussed simultaneously and in organic connection. ... Possible accords could be implemented only after the SALT II Treaty comes into force.[12]

In terms of the subject matter of the negotiations, therefore, the key difference between the two sides is the FBS question. The other long-standing Soviet objective, the inclusion of British and French forces, appears to have been postponed for the moment as something more relevant to SALT III proper.

The Forces

The two sides possess systems with such variations in numbers, types and quality, and held for such disparate purposes as to render a proper comparison extremely difficult. There is a boundary problem as LRTNFs blend into battlefield nuclear forces, and as wholly nuclear systems make way for dual-capable tactical air or anti-ship systems. At the other end there are the 400 Poseidon warheads that have officially been assigned to the Supreme Allied Commander Europe (SACEUR) and the Soviet SS-11 intercontinental ballistic missiles (ICBMs) in MR/IRBM fields, which are already SALT accountable. Any drawing of boundary lines must be quite arbitrary.

Missiles

NATO has not had any land-based missiles in Europe since the Thor and Jupiter, both of which were moved by the mid-1960s, apart from 18 French IRBMs and 180 short-range Pershing I missiles (72 of which are operated by West Germany under a dual-key system).[13] The two new systems—Pershing II and the Tomahawk GLCM—will not become operational until December 1983 at the earliest.[14] By mid-1986 the 108 Pershing IIs should all be deployed, but only 160 of the Tomahawks will be deployed. It will take until 1989 for the full force of 464 to be ready.[15]

The Soviet missile force reached a peak in 1964 with 733 SS-4s and SS-5s (598 in soft launchers and 135 in hardened launchers). By the late 1960s a few had been withdrawn, but the major move came in 1968 when about a quarter were removed to positions in the Far East facing China. At the time the Soviet Union was attempting to develop a solid-fuelled (for quicker reaction) follow-on system. A missile, designated the SS-14 by NATO, was tested using the last two stages of the SS-13 ICBM.[16] The SS-13 did not turn out to be a very popular ICBM and its low rating appears to have reflected on the SS-14, which was never deployed.[17]

From 1969 to 1971, perhaps as a result of this failure, 120 SS-11 ICBMs were deployed as an expedient in MR/IRBM fields. These SS-11s were unambiguously linked to the theatre tasks, though they were counted in the SALT totals.[18] They are still deployed in this mode.[19]

The SS-20 IRBM first became operational in 1976. It is of 4400 km in range and is derived from the first two stages of the SS-16, the latest, and unsuccessful, attempt to develop an efficient, solid-fuelled ICBM. The most recent estimate is that 160 SS-20s are now in place. They are found in 23 sites in the broad areas in the Soviet Union, in the Western Military Districts facing NATO, in the Far East facing China and in the centre of the country in swing sites, capable of being directed against Europe or China. This latter location is significant for any attempt to design a nuclear free zone for Europe or in assessing proposals to cut back on missiles located in the "Western parts of the Soviet Union." Estimates suggest that two-thirds of the SS-20s will be capable of attacking Europe. It is commonly assumed that the planned force level will be 250 SS-20s but that is only an assumption.[20]

There are 440 of the older missiles still deployed (380 SS-4 and 60 SS-5). It would appear that these are now wholly concentrated on Europe and the installations in the Far Eastern Districts are no longer operational.[21]

No SS-4s (but some SS-5s) have been dismantled since autumn 1979. This may reflect a desire to maintain a bargaining card for negotiations. At the June 1980 meeting of NATO's Nuclear Planning Group, Ministers "noted with concern the continued retention of Soviet SS-4 and SS-5 missile launchers. This, coupled with the continuing deployment of SS-20 missiles could lead to an even larger superiority in LRTNF in the mid-eighties than previously anticipated."[22] The previous estimate had been 50 SS-4s by the mid-1980s.[23]

Aircraft

Many aircraft that the Soviet Union appears to include under the FBS heading, the carrier-based A-6s and A-7s, and the F-4s, are dual-capable and generally unsuitable for strikes into Soviet territory. Technically, it would be very difficult to include them in an agreement. For example, if the carrier-based aircraft were included there would have to be some rules as to the patrols and composition of the American 6th Fleet in the Mediterranean.

Moreover, if these systems were included then the United States would probably demand the inclusion of comparable Soviet systems and this in turn would lead to great confusion. Although one can vary the picture according to the range/combat radius level chosen, the Soviet position at every point looks the strongest. Furthermore, as shorter-range aircraft are held in much larger numbers, they soon come to dwarf the longer-range and more unambiguously strategic TNFs. The position can at least be simplified by identifying the key long-range systems. It is best not to attempt to do this solely by range, because this is by no means a fixed measure.

On the American side it is very difficult to exclude the 170 F-111s based in Great Britain, which are included in official descriptions of NATO long-range TNFs.

On the Soviet side the most obvious systems would be the medium-range bombers under the control of the Soviet Long-Range Aviation Force. This would involve the Tu-16 Badger, the Tu-22 Blinder and the Tu-22M Backfire. At the moment the Soviet inventory includes 288 Badgers, 125 Blinders and 75 Backfires.[24] The Backfires are expected to increase at the rate of 30 per year, the maximum permitted following Brezhnev's June 1979 undertaking to Carter. The Badgers, which are now well over 20 years old, will be gradually phased out, though the Blinders, which were first deployed in 1962, will probably linger on for some time. As with the missiles, about a third of these aircraft are based close to China.[25]

When it comes to negotiating, a further problem could be those Soviet systems assigned to the Naval Air Force—in mid-1980s, 280 Tu-16, 40 Tu-22 and 70 Tu-22M. At the moment on the NATO side, American systems are supported by 57 British Vulcans and 33 French Mirage IVs. However, these will all be phased out by 1985.

Prospects for Agreement

From all this, what are the prospects for an agreement? On the basis of past experience with SALT it must be considered unlikely that the condition of parity, required as one of NATO's five principles, can be created through arms control unless it virtually exists already. To construct equality out of inequality requires the stronger to make extra concessions. This has not been the habit in the past.

There are four possible points of parity. First, by about 1985 there may be an equality in land-based missiles, at a level of some 250 missiles, halfway through the NATO programme and at the probable conclusion of the SS-20 programme. Codifying parity at this point might well appeal to the Soviet Union, even without FBS, but for the United States it would mean cancelling half her planned force without any compensating concessions on the Soviet side. It would also mean disregarding any older SS-4s and SS-5s that had not been retired. Providing an incentive to the Soviet Union to double her planned SS-20 production by fixing a ceiling to accommodate most, if not all, of the TNF programme is equally unappealing. From the NATO point of view another key objection is that no allowance is made for the three warheads on each SS-20 as against the single warhead Tomahawks and Pershings.

There are ways of addressing this last point. The first is through counting launchers rather than missiles, which has been normal in SALT. Strictly speaking the Tomahawks are carried and launched in batches of four on a GLCM transporter–erector–launcher (TEL). The problem with trying to create a parity on this measure is that the programmes of each side hardly begin to meet at the end of the decade. To achieve a ceiling at say 160 launchers, the NATO programme would be cut by 64 launchers, half of which might well be Pershings. The Soviet Union would have to hold her SS-20 force at about the current level, and dismantle all her SS-4s and SS-5s.

Current NATO studies prefer to count warheads rather than launchers, which would be more innovatory in SALT terms.[26] However, the required adjustments to existing programmes are similar to those of a launcher ceiling. If, as expected, the Soviet Union continues to build up her SS-20s for the duration of the talks, then adoption of either of these ceilings will require her to accept dismantling her own modern systems while American forces are continuing to grow.

A final approach to parity recognises Soviet concern over FBS by including the F-111 but in return draws in larger numbers in the Soviet Long-Range Aviation Force. If both programmes are allowed to run their course (and the Soviet Union does not drastically increase SS-20 production or hold on to obsolescent systems) then a form of parity might arrive naturally by 1989. An arms-control agreement which merely acknowledged this fact would hardly be taken seriously. A ceiling of 450 for delivery vehicles might prove attractive to the Soviet Union only in her older systems. For the same reason it would not appeal to NATO.

There is another obvious difficulty. About one third of all Soviet medium-range forces are facing China. To bring them into an agreement would seriously upset the calculations. From the Soviet perspective these systems have nothing to do with the European theatre: from the NATO perspective they could be turned against it, either as a result of a Sino-Soviet *rapprochement* or just through reinforcement measures in an emergency. It has been suggested that SS-20s could be transported to new sites by air. Whether or not NATO would feel able to exclude them might depend on how well such a movement could be observed. Certainly the movement of aircraft from one theatre to another would present no problems. The movement of aircraft based in the United States which could be sent to Europe in a crisis might be relevant to a compromise here.[27]

It is possible that an unequal deal would be accepted, despite offending the principle of parity, by which the Soviet Union would accept a firm limit on her force at, say, 200 SS-20s with a rapid run-down of SS-4s and SS-5s in return for a significant cut by NATO, for example the 108 Pershing II missiles which apparently alarm the Soviet Union more than cruise missiles. Their combination of accuracy and speed reduces warning time. Such a deal would have a number of problems: Pershing is generally considered a better missile than Tomahawk within NATO[28]; a deal of this sort would not seem very radical in Europe, merely confirming that arms control legitimises military programmes rather than restrains them; and it would not address the generalised hostility to cruise missiles in Europe. Another possibility might be to include the American F-111s but no Soviet aircraft. The Soviet Union may have something like this in mind but given the fuss about Backfire in SALT II it would seem a non-starter.

With all the options discussed thus far it is difficult to identify the mixture of incentives that might bring the negotiations to a successful conclusion. The Soviet objectives of severe limitation and cancellation of NATO's LRTNF programme and acceptance of the principles that account should be taken of American FBS are not compatible with the NATO position of only considering missiles and tolerating restraints in its own programme only to the extent that these are fully reciprocated by the Soviets. Furthermore, the Soviet Union is in a strong bargaining position, with her modernisation programme well advanced (and her final goal uncertain), while NATO's programme remains politically controversial.

This all suggests that without a substantial concession by one side or the other, there can be no agreement based on the principle of parity at the theatre nuclear level and it is even difficult to find one that accepts some permanent disparity. Politically this creates the prospect of East–West acrimony rather than comity, with both sides using the occasion for publicly shifting the blame to the other for a new arms race, adding to arms control's bad name and disappointing those in Europe anxious to see both the SS-20 and TNF programmes either abandoned or substantially reduced through mutual agreement.

The Integrative Approach

There is an alternative possibility: to merge discussions on theatre systems with those on central systems. This approach has some appeal in Europe but has found little favour in Washington where it has been felt that the main business of SALT III will be to achieve "deep cuts" in the ceilings agreed in SALT II, and that the inclusion of peripheral matters, such as LRTNF, will only complicate and delay. However, as this now provides the sole motor for SALT, the hierarchy of concerns no longer seems appropriate. Moreover, as we have seen, the prospects for a contained negotiation on theatre systems are not encouraging.

For NATO there are good arguments for a merger. It is profoundly "coupling" in recognising the strategic unity of the alliance and emphasising that American nuclear weapons provide a continuum of deterrence. A deal achieved solely within a European framework would inevitably encourage notions of a separate theatre balance, even if reached and enforced within some specially broadened SALT context.

One objection is that it will force difficult choices on Washington between the two types of systems and the competing demands of domestic and alliance constituencies. However, the existence of separate negotiations will not prevent links developing between the two, with concessions in the different negotiations being traded. Suspicions will inevitably grow within NATO that the United States is assigning top priority to central systems and will not squander precious negotiating capital on the secondary matter of theatre systems. If an all-inclusive ceiling can be achieved within SALT III with a freedom-to-mix arrangement, then the allies can sort out the proper balance between central and theatre systems among themselves, without the Soviet Union sitting in, acting almost as an arbiter. Lastly, at least for Washington, all these systems can reach the Soviet Union, while Moscow would have to make far more serious choices among its different adversaries.

Nevertheless, the Soviet Union ought not to have fundamental objections. It fits in with previous demands for FBS to be brought into SALT. It need not offend the principle of no amendment to SALT II until the Treaty is ratified.

The proposal is to add 400, either to the eventual ceiling for central systems under SALT II 2250 or to a lower figure if further cuts in central system levels are agreed. Into this raised ceiling can be included theatre missiles and aircraft as already specified, but to levels either less or more than 400 depending on the priorities of each side. The United States, for example, might still be interested in going beyond 400 in theatre systems to take up some of the gap between the SALT II ceiling and the currently planned force levels.[29]

Because this proposal is based on launchers rather than warheads it does not take account of the SS-20's three warheads.[30] This could be remedied by including the SS-20 under one of the existing sub-ceilings for missiles with multiple independently targetable re-entry vehicles (MIRV); the most likely would be the 1320 ceiling which currently gives the United States a 120 credit for bombers carrying air-launched cruise missiles (ALCMs). An equivalent 120 credit for MIRVed medium-range missiles could well seem appropriate.[31]

Within these sort of totals the 66 US FB-111s could not be excluded. The more politically awkward problems of the Soviet forces facing China and the British and French forces will remain. They may become linked, so that they are either brought in together or kept out together.[32]

It has not been possible to suggest anything more than expedients to many of the obvious difficulties over what to exclude and include, to be faced whatever the conceptual framework. The problems already identified will not evaporate by placing the negotiations in a different setting, and some new ones will emerge. There is a major mitigating factor: what appears as 20 per cent of some theatre balance will only represent a couple of per cent of the overall balance.

Arms control always involves a compromise between strategic logic and political convenience, and this case is no exception. The proposals outlined above are designed more to respond to a developing political situation and make no pretence to manufacture some new strategic stability—an elusive concept in these circumstances. The only real strategic benefit would be to couple theatre with central systems in an overall SALT ceiling, serving to reinforce an important piece of NATO symbolism. Nor is it claimed that major political benefits can be gained. The current requirement is to avert a political crisis.

This analysis warns of some pitfalls in the coming talks, and, in suggesting integration with the main body of SALT, also cautions against attempting to rescue SALT by persevering with European negotiations in isolation. It illustrates the general difficulties inherent in this sort of arms control, guided by the dubious precept of parity and dominated by the imperfect science of weapon counting. In offering any directions in this area the old *Punch* cartoon comes to mind—"to get there, I wouldn't be starting from here."

Notes

1. The present writer's scepticism as to the approach to arms control informing these talks is expressed in Lawrence Freedman, "Time for a Reappraisal," *Survival* 21, no. 5 (1979): 198.
2. Uwe Nerlich observed: "To some West Europeans, FBS had taken on symbolic functions, and American handling of the FBS issue was seen as the one indicator of future accountability to European interests of American strategic power in Europe." Uwe Nerlich, *The Alliance and Europe: Part V. Nuclear Weapons and East-West Negotiations*, Adelphi Papers no. 120 (London: International Institute for Strategic Studies, 1976): 4.
3. See for example, Klass De Vries, "Responding to the SS-20: An Alternative Approach," *Survival* 21, no. 6 (1979): 251–255. A key objection to this conciliatory approach was that it was unlikely that the US Congress would authorise funds to be spent on systems when there was a chance that they would never be deployed.
4. The Netherlands' position is that she will only consider opting into the programme in mid-1981 depending on the state of arms-control negotiations. Belgium apparently took a firmer stand in December 1979, implying that she would confirm participation after 6 months. However, the continuing political crisis in Belgium meant that a decision was postponed. In September 1980 the Belgium Cabinet announced that it would review the

position every six months, monitoring the negotiations, but could on balance be expected to accept the cruise missiles: "In the event of negotiations between the United States and the Soviet Union not reaching a conclusion, Belgium, in solidarity with its allies, will take all the measures agreed between the NATO partners." *International Herald Tribune*, 20–21 September 1980.

5. The preparation of the NATO arms-control position was the responsibility of the Special Group, up to December 1979. They operated from April 1979 in parallel to the High Level Group, which was responsible for the actual force plans. In 1980 the Special Group was reconstituted as the Special Consultative Group. It has met a number of times to refine the NATO proposal. Now that talks are to begin it will act as the main consultative mechanism to enable the American negotiators to be aware of the interests of all NATO members.
6. NATO *Communiqué, Special Meeting of Foreign and Defence Ministers* (Brussels: NATO, 12 December 1979).
7. This desire to push TNF to one side is reflected in one of the few published detailed studies on US strategies for SALT III, William E. Hoehn Jr, *Outlasting SALT II and Preparing for SALT III* (Santa Monica, CA: RAND, 1979): x; "SALT III negotiations must emphasize strategic ceilings and missile verification measures. Other important issues—the resolution of many protocol items, 'grey area' and theater systems, et cetera—must be the core issues for 'SALT IV' conducted in parallel but separately."
8. In a unilateral statement after the signing of the Interim Agreement on Offensive Arms in May 1972, the Soviet Union argued that she was entitled to increase the number of her nuclear ballistic missile-firing submarines (SSBNs) in line with any increases in those of American allies, implying that the extra numbers she was allowed under the submarine-launched ballistic missile (SLBM) ceilings took account of British and French forces. The United States refused to accept this argument and she has not been raised again since, even when France actually brought an extra submarine into operation.
9. "The progress that has been made in recent months is that the Soviet Union gradually gave up asking for compensations for the forward-based systems party because most of the forward-based systems, or I would say all of them, are not suitable for a significant attack on the Soviet Union," Press Conference of Secretary of State Henry Kissinger, Vladivostok, 24 November 1974. To be found, with a rich collection of source material, in Roger P. Labrie ed., *SALT Handbook: Key Documents and Issues, 1972–1979* (Washington, DC: American Enterprise Institute for Public Policy Research, 1979): 285.
10. Leonid Brezhnev, speech to a meeting of the *Baumansky* constituency in Moscow, 2 March 1979; "[The USSR] has already repeatedly said that it

stands not for the accumulation but the restriction of nuclear missile and other weapons by agreements based on full reciprocity. The same applies to medium-range weapons in Europe, taking into account the presence of American military bases there, of course."

11. There were only hints from Eastern European sources that SS-20 production might cease if the NATO programme was abandoned. See Milton Leitenberg, "NATO and WTO Long-Range Theatre Nuclear Forces," in *Arms Control in Europe; Problems and Prospects,* ed. Karl E. Birnbaum (Laxenburg: Austrian Institute for International Affairs, 1980): 143.
12. *Pravda*, 7 July 1980. The Soviet Union has argued that this was an alternative to previous proposals for the reduction of her medium-range weapons if no additional US weapons were deployed in Western Europe, or else a discussion in the framework of SALT III after the SALT II Treaty enters force. See *Pravda*, 15 July 1980.
13. Sea-based missiles include 64 British and 80 French SLBMs and the 400 Poseidon warheads (essentially the contents of two SSBNs) assigned to SACEUR by the United States.
14. It was reported in October 1980 that delays in the GLCM development programme, particularly with regard to the software and hardware for the control systems, have used up all the spare time built into the programme, so that "if further delays occur they would affect the operational commitment to NATO" in "Cruise Missile Project Delayed," *Aviation Week and Space Technology*, 20 October 1980, 24–25.
15. The current intention is for the 108 Pershing II to be based in West Germany, with Great Britain, Italy, Belgium and Holland respectively taking 96, 160, 112, 48 and 48 cruise missiles each.
16. There was also some work on a version using the first and third SS-15 stages to produce a longer-range missile—3000 as against 1100 nautical miles.
17. There may have been a sighting in the Far Eastern region close to the Mongolian border. If true, this confirms the importance of the "China threat" in stimulating Soviet MR/IRBM development.
18. In the now declassified presentations of the US Secretary of Defense to Congress (from which much of the above information is taken), these SS-11s were first discussed separately from the rest of the SS-11s and other ICBMs. However, in public presentations of the "threat" they were brought together. See Lawrence Freedman, *US Intelligence and the Soviet Strategic Threat* (London: Wiley, 1977): 158–159.
19. The relevant SS-11s, at the Derazhnya and Pervomaysk sites, are deployed in their silos at such an angle that they are clearly aimed at European targets and could not be retargeted against the United States. They have now been supplemented by 60 SS-19s at these sites and might eventually be replaced by more of this particular missile. There are reports of training

exercises with the SS-19 that show it being prepared for theatre use (*Armed Forces Journal International*, December 1979). The point is that the SS-I9s are in vertical silos and can therefore be targeted against the United States or Europe.

20. The 23 bases already operational or under construction are capable of supporting over 200 SS-20s.
21. See Harold Brown, *Department of Defense Annual Report, Fiscal Year 1981* (Washington, DC: US Department of Defense, 1980): 93.
22. NATO Nuclear Planning Group, *Final Communiqué*, M-N PG-1, 80 (Bodö, 4 June 1980). As the SS-4s and SS-5s are in different sites there is no necessity for a one-to-one replacement. The most relevant considerations are manpower and maintenance requirements.
23. See Brown, *Annual Report*, 93. He notes that this decline "is based upon current trends. It is possible, however, that the Soviet Union may wish to retain a larger proportion of the current force, perhaps for use as a bargaining chip in future arms control negotiations."
24. IISS, *The Military Balance, 1980–1981* (London: International Institute for Strategic Studies, 1980). This excludes 280 Badgers, 40 Blinders and 70 Backfires attached to the Naval Air Forces.
25. If the F-111 is included then there is a case for including the 370 Su-19 Fencers. This is the closest Soviet aircraft to the F-111. Furthermore, there has been a suggestion that it might be a medium bomber rather than a close support aircraft. However, while its combat radius is not dissimilar from the F-111, its weapons load at 8000 lb compares unfavourably with the F-111 at 28,000 lb and certainly does not warrant medium-bomber status (IISS, *The Military Balance, 1980–1981*, 90–91). Furthermore, once one starts including aircraft such as the Su-19 then it becomes difficult to exclude others essentially designed for combat support.
26. However, the fractionation limit in SALT involves a move in this direction. Counting the GLCM-TEL as a single launcher might well seem contrived. The normal inclination would be to count launch tubes (as in a submarine).
27. Potentially available in the United States are 44 F-111 E/F and 237 of the much older and less capable F-111 A/D. In addition there are 66 long-range FB-111s, which are part of the Strategic Air Command.
28. Furthermore, if the problem it creates for the Soviet Union is reduced warning time then the only significance this might have would be in confusing any plans for launch on warning. Persuading her of the impracticality of such a dangerous plan would seem a wholly desirable objective—stabilising in the traditional arms control sense.
29. One American problem is that all new missiles are likely to be MIRVed so that as old systems are phased out they will be unable to be replaced with

new missiles because they will come up against the sub-ceilings for MIRVed missiles and there are no new bombers coming into production. There are plans to "stretch" 66 FB-111s and 89 F-111Ds between 1985 and 1986 and to turn them into *Ersatz* long-range bombers. ("FB-111 Bombers Playing Crucial Role," *Aviation Week and Space Technology*, 16 June 1980, 144–145.)

30. It is assumed here that GLCMs will be counted in individual tubes rather than in batches of four on TELs.
31. It would obviously be very attractive for the United States if the SS-20 could be used to cut into the Soviet MIRVed ICBM (820) or even MIRVed ICBM and SLBM (1200) sub-ceilings.
32. Differentiating between Soviet bombers and missiles, as discussed earlier, might be matched by differentiating between French (which are not assigned to NATO) and British (which are) forces, particularly as it is now known that Great Britain will now have no more than five SSBNs until well into the next century. Neither Great Britain nor France could prevent the United States allowing the Soviet Union a credit for their systems.

Sir Lawrence Freedman is Emeritus Professor of War Studies at King's College London. He made distinguished contributions in the defence and international security fields. He has written extensively on nuclear strategy and the Cold War, as well as commentating regularly on contemporary security issues. He is a member of the 2009 United Kingdom's Iraq War Inquiry and held positions at the International Institute for Strategic Studies and Royal Institute of International Affairs (Chatham House).

CHAPTER 10

Nuclear Arms Control: Obstacles to Agreement

George Bunn

Why Have We Made So Little Progress in Negotiating?

Distrust, Asymmetries and Lack of National Consensus

The first major barrier is distrust among nations. For example, the United States fears a Soviet goal of world dominance, and the Soviets see the United States in much the same way.

Americans fear Soviet acquisition of more and more territory from Eastern Europe to Afghanistan. They fear Soviet-influenced Communist takeovers from Angola to Cuba to Vietnam and Kampuchea, naming just a few. They fear that superior military power in Soviet hands could eventually give them control over most of the world. Some Americans even fear Soviet subversion of American peace movements.

Soviets fear encirclement by capitalists on their western boundaries and Chinese on their eastern ones. They fear loss of control over Poland and other East European countries to independent mass movements

---

Originally published in David Carlton and Carlo Schaerf, eds., *Reassessing Arms Control* (London: Macmillan and New York: St. Martin's Press, 1984): 81–89.

G. Bunn (✉)
University of Wisconsin-Madison, Madison, WI, USA

P. Foradori et al. (eds.), *Arms Control and Disarmament*,
https://doi.org/10.1007/978-3-319-62259-0_10

"subverted" (as they see it) by the United States; rising American influence in the Middle East; a reindustrialised, remilitarised Germany with possible access to American nuclear weapons; and a reinvigorated China improving her ties with Washington. Soviets fear the recklessness with which Americans sometimes talk about nuclear weapons and even threaten their use. Twice at least, in Iran in 1945 and in Cuba in 1962, the United States faced down the Soviet Union with superior nuclear might.

The mutual distrust of nations, which see themselves as potential enemies, has probably been the greatest single obstacle to successful arms control negotiations, according to American historian Barbara Tuchman. Why? She quotes Salvador de Madariaga, Chairman of the League of Nations Disarmament Commission and Conference:

> The trouble with disarmament," he wrote in 1973, "was (and still is) that the problem of war is tackled … at the wrong end. … Nations don't distrust each other because they are armed; they are armed because they distrust each other. And therefore to want disarmament before a minimum of common agreement on fundamentals is assured is to want people to go undressed in winter. Let the weather be warm, and people will discard their clothes readily.[1]

What can be done to reduce distrust? Trade and cultural exchanges between East and West help. Americans need to be reminded that they were once at war with what is now Canada, and that a successful arms control agreement now governs fortification of the common boundary. East and West must come to see that their common interests in survival outweigh their differences. Somehow we must all take seriously the command of the United Nations Charter not to use or threaten force in the settlement of disputes, and not to interfere in the political independence of other countries. The conflicts relating to Afghanistan and Vietnam both made arms control negotiations much more difficult.

Mistrust that the other side will violate arms control agreement is perhaps easier to deal with than the fear each side has of world dominance by the other. Verification is of course the answer. But many identify Western demands for verification, or Soviet refusals of on-site inspection, as obstacles to agreement. They have been, though not as important in the present writer's view as the underlying mistrust each side has of the intentions of the other. The Strategic Arms Limitation Talks (SALT) agreements could probably be verified effectively, with satellite and electronic detectors, with exchange of data in the consultative commission, but without any on-site inspections. The barrier to agreement here is not verification but more basic distrust.

After distrust, the second major barrier to agreement is asymmetry. There are differences between East and West in numbers of weapons and choices of kinds of weapon systems, in what those weapons are intended to defend (including alliances), in distances between likely battlegrounds and home bases, in the degree to which population and industry are concentrated in each country, in availability of seaports and vulnerability of coastal populations and so forth. It is simplistic, for example, to talk of seeking equality in numbers of inter-continental missiles without looking at what those missiles are supposed to do, and at the differences in geography, in population and industrial concentrations and in other weapons available. Each side has chosen its weapons constrained largely by budgets, technology, its own geography and military strategy, not by arms control agreements, except for a few SALT limitations.

Usually there are good reasons of national choice why one side has emphasised a particular weapon system more than the other side has. But the perceived differences often prevent agreement. In 1964, the United States proposed a freeze in long-range missile deployment and testing much like the freeze we are talking about today. But the Soviets refused to consider the Americans' proposal seriously, in part, perhaps, because they demanded on-site inspections, but mostly, in the present writer's opinion, because they thought they were behind. Now the United States rejects a nuclear missile freeze because it is thought to be behind.

Will there ever be a time when each side will perceive that it is equal to the other? Certainly the asymmetries, real and perceived, have long been major barriers to agreement.

A third barrier is the lack of national consensus to negotiate. It is related to the first two. The present writer can speak here only of his own country, the United States. When the main leadership groups there have achieved some sort of consensus favouring arms control and when the government agrees, arms control agreements have been possible. We do not now have either the consensus or a pro-arms-control administration in the United States. But the picture is beginning to change.

Certainly in the United States, at least, a government needs to have the strong support of a very large majority of the people to overcome the mistrust and the concern that the Americans are somehow behind. There may always be many Americans who distrust the Soviets excessively and perceive American weakness in the face of Soviet strength. Hence if leadership groups remain apathetic, agreement will be unlikely.

Even Barbara Tuchman's otherwise gloomy review of the history of arms control negotiations sees hope in an awakened public: "There can be no real progress toward arms control until public tolerance of existing policy ends. When public tolerance stops, piling up overkill must stop."[2]

Preparations and Framework for Negotiations

When the press reports that negotiators have arrived in Geneva for disarmament talks, we on the outside see only the tip of the iceberg. Preparations and internal negotiations have probably been going on for years. Moreover, the negotiator is typically only a message carrier. Almost never does he have authority to compromise from his initial position without first reporting back to his capital.

The difficult preparations for the conference, and the limitations in the negotiator's instructions, both brake the speed of negotiations. But they seem inherent in the process of at least the American system of government. The same is probably true for many other governments.

How do the Americans prepare for negotiations? Even if all the staff work is done in all of the agencies concerned, a long process of negotiation within the government is necessary to achieve an American position. The Arms Control Agency, the State Department, the Defense Department, the Joint Chiefs of Staff, the three services and the White House all have arms control experts. So do the relevant Congressional committees. The National Security Council usually coordinates the views of the experts from the executive agencies to produce a single position. Negotiations among the agencies can take many months. When administrations change, as at the beginning of 1981, the whole process must start over again. The decision-making process is cumbersome and slow.

Ratification of an arms control treaty requires a two-thirds majority of the United States Senate. There are almost always ten or more votes in the 100-member Senate against a significant arms control treaty. The opposition of a single important agency can raise that number quickly to 34 votes, unless there is strong popular opinion favouring negotiations. Is it therefore any wonder that US arms control positions tend to reflect the lowest common denominator among the agencies—the position which offers the least compromise to the other side? Only a strong

President interested in progress and backed by a strong public demand for progress is likely to face down an important agency urging an unrealistic position.

Once the internal negotiations are concluded by the Executive Branch, there are likely to be further negotiations with key Senate committees and with affected US allies. These can also take time and water down the American position. By the time the American negotiator reaches Geneva, he will probably have little to offer which appeals to the other side at the beginning of the talks. Because a similar clearance process apparently takes place on the other side, the American negotiator may wait a month or so for a Soviet response to the initial American offer. Then, unless the American negotiator was given a "fall-back" position in the inter-agency and inter-allied negotiations, he must seek new instructions before adopting a new position. Often the elaborate clearance process must be repeated to produce a new offer. Thus, the American negotiator is typically only a message carrier, and the clearance process can delay serious talks for months and years even when there is a national consensus for agreement.

Governmental Negotiating Objectives

The objectives for arms control negotiations, which are usually given by governments, are to reduce the risk of a nuclear exchange and the damage which would occur if an exchange took place. There are usually other objectives—side effects of the negotiations—which may be very important:

1. Maintaining negotiating contact with the other side—just to "stay in touch."
2. Securing intelligence information about the other side's plans or capabilities. One does find out a lot about the other side during negotiations.
3. Improving the nation's propaganda position. The United States finally produced a plan for General and Complete Disarmament in 1962 to counter the Soviet plan because, like the Salvation Army Band, the Americans were tired of "letting the devil have all the good tunes." As one of the drafters of that plan, the present writer can confidently assert that no one had any hope of reaching agreement on General and Complete Disarmament soon. Many, how-

ever, hoped that the weapon production freeze and the 30 per cent "across-the-board" reductions proposed for the first stage in the American plan could be the basis for serious negotiations with the Soviets. We were disappointed.

4. Finally, arms control negotiations can be used, wittingly or unwillingly, to legitimise the military status quo or even a build-up.

For example, the Mutual Balanced Force Reduction (MBFR) negotiations began soon after sentiment developed in the US Congress to bring some American troops home from Europe. The Richard Nixon and Jimmy Carter administrations feared that Congress would cut American troop levels in Europe *unilaterally* without any corresponding reduction of the Soviet troops on the other side of the East-West boundary. A stated purpose of the negotiations was to reach an agreement between the North Atlantic Treaty Organization (NATO) and the Warsaw Pact on the level of troops and tanks on each side of the boundary. The talks have gone on for almost ten years with considerable progress but no agreement being reported. In the absence of agreement, the Americans are still in Europe as they were in 1973. The negotiations have legitimised US troop levels in Europe, which might otherwise have been reduced. Similarly, the Long-Range Theatre Nuclear Force (LRTNF) or Intermediate Nuclear Force (INF) negotiations began because NATO allies insisted on an attempt to reduce the new Soviet land-based missiles aimed at Europe by negotiations *before* the Americans deployed new missiles in Europe to counter these Soviet missiles.

But European peace groups, particularly in Holland and Federal Germany, were adamantly opposed to the deployment of new American missiles. Hence NATO adopted a so-called two-track decision during the Carter administration. One track is the arms control negotiations and the other is the missile build-up. Without the negotiations, the new American missiles would probably be impossible politically for the Dutch and the West Germans to accept. As perceived by American negotiators, the new American missiles are a "bargaining chip" to strengthen the negotiator's hand with the Soviets. But we need something to offer if we expect the Soviets to give up their missiles. Whether the Soviets will trade our plans for their deployed missiles is another matter. Probably the TNF negotiations are largely to legitimise a military build-up, which the Americans perceive as necessary. In the negotiations little progress has been reported.

Again, the Strategic Arms Reduction Talks (START) can be looked at the same way. The United States is asking the Soviets to give up more existing long-range land-based missiles than the Americans would relinquish. At the same time, the Americans propose no limits on their proposed MX or cruise missiles, or on the B-1 or Stealth bombers. If Moscow were to accept the Americans' START proposal, the building of new American missiles and bombers would be legitimised. The Soviets will not of course do that. But the same legitimisation results if the negotiations go on and on without agreement while the new weapons are produced. American and Soviet negotiators began talks in 1982. Little progress has been reported. If the Americans are to "force" Soviet concessions, say the American negotiators, they must negotiate from a "position of strength." That means they must have many "bargaining chips" on their side to offer to trade for Soviet concessions. Hence the dilemma is real. For some arms control advocates believe that negotiations as presently conducted can do little more than legitimise military build-ups.

Negotiating Tactics

How have negotiators achieved any significant agreements in the face of the obstacles just described? What tactics if any reduce these obstacles?

First, it is impossible to prevent a government from seeking, as its primary goal, some of the "side effects" objectives of negotiations, namely: maintaining contact, intelligence and propaganda. Indeed, these side effects are useful to governments and probably inevitable.

Secondly, preventing the use of negotiations to legitimise arms build-ups, while essential, is difficult. A watchful public can help. But outsiders are likely to hear little except the propaganda handed out by each side.

Arms control negotiations seem inevitably to be conducted in two rooms—one to which the media is admitted, and one to which it is not. At the Geneva Disarmament Conference, there is a public forum for speeches. Often the speeches are little but propaganda. But they are all that outsiders are likely to hear about until the results of serious negotiations carried on in private rooms are announced. The public propaganda is a cover for the secret negotiations.

For some negotiations, as with START, there is no regular public forum. But heads of government or ministers of state make speeches outlining their governments' positions, as President Reagan and Chairman Leonid Brezhnev did. The negotiators occasionally meet the press to hand out titbits of information. But the serious negotiations go on in private. This has the virtue of permitting governments to accomplish their propaganda objectives with minimal effect on the negotiations. During the talks on the Non-Proliferation Treaty (NPT), the Soviets made set speeches attacking American intervention in Vietnam. The Americans replied. Meanwhile private negotiations went on seriously and without interruption. Though many reporters asked many questions, the negotiators were able to keep progress secret for months at a time.

The only way known to this writer to stop negotiations from legitimising an arms build-up is to reach prompt agreement on a freeze of the production, testing and deployment of the nuclear weapons systems on which agreement is to be sought—pending the outcome of negotiations. This is the fundamental reason why the Forsberg-Hatfield-Kennedy freeze proposal is designed as it is. The model for it is the moratorium on nuclear weapon tests observed by both sides without formal agreement from 1958 to 1961 during the test ban negotiations. The moratorium was both mutual and verifiable during that period. Though that moratorium did not produce a treaty banning all tests, it did halt tests on both sides for three years, thus effectively slowing development of nuclear weapons during the negotiations.

Thirdly, a major objection to present negotiating tactics is that they constitute what Roger Fisher and William Ury call "positional bargaining."[3] Each side comes to the negotiations with a fixed public position. The goal is victory over an adversary—the other side at the negotiating table. Concessions are demanded as a condition even of talking. Insults are exchanged freely, at least in the public, propaganda channel. The negotiators in "positional bargaining" haggle interminably, exchanging minor concessions only at the cost of public criticism from hostile critics at home. For example, the United States started with a cleared position of 25 on-site inspections per year for a Comprehensive Test-ban Treaty in the late fifties. The Soviets started at none. After many years, the Americans got down to seven (paying a terrible price to critics at home). But the Soviets did not go above three. The result was not a Comprehensive Test-ban Treaty but a limited one which does not prohibit underground tests but can be verified without on-site inspections.

Fisher and Ury suggest: (1) that the negotiators attend the conference without any fixed position; (2) that they look upon themselves and the other side as problem-solvers, not adversaries; (3) that, rather than insulting each other, they discuss what the problem is that they both have an interest in solving; (4) that, rather than demanding concessions to begin the talks, both sides focus on a common interest such as avoiding nuclear war; (5) that they develop common objective principles for agreement, such as reducing the risk of a nuclear exchange; (6) that they list various options which will be mutually beneficial in that each option will serve this common objective; and (7) that they seek agreements based rationally on these principles of mutual interest, rather than trying to gain a strategic advantage over the other side.[4]

Cynics will see these ideas as far too idealistic for governments. And to send a negotiator off to reach an agreement without fixed instructions would be unthinkable for the agencies in Washington, which must acquiesce in a common government position if a two-thirds majority is to be achieved in the United States Senate. But if negotiators were sent off without fixed positions to develop options in the Fisher–Ury manner, the agencies in Washington could then make a more informed choice among the options identified by both sides than is now possible when developing an initial position knowing less about what is negotiable. After Washington had chosen two or three options, the negotiators could be sent back with instructions based on these options. This would produce positions with some realistic chance for success, at least if other countries followed the same procedure.

In two successful negotiations in which the present writer was personally involved, tactics something like what Fisher and Ury propose were followed. Neither involved an initial position for negotiations, however. Both came about when the initial position had got nowhere and the negotiations were stalemated. In one case, Secretary of State Dean Rusk was present. We therefore had more flexibility than a negotiator usually has. We could explore options, report them to Washington for approval and expect prompt and sympathetic answers. In the other case, we had no flexibility at all and neither did the Soviets. But we agreed with the Soviet negotiators to recommend a treaty text to our government calling it the Soviet negotiators' proposal, while they agreed to recommend the same text to their government calling it the American negotiators' proposal.

In both cases, there had been lengthy discussions of our common interests and of the options available to achieve those interests. In both cases, each side recommended governmental agreement to a common position not within their instructions. In both cases, governmental approval and agreement were ultimately achieved on the text reported back, with very little change. Probably there would today be no NPT without such tactics.

Bargaining based on fixed but different Soviet and American instructions had only produced stalemates.

The NPT was of course perceived to be of less consequence than a SALT or START treaty. But there were nevertheless sharp disagreements among the interested agencies at home. Haggling from rigid positions got nowhere. Joint problem-solving achieved success.

Notes

1. Barbara Tuchman, *The Alternative to Arms Control* (Los Angeles: Center for International and Strategic Studies, University of California, 1982).
2. Ibid.
3. Roger Fisher and William L. Ury, *Getting to Yes: Negotiating Agreement without Giving In* (London: Penguin Group, 1981).
4. Ibid.

George Bunn (1925–2013) was an American diplomat, lawyer and non-proliferation expert. He served as Dean of the Law School at the University of Wisconsin-Madison and spent the last two decades of his career at the Center for International Security and Cooperation at Stanford University. He was one of the lead US negotiators of the nuclear Non-Proliferation Treaty and was appointed Ambassador to the Eighteen-Nation Committee on Disarmament.

CHAPTER 11

International Systemic Features Inhibiting Disarmament and Arms Control

David Carlton

Introduction

The term "disarmament" is often used imprecisely and it is therefore essential that we should distinguish among various possible forms. The most ambitious is that implying a world in which armaments would disappear altogether as a means of influencing the relations between communities. This is sometimes called General and Complete Disarmament (GCD). Another possibility would be a universal disarmament convention whereby all states would agree not to abolish armaments altogether but would perpetually freeze them at mutually agreed levels (presumably at much lower levels than at present). A third use of the term relates to limited measures covering only some countries or involving only particular weapons. This third form is probably most accurately described as arms control. We may conveniently examine each of these forms of "disarmament" in turn in order to explore how far the present international system serves to prevent agreement.

Originally published in David Carlton and Carlo Schaerf, eds., *Reassessing Arms Control* (London: Macmillan and New York: St. Martin's Press, 1984): 28–35.

D. Carlton (✉)
University of Warwick, Coventry, UK

P. Foradori et al. (eds.), *Arms Control and Disarmament*,
https://doi.org/10.1007/978-3-319-62259-0_11

General and Complete Disarmament

This idea was first advanced in the late twenties at the Preparatory Commission of the Disarmament Conference by the Soviet Union. It was at first not well received by most other states on the grounds that it was impractical. But when Nikita Khrushchev revived the approach, speaking to the United Nations General Assembly on 18 September 1959, the Western powers responded with declaratory support for it in principle while reserving their position about the methods for applying the scheme.

The principal objection to GCD, as canvassed by the Soviets, is that it does not give any guarantee that aggression would not continue to take place after it had been carried out. First, there is the possibility that sovereign states might attack a neighbour with internal police forces or the sheer weight of unarmed numbers. As long ago as 1928 a spokesman of the Netherlands pointed out that the internal police forces of the more populous states—and the Soviets have never suggested the abolition of such forces—could be converted, without any formal breach of a GCD Treaty, into armies larger than those of Charles XII, Gustavus Adolphus or even Napoleon.[1] Supposing, however, that this difficulty could be overcome, would not, say, the unarmed Chinese, possessing only stick and stones, find it relatively easy to overrun their unarmed North Korean neighbours, and would not, say, unarmed France easily be able to overrun unarmed Luxembourg? Then there is the problem of possible derogation by sovereign states from a GCD agreement. As the Italian delegate to the Preparatory Commission put it more than 50 years ago:

> I would point out that, supposing that complete disarmament had been carried out all over the world we should be faced with the following situation: there would be some countries, which owing to their wealth, the organisation of certain industries and the extent of their population could—if at any moment they so decided—arm again much more easily and much more rapidly than other poorer, smaller countries not so well endowed from the industrial point of view. I wonder, in such a case, what degree of security this second class could really count on.[2]

We are thus driven inexorably to the view that GCD will not come about in circumstances where some sovereign states would feel that their security would not thereby be enhanced. Yet how could we in every case overcome the doubts caused by the foregoing considerations? The brief answer would seem to be that no sovereign state must have the means to resist an all-powerful world peacekeeping authority in a world of GCD. But for that to be so the world authority must have powerful arms at its disposal, must be free from the great power veto system that currently applies to the United Nations Security Council and must have the confidence of all leading states as being likely to carry out its peacekeeping duties impartially on the basis of an agreed definition of aggression. Whether or not we choose so to call it, we are here talking about a world government elected by consensus among the present sovereign states. Existing deep divisions, ideological and power political, make this idea, however desirable, a utopian dream. The Soviet Union, for example, though for so long an advocate of GCD, has always rejected the idea of the United Nations having peacekeeping forces not subject to veto. The United States, by contrast, adopted a line in 1962 that suggested that she favoured GCD provided only that such a veto-free peacekeeping force were established. At that time, however, the Americans could count on a majority in the United Nations on most contentious issues. Now the picture has changed: they, like the Soviets before them, are in a minority. It should not, therefore, surprise us if the Americans would now be no less opposed than the Soviets to disarming and giving supreme power to a veto-free United Nations peacekeeping body.

The Soviets of course believed in the early days of their Revolution that the key to disarmament was world proletarian revolution. Vladimir I. Lenin, for example, wrote: "Only after the proletariat has disarmed the bourgeoisie can it, without betraying its world-historical task, throw to the scrap heap all kinds of armaments in general—and the proletariat will doubtless do it—but only then, by no means before."[3] It is rather improbable, however, that many in the Kremlin would now take so simplistic a view. For the "proletariat" of China, though it has overthrown its "bourgeoisie," does not seem to be on the best of terms with the "proletariat" in the Soviet Union. They show no signs of reaching agreement about the kind of non-armed border arrangements that have been the case—alas all too uniquely—over more than a century between those two capitalist neighbours, the United States and Canada. Would, therefore, a world of sovereign Communist states be any nearer than the present world to disarmament and permanent peace?

We may conclude that disarmament in the sense of GCD is in practice incompatible with the continued existence of sovereign states of whatever character. Sovereign states, however, irrespective of ideological similarities or differences, show no sign of withering away. On the contrary, there are more sovereign states than ever before. And far from most states moving towards merger of sovereignty, the tendency may be in the other direction. There is also the further complication of the growth of armed substate actors, one of which, the Palestine Liberation Organisation (PLO), has even been allowed, rightly or wrongly, to present its case at the United Nations and in other forums otherwise limited to sovereign states.

Multilateral Disarmament to Levels Consistent with National Safety

During the interwar disarmament negotiations this was the stated goal of all League of Nations members apart from the Soviets, who, as stated, supported the more radical idea of GCD. Between the wars, however, neither goal was achieved. Is the former idea today any more practical? The difficulty in this case is not that states would be asked to disarm totally and rely on a world authority for their security; it is simply that the perceived needs of states vary greatly and cannot be easily reconciled. Here we may cite Winston Churchill's famous fable:

> Once upon a time all the animals in the zoo decided that they would disarm, and they arranged to have a conference to arrange the matter. So the rhinoceros said when he opened the proceedings that the use of teeth was barbarous and horrible and ought to be strictly prohibited by general consent. Horns, which were mainly defensive weapons would, of course, have to be allowed. The buffalo, the stag, the porcupine, and even the little hedgehog all said they would vote with the rhino, but the lion and the tiger took a different view. They defended teeth and even claws, which they described as honourable weapons of unmemorable antiquity. The panther, the leopard, the puma and the whole tribe of small cats all supported the lion and the tiger. Then the bear spoke. He proposed that both teeth and horns should be banned and never used for fighting by any animal. It would be quite enough if animals were allowed to give each other a good hug when they quarrelled ... However, all the animals were very offended with the bear and the turkey fell in to a perfect panic. The discussion got so hot and angry, and all those animals began thinking so much about horns and teeth and hugging when they argued about the peaceful intentions that had brought them together that they began looking at one another in a very nasty way. Luckily the keepers were able to calm them down and persuade them to go back quietly to their cages, and they began to feel quite friendly with one another again.[4]

It is not necessary to share Churchill's view that even to seek disarmament is folly to appreciate that most states, like his animals, are of differing sizes, with varied interests to defend, and hence cannot easily, even with goodwill, reach agreement on arms ratios that are fair to all. And the lower the level of armaments proposed, the greater will be the difficulty, for irreducible differences of geography, natural resources and population will come to have ever more importance in assessing the implications for each individual nation's security.

Possibly if there were only a handful of sovereign states, an acceptable formula for far-reaching disarmament measures could be arrived at. But, with the best will in the world, those required to draft an agreement for over 150 sovereign states face a daunting task. And it is not much less formidable if we confine our attention to the world's two dozen leading military states.

Perhaps we are again driven back to the view that the key to disarmament and permanent peace is the abolition of sovereign states, difficult though that would be to accomplish.

Arms Control

In practice it is to more limited measures that the world's negotiators must look in the foreseeable future. These will certainly not guarantee an era of assured peace or bring an end to reliance by many states on the notion of deterrence or enable vast resources to be rapidly transferred to the task of developing the poorer parts of the planet. But some arms control measures have already brought some limited benefits and hence this approach should not be too lightly dismissed. For example, between the wars the leading naval powers were able to reach substantial agreement on limiting the major categories of naval armaments. And in the last two decades several measures have been arrived at, such as the partial nuclear test ban and the Non-Proliferation Treaty (NPT). True, not all the relevant powers have yet subscribed to these agreements but the fact that some have done so is arguably of great value in that there have been at least some limits set to the scale of the arms race.

Most spectacular of all these partial measures of arms control has surely been the Strategic Arms Limitation Talks I (SALT I). Even those who are most sceptical about arms control cannot deny that the superpowers have placed firm limits on the numbers of their land-based intercontinental ballistic missiles (ICBMs), their missile-carrying submarines, and anti-ballistic missiles (ABMs). These levels are of course very high (except in the case of ABMs) and would still permit either superpower to inflict enormous damage throughout the world. And certainly non-nuclear signatories of the NPT can justly complain that the superpowers have not wholly honoured the spirit of their undertakings given in 1968. Yet the SALT I agreements have brought great advantages to both Moscow and Washington in recent years. First, each superpower knows that for the present at least it has an invulnerable second-strike capability and hence need not fear a rationally calculated pre-emptive strike. This is the so-called Mutually Assured Destruction relationship, which is not as indefensible as its acronym would suggest. Second, neither superpower need engage in a financially crippling race to acquire more of the items covered by SALT I.

What we have to seek are ways of extending the SALT I principle of limitation, to the *qualitative* sphere, to other kinds of weapons and to many more states. Unfortunately, however, progress will not be easy. This will be the case even if goodwill is shown in all quarters. For once again the existence of sovereign states limits what can be achieved. Above all, the difficulty lies in ensuring adequate verification of any proposed measures. The principal arms control arrangements made so far—the interwar naval agreements, the partial nuclear test ban and SALT I—have in common that cheating would be almost impossible. In each case national means of verification existed. In the 1920s, for example, no state could build a battleship without the espionage services of another state easily being able to discover it. Again blatant breaches of the partial test ban can be discerned by seismologists throughout the world. And satellite observation permits the superpowers to assure themselves that significant cheating by their rival under the SALT I agreement is not taking place.

More far-reaching measures cannot be verified in this way. For example, serious doubts about the verifiability of aspects of the SALT II agreements were at least as important as the worsening of the relations between Washington and Moscow arising out of the Soviet invasion of Afghanistan in causing the non-ratification of those agreements by the United States Senate. And the verification problems would be even more severe in any far-reaching arms control agreement on a multilateral basis covering conventional weapons. Intrusive inspection on a vast scale would be needed, for example, to verify an agreement to ban tanks—something many states would refuse to permit on the grounds that it would constitute an excuse for generalised espionage and interference in internal affairs. And in some respects no amount of inspection, even if acceptable, would suffice. Most importantly, fissile material for making nuclear bombs is now unfortunately possessed by many states in unknown quantities and it could be stored away in such a small space that no certainty could ever exist that a given state had actually renounced the possession of such material. Some will retort that all that is needed is trust. But it is precisely the absence of mutual trust that has led most sovereign states to acquire the means of waging war in the first place. We are thus again driven back to the root cause of the arms race: the existence of sovereign states.

Another unfortunate fact is that we have no way of preventing scientists from presenting their governments with technological innovations. And alas the next years may see developments that will undermine the value of such arms control agreements as we already have. In particular, the cruise missile represents a terrifying threat. The Americans will soon be able to give them a range and an accuracy that will enable them to deliver nuclear weapons anywhere by these means. And over time many other states will also be able to acquire this cheap means of delivery. The most tragic aspect, however, is that these vehicles are so small that once in mass production no agreement to ban or limit them could in practice be adequately verified. This is nobody's fault: it is simply a technological breakthrough that can be no more prevented in a world of sovereign states than any other advance in knowledge. But it does mean that the SALT I agreements limiting those means of delivery that can be seen from overhead satellites will become increasingly irrelevant. It is, of course, open to the American President in a unilateral gesture to slow down or halt production of cruise missiles. But even if he said he would do so, how could the Soviets—and others—be sure that the Americans would not secretly cheat? And even if he were sincere, how could the American President know that the Soviets—and others would not use the period of unilateral American restraint to push ahead with research designed to catch up? Experience suggests that no sovereign state, having a perceived need for nuclear weapons, could in elementary prudence forgo the possibility of acquiring the most up-to-date means of delivery particularly when unsure whether its adversary might be planning to acquire the same means.

Conclusion

We are thus required to live in a world where technological innovation conspires with the existence of the sovereign state to present arms controllers with formidable and often intractable problems. In these circumstances it ill-becomes any government to make sterile propaganda attacks on its peers. It is also no service for governments to raise false hopes by arguing that easy solutions are at hand. Each government has a duty not merely to avoid such shallow rhetoric but, on the contrary, to admit the existence of almost insuperable difficulties. Only from the basis of having made an honest diagnosis of the essential elements of the situation facing mankind will it be possible to make even that minimum of progress which is unfortunately the most that can realistically be expected.

Notes

1. *Official Journal of the League of Nations* 9 (May 1928).
2. Ibid.
3. Nikolaj Lenin, "The Disarmament," first published in *Sbornik Sotsial-Demokrata* 2 (December 1916) reprinted in *The Collected Works of Vladimir I. Lenin*, vol. 19, 2nd Russian edition (Moscow: Progress Publishers 1964): 326.
4. Quoted in Henry W. Forbes, *The Strategy of Disarmament* (Washington, DC: Public Affairs Press, 1962): 47–48.

David Carlton is former Senior Lecturer in International Studies at the University of Warwick. He previously held similar positions at the Open University and at what is now the London Metropolitan University (where he is a professorial research fellow). His publications include (1) *Anthony Eden: A Biography* and (2) *The West's Road to 9/11* (London: Palgrave Macmillan).

CHAPTER 12

The Problem of the Nuclear First-Use Option

Cui Liru

> The real problem at issue ... is the problem of our attitude towards weapons of mass destruction in general, and the role, which we allot to these weapons in our own military planning. Here, the critical question is: Are we to rely upon weapons of mass destruction as an integral and vitally important component of our military strength, which we would expect to employ deliberately, immediately, and unhesitatingly in the event that we become involved in a military conflict with the Soviet Union? Or are we to retain such weapons in our national arsenal only as a deterrent to the use of similar weapons against ourselves or our allies and as a possible means of retaliation in case they are used? According to the way this question is answered, a whole series of decisions are influenced.[1]

These words were written in 1950 by George Kennan, but they still apply. The question touches the heart of the issue of the nuclear danger and the arms race today. The way this fundamental question has been answered in the past years has already cost us a great deal, but it has also taught us a lot and spurred us on to find a better way to answer it.

Originally published in David Carlton and Carlo Schaerf, eds., *The Arms Race in an Era of Negotiations* (London: Macmillan and New York: St. Martin's Press, 1991): 20–25.

C. Liru (✉)
China Institutes of Contemporary International Relations,
Beijing, The People's Republic of China

P. Foradori et al. (eds.), *Arms Control and Disarmament*,
https://doi.org/10.1007/978-3-319-62259-0_12

It is true that no other single concept has dominated international strategic theory during the last three decades so much as that of nuclear deterrence. And it is equally true that we can hardly discuss this subject meaningfully without discussing the relationship between the United States and the Soviet Union, who are the major players in this game. The calculus of deterrence is intimately connected to advanced weapons technology, and this technology constantly changes. The calculus of deterrence is also intimately connected to perception and assessments of adversaries about one another, and perception and assessments are subject to the influence of strategy, political intention, international events, ideology, historical legacy and established structures. When adversaries have developed roughly equal capabilities in technological progress and a strategic parity is achieved, the factors of perception and assessment become more important in the workings of nuclear deterrence. For they could either positively help to create a stable environment in which deterrence holds with maximum certainty, which would in turn reduce the possibility of nuclear war to minimum, or negatively help to create circumstances where nuclear deterrence holds with minimum certainty, which would in turn increase to the greatest degree the possibility of nuclear war.

The object of nuclear deterrence is to deter an opponent from launching a first nuclear strike by holding nuclear weapons capable of confronting such an aggressor with the threat of an unacceptable level of nuclear retaliation. Deterrence with maximum certainty implies the acceptance of the notion by nuclear powers that the only utility of nuclear weapons is to deter use of their counterparts and thus encourage the end of the arms race in nuclear weapons, given that more weapons contribute nothing to deterrence. If this first stage could be successfully reached, then, following the same logic, it might be possible to go on to real reductions and finally to the elimination of nuclear weapons. For retaining them would be meaningless if the possibility of their being used had vanished. Though reasonable in theory, the prospect of the realisation of this programme seems hopelessly remote. For the development of technology makes the calculus of deterrence more and more complicated, and mutual misperception and distrust mean that deterrence is based on minimum certainty. Moreover, the development of so-called extended deterrence has contributed to both of these aspects of the central problem.

Extended deterrence is intended to deter conventional military attacks by posing a risk that nuclear weapons will be introduced into any conflict.

This differs from the original purpose of nuclear deterrence. The rationality of nuclear deterrence is thus undermined and the chances of a nuclear war are thereby increased. Since 1949, when the Soviet Union tested its first atomic bomb, the United States has been seeking to re-establish her nuclear superiority, and to regain the diplomatic initiative by marrying the concept of nuclear deterrence to the concept of nuclear first use, a pernicious theory which, as Kennan has argued, "has lain at the heart not only of the nuclear weapons race … but also of the proliferation of nuclear weapons across the globe."[2]

There are not many who disagree in principle with the concept that nuclear weapons are not like other weapons, and similarly there are not many who reject the concept of nuclear deterrence. But the issue of nuclear first use is more contentious. As Kennan noted in 1950:

> it is not questioned that *some* weapons of mass destruction must be retained in the national arsenal for purpose of deterrence and retaliation. The problem is: for what purpose and against the background of what subjective attitude, are we to develop such weapons and train our forces in their use.[3]

To put it plainly, should or should not such weapons be included in a military strategy that predicates and is dependent upon their use? Retention of the first-use option obviously means "yes," although its advocates try to rationalise it with talk about the conception of deterrence.

One of the arguments for the "first-use" option is based on the belief that in the nuclear age the choice before us is no longer peace and war, but peace and extinction. Accordingly, any major military conflict between the United States and the Soviet Union, or any conflict involving their vital interests, will in all likelihood escalate to nuclear war. Therefore, in order to prevent nuclear war, it is necessary that there should be no conventional warfare between the United States (and its allies) and the Soviet Union. "First use" is intended to make the escalation inexorable, so that conventional attack from the other side will be deterred. But the "first use" advocates here forget, or rather avoid, an essential difference between the concept of "first use" and that of deterrence: deterrence is based on the defensive intention to use nuclear weapons for retaliation after one has been attacked by nuclear weapons; "extended deterrence" is based on the offensive intention of initiating the use of nuclear weapons to assist others being attacked by conventional weapons.

The rationality of deterrence in essence rests on the belief that nuclear weapons are not like conventional weapons. They are there to deter their use. Mutual Assured Destruction (MAD) is credible not only because no one wants it, but also because it creates no misperception of the other's intentions, and the two parties expect to be treated according to the same standard. MAD is defensive in nature, posing no threat to the other side. Instead it serves to deter either party from being tempted when a crisis arises to seek the advantage of initiating a first strike. This deterrent arrangement cannot, however, be simply extended to provide a deterrence against the other party attacking a third party with conventional forces. Efforts to extend deterrence in this way lead only to its credibility being questioned. For the rationality of mutual nuclear deterrence is thereby undermined. All concerned face a vitally challenging question: would the United States be willing for someone else to trigger off a nuclear war at the cost of her own survival? In other words, extended deterrence is not credible unless a case can be presented to convince both allies and adversary that the first-use policy is plausible. This has led to the development of a war-fighting theory, which argues that nuclear war could be fought in a controllable way. The initiation of the use of nuclear weapons at a low level, it is argued, would not necessarily lead to an all-out nuclear exchange, especially if one has adequate capability and means to convince the adversary that escalation could only make his situation even worse while giving him no military or political advantage. But such an attempt to rationalise a nuclear war to make extended deterrence seem more credible not only puts the European countries in an awkward position, but it also has some other serious consequences.

Once a strategy prepared for the possibility of waging a nuclear war is adopted by one side, the opposite side cannot but feel itself subject to at least potential threats. In response the adversary would always feel it necessary to treat such a strategy as an offensive one even if it is claimed to be merely intended for defensive purposes. Despite the fact that there is no evidence to prove the Soviet Union has ever planned or even wanted a war against Western Europe, despite the fact that most policymakers in the West deem it most unlikely that the Soviet Union will launch a surprise attack against countries of the North Atlantic Treaty Organization (NATO) in any imaginable circumstances, the Soviet Union is treated in the NATO strategy as one prepared for an offensive conventional war, largely because of her capability to wage such a war. Similarly, therefore, when the Soviet Union adopts, as she does, the same attitude towards

NATO's nuclear "first-use" option and the strategy based on it, the possibility of fighting a nuclear war becomes more thinkable. This consequently contributes to the tension and distrust between the two sides. Furthermore, as noted earlier, to fight a nuclear war in a controllable way actually calls for adequate capabilities at every level of nuclear forces so as to discourage the adversary from escalating the war to a higher level, and thus hopefully to end the war before it gets out of control. Such adequate capabilities obviously mean a kind of superiority. Though it is fundamentally doubtful whether any meaningful superiority could be achieved, adoption of such a strategy automatically inclines the other side to interpret whatever efforts are made to strengthen nuclear forces as an attempt so to do. Given the adversarial nature of the NATO-Warsaw Pact relationship, therefore, extended deterrence based on a "first-use" option, while allowing the European allies to develop their conventional forces at a less rapid pace, has by and large promoted the arms race, especially at the nuclear level.

The "first-use" option is naturally based on the assumption that the United States might have to fight with nuclear weapons and that she might be the one to initiate their use. So she must be prepared for this possibility. Consequently, nearly all aspects of the training and equipment of her armed forces, not to mention the strategy and tactics underlying their operations, have been affected by the assumption. This cannot but enhance the hostility and increase the possibility of these weapons being used.

Some believe, however, that extended deterrence could be made to appear more credible if the United States seems to be closely bound to her allies. For it is claimed that the inexorability of escalation would happen regardless of interests. Thus all eggs are put in the basket of "first use." There is a certain logic here in that a premeditated war may thus be deterred. But is there any evidence that Europe has ever been threatened by a premeditated war since the end of the Second World War? Wars do not always arise from acts of outright aggression; they are more apt to proceed, as history shows, from a confused situation arising out of a background of extreme political tension. In this sense NATO's strategy could play a negative and harmful role. For it surely increases the possibility of nuclear war by deliberately lowering the nuclear threshold and by making escalation inexorable.

When extended deterrence was first adopted, however, it seemed credible because these problems were obscured by the overwhelming American nuclear superiority. The United States seemed able to fulfil her commitments in Europe at no risk of being destroyed. But as soon as the Soviet Union achieved nuclear parity with the United States, the question of the credibility of the American nuclear guarantee became an acute one. Strategies involving limited war and flexible response represent American efforts to make it seem that the United States heartland could escape the effects of nuclear war while permitting the country to fulfil its commitments. But these efforts have made things even worse in the view of many Europeans. For to think about the use of nuclear weapons makes the use of these weapons more thinkable. One cannot consider the possibility that deterrence may fail without contributing to the likelihood of its failure. Thus the "first-use" option is not a natural and indispensable part of deterrence as some argue but a dangerous option adopted out of expediency. In the search for disarmament, abolishing this "first-use" option would be by itself a positive step, if only because "first-use" has long been a major reason and excuse for the war-fighting advocates to ask for the building-up of an "efficient" and more "credible" nuclear arsenal, which has been one of the major causes of the arms race. Finally, "no first use" is not just words. It is also, as with "first use," a policy. It is rejected because it requires fundamental changes in the established political, economic and military arrangements of the Western Alliance.

The insistence on "first use" largely reflects the negative and hopeless quality of Cold War policies, which include the endless series of distortions and oversimplifications, the reckless application of double standards to judgements about Soviet conduct and the United States' own, and the corresponding tendency to view all aspects of the relationship in terms of a supposed total and irreconcilable conflict of concerns and of aims. What has resulted is a kind of rigidity and traditionalism, causing the governments in Western countries to ignore the fundamental distinction between conventional weapons and weapons of mass destruction and preventing them from finding, or even seriously seeking ways of escape from the fearful trap into which the emphasis on nuclear weapons is leading them.

The commitments that have woven the nuclear arms race so tightly into the fabric of American foreign policy since Hiroshima have convinced many that the arms negotiations, at least until recently, have been "an integral part" of the arms race. To break the stalemate requires fundamental and extensive changes in both outlook and in approaches to these problems. But such changes cannot be made without some kind of transformation of the *status quo*, namely, the established bipolar system, from which these problems largely derive.

Notes

1. Reprinted in George F. Kennan, *The Nuclear Delusion: Soviet-American Relations in the Atomic Age* (New York: Pantheon Books, 1982): 3–4.
2. Ibid., 106.
3. Ibid., 4.

Cui Liru, professor, is a senior advisor to the China Institutes of Contemporary International Relations, a think-tank in China known for its comprehensive studies on current international affairs and prominent role in providing consulting services to the Chinese government. He is a member of the Committee of Foreign Affairs of the Chinese Peoples' Political Consultative Conference and also serves as director of board and/or senior adviser to multiple institutions for the study of national security and foreign relations. As a senior researcher, his studies cover US foreign policy, US–China relations, international security issues and Chinese foreign policy.

CHAPTER 13

The Problem of Extended Deterrence in NATO

Jane M.O. Sharp

Introduction

The reliability of the American security guarantee can be analysed in three dimensions: how credible a deterrent to the Soviet Union as the presumed adversary, how reassuring to the European allies who are being offered protection and how convincing to the United States herself as a basis for managing alliance relations. Denis Healey, a former British Defence Minister, captured the difference in allied and adversary perceptions when he noted that it only required a 5 per cent chance of nuclear retaliation to deter a Soviet attack on Western Europe, whereas a 95 per cent chance might not be enough to reassure the more nervous and dependent allies in the North Atlantic Treaty Organisation (NATO).

Originally published in David Carlton and Carlo Schaerf, eds., *The Arms Race in an Era of Negotiations* (London: Macmillan and New York: St. Martin's Press, 1991): 60–70.

J.M.O. Sharp (✉)
Centre for Defence Studies, King's College London, London, UK

P. Foradori et al. (eds.), *Arms Control and Disarmament*,
https://doi.org/10.1007/978-3-319-62259-0_13

Deterring the Soviet Union

The bipolar balance of power that emerged after the Second World War would probably have been enough to deter each superpower from directly threatening the interests of the other, even if nuclear weapons had never been developed and if NATO had never been established. For Soviet leaders who had twice seen the United States come to the aid of Western Europe in the absence of any alliance commitment to do so, the integration of American and West European forces in NATO, and the forward deployment of American forces and equipment, must seem like icing on the cake of an already firm commitment based on manifest American interests. While nuclear weapons may induce additional caution in superpower relations, the exact nature of the NATO guarantee and the precise level of American nuclear forces do not change in any fundamental way the existential deterrence of Soviet military adventures in Western Europe, or for that matter NATO intrusions into Eastern Europe, that rests on the allocation of political, economic and non-nuclear military power in the international system.

We cannot prove that nuclear deterrence has deterred the Soviet Union from invading Western Europe, because we cannot prove that the Soviets ever had any interest in doing so. Indeed, apart from some pressure on West Berlin in the late fifties and early sixties, there is little evidence to suggest that Soviet leaders had either the opportunity or the urge for any territorial aggrandisement westwards. They appear to have been preoccupied with the effort to consolidate Eastern Europe as both a security and an ideological buffer against military threats and destabilising influences from the West. Keeping the Soviets out of Western Europe may well be a much less demanding task than NATO planners assume. For a nation as cautious and risk-averse as the Soviet Union has proved to be, any military move westwards would have to be a last-ditch effort to avert the imminent collapse of Soviet control over Eastern Europe or an imminent attack from the West. But even assuming such a motive, in addition to the prospect of nuclear retaliation, Soviet planners would have to weigh both NATO's conventional military capability and the American resolve to defend its interest in the political and territorial independence of Western Europe.

Thus, Soviet perceptions of American interests are as important as their assessments of American military capability in deterring Soviet aggression against Western Europe.

Reassuring the Allies

Perceptions of American interests are also a crucial aspect of reassurance for the NATO allies, who are constantly seeking evidence that American interests are unambiguously coupled with those of Western Europe. This coupling is most reassuring in its economic and political dimensions as exemplified by the Marshall Plan and the North Atlantic Treaty. To the extent that the NATO guarantee is measured by the American military presence in Western Europe, however, the process of reassurance becomes hostage to the East-West arms control process, since the prospect of limits on American forces can unsettle dependent allies.

The North Atlantic Treaty, signed in 1949, was intended to communicate American intent to the Soviet Union and did not initially involve the forward deployment of American forces on a permanent basis. In 1950, however, the outbreak of the Korean War called into question the value of a purely political guarantee, so President Harry S. Truman sent four divisions to Western Europe to join the two divisions already there on occupation duty. A second, equally important, purpose of these American troops was to make West German rearmament more widely acceptable in Western Europe, especially in France.[1] For many West Europeans, NATO's role of binding West Germany firmly to the western democracies remains as important as its role in deterring the Soviet Union, a point often forgotten in the United States.

In the early fifties, Great Britain and the United States both adopted military strategies based on the assumption that nuclear weapons could substitute for inadequate conventional forces. It should be noted, however, that intelligence assessments in the late forties (as opposed to government declarations) suggest that Western conventional forces were never seriously inferior to those of the Soviet bloc.[2]

The initial American nuclear guarantee to NATO stems from an October 1953 National Security Council Paper (NSC 162/2) on Basic National Security Policy, and from the "massive retaliation" speech that Secretary of State John Foster Dulles made to the Council on Foreign Relations in January 1954, later published in *Foreign Affairs*.[3] Both documents posit a massive retaliation with American nuclear weapons to any kind of Soviet encroachment on Western Europe; a policy that was endorsed by NATO as a whole at the North Atlantic Council meeting in December 1954.

Lawrence Freedman argues that NATO policy was always "'flexible retaliation' with 'massive retaliation' as just one option among many,"[4] but any use of American nuclear weapons on Europe's behalf became less credible (to the allies) and more risky (to the United States) as the Soviet Union acquired her own nuclear weapons, and especially when she acquired intercontinental-range nuclear delivery vehicles. The strategy was thus modified (though not without a long and difficult intra-alliance debate) to "flexible response," which posited a NATO response appropriate to the level of aggression, from non-nuclear, through tactical nuclear to strategic nuclear. The "flexible response" strategy is a conscious compromise that can imply one thing to the Soviet Union, another to the United States and yet another to the West Europeans. As long as the concept is left fuzzy it has been generally reassuring to the allies, but problems arise whenever officials attempt to clarify the implications of executing the strategy.

NATO's Security Dilemma

States join alliances to gain protection by pooling resources with others facing a common threat. These benefits are nevertheless offset by the costs associated with dependence and reduced freedom of action. Thus each ally oscillates between fear of abandonment in a crisis (the cost of dependence) and fear of entrapment in a conflict not of its own choosing (the cost of reduced freedom of action).

As Glenn Snyder argues, the alliance security dilemma rests on the fact that these two costs are themselves in conflict, since the steps that a security guarantor takes to alleviate one alliance fear tend to trigger the other.[5] Thus, when the United States responds to European fears of abandonment with a new generation of nuclear weapons, this quickly produces fear of entrapment in a Soviet-American nuclear exchange. NATO's cycle of anxiety, whereby European calls for reassurance generate American hardware responses that in turn generate new anxieties, has been described in detail elsewhere.[6] Suffice to note here that two of the more significant cases were the December 1956 decision to distribute battlefield nuclear weapons to the European allies after the trauma of Suez, and the December 1979 decision to deploy Cruise and Pershing II missiles in Western Europe after the trauma of the Strategic Arms Limitation Talks (SALT) II. Plans for a new generation of short-range nuclear weapons in Europe emerged from the trauma of the 1986 Reykjavik summit, but were shelved in the new détente of the nineties.

The US Dilemma

For the United States the dominant NATO risk is clearly entrapment in a nuclear conflict on behalf of her European allies. American leaders thus prefer to reduce the risks associated with alliance by adopting a military doctrine that delays as long as possible the need to use nuclear weapons. In the American view, this requires the West European allies to acquire robust conventional forces to hold out as long as possible in the event of an attack from the Soviet Union and her Warsaw Pact allies. The move away from a firm American commitment to nuclear retaliation on behalf of Europe began during the John F. Kennedy administration, with the move to "flexible response," and has continued with pressure to move from "flexible response" to "no first use," with President Ronald Reagan's interest in strategic defences and deep cuts in European-based American nuclear weapons, and with numerous proposals to withdraw substantial numbers of American troops from Europe and to boost European conventional forces. These moves increase European fears of nuclear abandonment, which are further exacerbated by public statements of former government officials confirming that no American President could risk nuclear holocaust on behalf of the allies.

Henry Kissinger, for example, told a West European gathering in September 1979 that it was absurd to base the strategy of the West on the credibility of mutual suicide. He chastised the West Europeans for "asking us to multiply strategic assurances that we could not possibly mean, or if we do mean we should not want to execute because if we execute, we risk the destruction of civilization."[7]

More recently, in a BBC interview, former Secretary of Defense James Schlesinger claimed that American nuclear weapons in Western Europe had no military significance, since American strategic forces represent the retaliatory capability of the alliance and are more than adequate to handle any threat from the Soviet Union. Schlesinger claimed that Cruise and Pershing II were deployed in Western Europe primarily to reassure Europeans who were concerned that "President Carter was a weakling, a wimp in the White House."[8]

How decision-makers assess the requirements for credible deterrence of an adversary, and reassurance of an ally, depends on beliefs about the nature of the threat, the relative value of military force and diplomacy in enhancing security, and the utility of different kinds of force postures and bargaining tactics towards these ends. Belief systems vary both within and between governments. Discounting Manichean thinkers at one end of the spectrum and pacifist "accommodationists" at the other, we can identify at least three different mindset in the political culture of most countries: confrontational, competitive and cooperative.

Confrontational decision-makers in the United States tend to see the Soviet Union as inherently expansionist and aggressive, and believe that unambiguous American superiority in all categories of military force is the only safe and feasible option for the West. They tend to emphasise the need to acquire a capability to fight and win nuclear wars, over the need for cooperative arms control regimes with the Soviet Union. They favour a unilateral interventionist foreign policy and tend to see NATO as an unnecessary drain on American resources and a restraint on unilateral American interventions in Third World trouble spots.[9] This has been a prevalent view at the Heritage Foundation, the Committee on the Present Danger and among key figures in the first Reagan administration. Richard Perle was the prime example and a source of considerable anxiety among those West Europeans whose dominant alliance fear is of entrapment in a Soviet-American conflict.

Competitive, rather than confrontational, mindsets dominated the Jimmy Carter, Gerald Ford and Richard Nixon administrations. They viewed the Soviet Union not as inherently expansionist but as a formidable adversary whose military power and political influence must be contained rather than rolled back. They saw Soviet-American nuclear parity as the minimum necessary for stable deterrence, but wanted to maintain American technological superiority and the capacity for limited nuclear options, to maintain at least a theoretical capability to control escalation in a nuclear exchange. They also tended to discount the contribution that NATO nuclear forces make to the Western deterrent, and to be unduly pessimistic about the balance in non-nuclear forces.

Cooperative decision-makers tend to see the Soviet Union as a basically status quo defensive power, opportunistic rather than aggressive in its foreign policy. Robert McNamara, Cyrus Vance and the early Carter are good examples: advocates of more modest deterrent forces, who saw the invulnerability of the deterrent as more important than sheer numbers, and who took a more sanguine view of the East-West conventional balance. The irony is that American policy-makers who advocate cooperative behaviour towards the Soviet Union will often be the least reassuring to conservative West European government officials who fear abandonment.

Have NATO Perspectives Changed over Time?

Edward Luttwak and others have argued that we are now in the post-nuclear era, by which they mean that, although possession of nuclear weapons can still deter direct nuclear attack, the credibility of deterring non-nuclear attacks against third parties has eroded.[10]

But the fact is that West European leaders have always worried about the credibility of extended nuclear deterrence, especially since Sputnik, when it was clear that the Soviet Union would acquire the capability to strike American territory. It was clear from that moment on that the risk of American retaliation on Europe's behalf was too great. In articles and interviews, former West German Chancellor Helmut Schmidt claims that "flexible response" was never credible to him, and he resolved on becoming Minister of Defence in 1969 that in the unlikely event of a Soviet attack, he would do nothing to aid escalation by the West into a nuclear war. More significantly, Schmidt now sees reliance on American nuclear weapons as undermining West Germany's will to defend herself.[11]

France and Great Britain never showed much confidence in the American nuclear guarantee and acquired their own nuclear deterrent forces as soon as possible after the Second World War, despite various US efforts to maintain central control of the Western deterrent. When France embarked on her nuclear weapons programme, President Charles de Gaulle asserted that nuclear weapons could be credibly used by a nation state only in the direct defence of its own territory, thereby denying the plausibility of one nuclear power extending deterrence over another. The French deterrent is thus an expression of French independence from the United States. The British motivation to acquire nuclear weapons was different, namely, to restore the special Anglo-American relationship that had prevailed through the Second World War under President Franklin Roosevelt and Prime Minister Winston Churchill. The British sought interdependence and shared responsibility—a seat at the high table—rather than independence.

For successive West German governments, denied an independent nuclear capability as a condition of joining NATO in the mid-fifties, the nuclear arsenals of Great Britain and France (be they interdependent or fully independent) serve primarily to highlight the West German sense of singularity within the alliance, and its emasculating dependence on the US security guarantee. This partly explains Chancellor Helmut Kohl's resistance, through the summer of 1987, to giving up his 72 American nuclear-armed (but obsolescent) Pershing IA missiles, even at the risk of sabotaging a Soviet-American Treaty on Intermediate-Range Nuclear Forces (INF).

The degree of reassurance that West Europeans need depends on many factors: perceptions of the East-West military balance, geography, historical experience of support in crises, images of the adversary and the confidence of its political leadership. In general, the less confident fear abandonment more than entrapment and require more military reassurance from the United States, while the more confident fear entrapment more than abandonment and are more reassured with a minimal deterrent capability.

Fear of abandonment is most prevalent among West German leaders who are on the forward edge of the hypothetical battle area, and are less sure of their political identity than other West Europeans. No other NATO ally seems to fear abandonment to the same degree, but the prospect of fighting any kind of war, either nuclear or non-nuclear, is abhorrent to all Europeans. So West European governments often seem more comfortable with a defence policy that advocates early use of nuclear weapons as a means of deterring the Soviet Union; not so much because they fear a Soviet attack as because they do not want to contemplate either preparing for, or fighting, another non-nuclear war. Thus the conservative governments in Great Britain, France and West Germany were not at all enthusiastic about the double-zero INF agreement forged by the United States and the Soviet Union, since they believed this would generate pressure to build up conventional forces in Western Europe.

Opposition leaders in Great Britain, France and West Germany, largely but not exclusively of the centre and centre-left, were generally in favour of the double-zero INF treaty signed in December 1987, and would prefer to explore Mikhail Gorbachev's initiatives on comprehensive European arms control than to compensate for INF limits with a further build-up of conventional forces.[12] It should be noted, however, that West European leaders who advocate cooperative arms control diplomacy with the Soviet Union while in opposition tend to become more confrontational and competitive when in government; Carter, Schmidt and François Mitterrand are good examples of this phenomenon. Thus, Socialists and Social Democrats, now mainly in opposition in Western Europe, can be expected to toughen their positions considerably if and when they are called upon to govern.

The Problem of Political Confidence

Despite the inherent incredibility of a nuclear guarantee, West European expressions of confidence in the American nuclear umbrella have been considered important politically, as a means of complicating the calculations of Soviet leaders who might be contemplating an attack, of coupling American and West European security interests and, most important economically, of rationalising smaller non-nuclear forces in the West European NATO countries than might otherwise have been necessary. West European leaders may keep up the fiction of the efficacy of nuclear weapons, but the two most important components of deterrence have always been the presence of American troops in Europe, and manifest American interest in the political and territorial independence of Western Europe. Thus, political confidence in United States leadership counts far more than details of the force posture, and to the extent that the force posture does matter, it is conventional forces that matter more than nuclear arsenals.

West European confidence in the United States can be shaken by many different kinds of activity. NATO leaders more prone to fears of entrapment tend to lose confidence when American leaders refuse to negotiate with the Soviet Union, or when they intervene in Third World trouble spots like Nicaragua, Grenada, Libya, Lebanon and the Gulf. All this suggests a trigger-happy foreign policy that might drag Europe into a shooting war that would be hard to control. For West Europeans most prone to fears of abandonment, any unilateral moves that the United States makes in an East-West context can trigger anxieties. Moves towards radical arms-reduction agreements with the Soviet Union made without intra-alliance consultations, as at Reykjavik in late 1986, have often seemed particularly traumatic for West German leaders; though not, according to recent polls, for West German publics. As we saw in the late seventies this fear of abandonment can generate calls for reassurance and reaffirmation of NATO security guarantees that can be very destabilising. Any hint that the United States is putting her other global commitments ahead of her NATO commitments also generates anxiety. In the sixties, for example, the Vietnam War drained off thousands of troops from the US 7th Army based in Europe. There have been no significant withdrawals since that time, but through the seventies and eighties the American Congress repeatedly complained that the West Europeans were not pulling their weight in the alliance, and suggested it was time to reduce overseas manpower deployments.[13]

It is this lack of West European confidence in the continued presence of American conventional forces that most threatens NATO cohesion, since most of the potential solutions to the problem imply greater West European independence from Washington, if not necessarily closer rapprochement with the East. More West Europeans are adopting a Gaullist attitude towards the United States. Schmidt, now free of the responsibility of government, is one of the most articulate of these, and in a series of speeches and articles has recently been emphasising the debilitating effects of long-term dependence on the United States. Government leaders are not as outspoken as Schmidt but many are looking for ways to make West European policies more independent of Washington. For some, like Great Britain's Geoffrey Howe, this means building a stronger European pillar in NATO, but in France and West Germany there is talk of a much more independent approach.

In its more confrontational variant, this could lead to a Western defence entity based on a nuclear-armed West European Union that would look to Paris, rather than to Washington, for leadership. The establishment of a Franco-German brigade and a Franco-German Defence Council point in this direction as do the discussions (albeit vague and inconclusive) about the extension of French nuclear deterrence over West Germany. It must be said, however, that few West Europeans, especially those currently plagued by fears of abandonment by the United States, want a new security arrangement that makes them dependent on France.

A nuclear-armed West European Defence Community is of course anathema to the Soviet Union.[14] Reformers who seek more independence from the United States, while at the same time wanting to avoid a new military threat to the East, would restructure West European military forces into manifestly non-offensive defence postures, and pursue East-West arms control more energetically. This seems the more promising option if conducted in parallel with a cooperative Soviet-American arms control regime. In the absence of Soviet-American détente, however, radical restructuring of West European forces might look dangerously destabilising and could further isolate the United States from her NATO partners.

To sum up, American nuclear and conventional guarantees are perceived to be neither as credible nor as reliable, to West Europeans, as they used to be in the days of unambiguous American strategic superiority; but nor are they as necessary. Western Europe is militarily strong and economically prosperous and no longer in as much need of protection as it was when the NATO commitments were first made. On the other hand, most Europeans (Eastern and Western) see NATO as the best arrangement for preventing West-West conflicts and for keeping West Germany firmly anchored to the West. Moreover, geography alone makes the Soviet Union a serious challenge to the integrity of Western Europe, so the necessity of some form of collective Western security arrangement is widely accepted. West Europeans are currently confused, however, about how to deal with both superpowers.[15]

The most dependent conservatives feel more comfortable with a hierarchical alliance structure, strong US leadership and a high degree of East-West tension that keeps NATO cohesion tight. They prefer a structure in which alliance loyalty means standing firm with the United States against the Soviet Union, and they are thrown into near panic by "peace offensives" from the East. West European centrists who are unhappy with the current leadership in Washington, and feel emasculated by long dependence on the United States, are struggling to find some independence within the basic NATO structure. Some West Europeans on the left would prefer to explore a more cooperative relationship with the East even if this means loosening transatlantic ties. Others on the left set a high priority on maintaining NATO intact, primarily as a means of restraining confrontational and interventionist tendencies in Washington, precisely the aspect of NATO that American conservatives find most objectionable.

Much of the unease in Western Europe during the eighties stemmed directly from the Reagan administration's eccentric foreign policy, but any American administration must confront the basic dilemma of alliance leadership, namely, how to establish a security regime with the Soviet Union that also reassures Western Europe. Specifically, the United States must find ways to make détente and arms control with the Soviet Union acceptable to conservatives in Western Europe. Otherwise, if confrontational policy-makers are allowed free rein on both sides of the Atlantic, Washington risks not only the loss of Soviet-American détente, but also the political centre in Europe, and with it the bedrock of the alliance.

Notes

1. Timothy Ireland, "The Year of Departure: 1950," *The Fletcher Forum of World Affairs* 2 (Summer 1978): 114–138.
2. Matthew Evangelista, "Stalin's Postwar Army Reappraised," *International Security* 7 (Winter 1982–1983): 110–138.
3. John Foster Dulles, "Policy for Security and Peace," *Foreign Affairs* 32 (April 1954): 353–364.
4. Lawrence Freedman, *The Evolution of Nuclear Strategy* (London: Palgrave Macmillan, 1982): 76.
5. Glenn Snyder, "The Security Dilemma in Atlantic Politics," *World Politics* 36 (July 1984): 461–496.
6. Jane M. O. Sharp, "After Reykjavik: Arms Control and the Allies," *International Affairs* 63 (Spring 1987): 239–358.
7. Henry A. Kissinger, "NATO: The Next Thirty Years," *Survival* 21 (November–December 1979): 266.
8. James R. Schlesinger, interviewed at BBC radio, 15 July 1987.
9. Melvyn B. Krauss, *How NATO Weakens the West* (New York: Simon and Schuster, 1986).
10. Edward N. Luttwak, "Soviet Military Strategy in the Emerging Post-Nuclear Era," in *National Security Issues of the USSR*, ed. Murray Feshbach (Dordrecht: Nijhoff, 1987): 277–289.
11. *Die Zeit*, May 8, 1987.
12. Horst Ehmke, "A Second Phase of Détente," *World Policy Journal* 4 (Summer 1987): 363–382.
13. For a survey of Congressional proposals on troop cuts and equitable burden-sharing, see A. Maroni and M. Sawtelle, "Selected Congressional Actions and Proposals Concerning Defense Burdensharing," in Stanley R. Sloan, *Defense Burdensharing: US Relations with the NATO Allies and Japan* (Washington, DC: Library of Congress, Congressional Research Service, 1988) appendix 5. See also the debate on burden-sharing amendments in *The House of Representatives Congressional Record*, 29 April 1988, H2735–H2743.
14. *Pravda*, 23 June 1987; and numerous radio broadcasts on 23 and 24 June 1987 transcribed in *FBIS Daily Report*, 29 June 1987.
15. François Heisbourg, "Can the Atlantic Alliance Outlast the Century?," *International Affairs* 63 (Summer 1987): 413–423.

Jane M. O. Sharp held research fellowships at Harvard and Cornell Universities, at Stockholm International Peace Research Institute (SIPRI) and at King's College London. From 1998 to 2003 she served as the United Kingdom representative on the United Nations Secretary General's Advisory Board on Disarmament Matters, and she is a trustee of the Institute for War and Peace Reporting. She has published widely on arms control and disarmament diplomacy.

CHAPTER 14

Minimum Deterrence and International Security

Richard H. Ullman

Introduction: The Attraction of Minimum Deterrence

For something like three decades, "overkill"—the label connoting the fact that the United States and the Soviet Union each possesses enough nuclear weapons to destroy the other's society several times over—has been like the weather: nearly everyone has talked about it but by and large no one has done anything about it. Rather, as new and more sophisticated nuclear weapons and delivery systems were developed, the two superpowers have gone on adding more and more warheads to their stockpiles. That nuclear arsenals should grow was considered almost a natural process—again, the weather comes to mind.

Now, suddenly, it seems that everyone wants to do something about overkill. Leading the procession were the American and Soviet leaders,

Originally published in David Carlton and Carlo Schaerf, eds., *The Arms Race in an Era of Negotiations* (London: Macmillan and New York: St. Martin's Press, 1991): 73–99.

R.H. Ullman (✉)
Princeton University, Princeton, NJ, USA

P. Foradori et al. (eds.), *Arms Control and Disarmament*,
https://doi.org/10.1007/978-3-319-62259-0_14

Ronald Reagan and Mikhail Gorbachev. At their Reykjavik summit meeting in October 1986 they astonished the world by embracing the short-term objective of reducing by half the size of their nuclear arsenals and the long-term goal of eliminating completely middle- and long-range nuclear ballistic missiles.

After Reykjavik, Reagan, at least, was persuaded by his advisers to ignore, if not actually to disavow, the second objective. But the two leaders instructed their negotiators to pursue the first. The result has been the treaty now nearing completion that will codify something like 30 per cent reductions in the long-range nuclear weapons deployed by each side. This agreement will come on top of the 1987 treaty under which each gave up entirely its intermediate-range nuclear missiles.

Success in these negotiations will diminish but not eliminate overkill. The two superpowers will still have more than enough long- and short-range nuclear warheads remaining to cover the ambitious target lists that their military staffs have constructed over the years. Even if only military and industrial targets were selected (as distinguished from population centres as such), the collateral damage from all-out attacks would result in the near-total destruction of their societies.[1]

Awareness of these likely dire consequences has in recent years given rise to a number of proposals for reductions in US and Soviet nuclear weapons stockpiles, much more drastic than the 50 per cent goal embraced at Reykjavik. Robert S. McNamara, the former American Secretary of Defense, has called for reductions down to a level of about 500 warheads for each superpower.[2] Others (including the present author) have suggested that no more than 2000 warheads on each side would be ample to assure the superpowers' security.[3]

Notions like these have been labelled "minimum deterrence" (sometimes "finite deterrence"). They are based upon the premise that there are levels of nuclear stockpiles far lower than those at present that would nevertheless deter the use of nuclear weapons by other states and not expose the state making the reductions to a risk of being disarmed by an opponent's first strike. Thus, to extend the image referred to above, they would provide for "kill" rather than "overkill" or "underkill" (the latter being the condition of a state having too few nuclear weapons to deter the contingencies that cause it concern). For proponents of minimum deterrence, achieving it should be a central objective of American—and Soviet—state policy.

Why Not Abolition?

But should the superpowers not go further and seek the total abolition of nuclear weapons? Reagan's answers to this question seemed ambivalent. On occasion he embraced abolition as a distant goal. Simultaneously, he asserted that his Strategic Defense Initiative was intended to make nuclear weapons borne by ballistic missiles (if not all nuclear weapons) "impotent and obsolete." However, most of Reagan's advisers, and the leaders of the United States' major European allies, have made it clear that they think that abolition is not advisable for the foreseeable future.

Gorbachev's answer has been a resounding and unambiguous "yes." In a major statement on 16 January 1986, and on a number of occasions since then, he has called for doing away with all nuclear weapons.[4] Abolition has, indeed, been a stated objective of Soviet policy ever since the beginning of the nuclear era. There is good reason, however, to think that Gorbachev is more serious about it, and as a practical goal, than his predecessors have been. Like them, however, he has not spelled out just what arrangements, in a world of independent sovereign states, might provide the same central core of deterrence that has been the single (and scarcely negligible) positive contribution that nuclear weapons have made.

The main argument against abolition, of course, is that many states would feel less secure. International agreement might abolish nuclear weapons, but not the knowledge of how to make them. That knowledge is now widespread; it surely must exist among the scientists and engineers of most industrialised, and many developing, states. So long as it exists (and it could not easily be extirpated), states would have every reason to fear that rivals might evade an abolition regime and, either previous to or during a crisis, manage secretly to manufacture and hide away a stockpile of nuclear weapons. Such weapons need be neither numerous nor sophisticated to be potent instruments of blackmail. They might be made either by "advanced" states or "backward" ones, including "crazy" states like Muammar Gaddafi's Libya. For that reason, so long as the international system contains no effective overarching authority, nation-states would be likely to feel less secure in the absence of overt, acknowledged stockpiles of nuclear weapons than with them.

Indeed, the presence of acknowledged stockpiles introduces a pervasive note of caution into international relations, particularly relations between the superpowers. That would be the case even in a minimum-deterrence regime of small arsenals retained by states for the declared purpose only of deterring nuclear use by others. For declared policies can change in the crucible of a crisis. Because no one can say whether a government that had made such a declaration would nonetheless in desperation resort to the use of nuclear weapons to halt a conventional invasion, deterrence tends to be pervasive rather than nuclear-specific. That is what some analysts have termed "existential deterrence." The mere existence of nuclear weapons, regardless of the declared policies of possessing states, exerts a widespread deterrent effect.

Why Minimum Deterrence?

Understandably, the idea of global abolition has attracted a wide popular following. Rallies and demonstrations in support of abolition have filled many plazas in many of the world's great cities. Political parties in many countries (although not in the United States) have made it part of their platforms, and politicians who embrace it have won (and lost) elections. But the arguments for abolition put forward by serious analysts have not been strong ones. Nearly always they tend to make much of the probability that abolition will introduce new elements of trust and cooperation into international relations, and to minimise the probability and perhaps also the consequences of "breakout"—a sudden overt departure from a regime of constraints.

Minimum deterrence also has its proponents. Some, indeed, are governments that have made the concept their state policy. Great Britain, France and the People's Republic of China—the three states other than the superpowers that have acknowledged nuclear forces—all adhere to the notion and implicitly promise never to allow their forces to expand beyond the point necessary to destroy at most a few key cities within a potential adversary's domain. Undoubtedly, they make a virtue of necessity. Who knows what sort of nuclear forces they would buy if they had more resources to spend? But their arguments should be considered on their merits. So should those of the many analysts who have declared themselves in favour of some version or another of minimum deterrence.

Three strands of argument stand out.

"Excess" Arguments

The first set of arguments stems directly from the notion of overkill. Too many nuclear weapons exist; if ever even a small proportion of the superpowers' arsenals is used, their effects would be catastrophic beyond most persons' capabilities for imagination. These effects would include those that have been known ever since the atomic bombings of Hiroshima and Nagasaki: first the primary effects caused by blast and fire, then the secondary and tertiary effects caused by radiation, the spread of disease in the absence of functioning medical and sanitation systems, and the breakdown of most elements of organised society. Depending upon which targets are struck, they might also include the effects posited a few years ago under the label of "nuclear winter"—starvation induced by the failure of crops owing to lack of adequate sunlight brought on by the sun's rays being deflected by pervasive layers of smoke sent into the atmosphere as a by-product of burning cities and oil fields.[5]

The larger the quantity of nuclear weapons that exist, the greater the probable extent of catastrophic damage should nuclear war actually occur. Some experts might hedge such a proposition with the qualification "not necessarily," but they also would concede that the extent of damage is highly dependent on war scenarios, and they would admit that the chances are considerable that once nuclear war begins, the fear of being preempted would cause governments to use their nuclear weapons rather than see them rendered impotent.

A second proposition follows the same line: the larger the quantity of nuclear weapons the greater the probability of an accident that might lead to an explosion, or of an unauthorised launch owing either to system malfunction or to human will. The more the weapons that exist, the greater the chance that one will suffer an accident while in storage or being transported, and the greater the chance that it will fall under the control of terrorist groups or even crazed individuals who might be able either to launch or at least to explode it. Specialists aware of the enormous amount of attention the current nuclear-weapon states pay to safety know that the probability of any such event occurring is extraordinarily low. In the mind of the public it is higher, however. In particular, it seems considerably less low to public in Europe, where the accident at the Chernobyl nuclear power-generating plant is a vivid and alarming memory, than it does in the United States.

"Stability" Arguments

A second set of arguments revolves around the notion of stability. The superpowers' current arsenals contain many nuclear weapons whose characteristics and mode of deployment invite early use. And both the US and the Soviet military establishments have been wedded to strategic doctrines that place a premium on pre-emption in order to destroy the other side's forces before they can be used, and on escalation to bring a conflict to a conclusion successful for the party initiating the escalation.

In this view, really drastic reductions in the size of superpower arsenals—down, perhaps, to 2000 or fewer warheads (both strategic and tactical)—would wring from the system its current built-in propensities for nuclear war-fighting. Whether this occurred, however, would depend upon what sorts of weapons remained and how they were deployed. The smaller a nuclear force the more important it is that it be as near as possible invulnerable, that its ratio of warheads to launch vehicles be as low as possible, and that its warheads be sufficiently dispersed so as to require an adversary to expend at least one warhead (better still, one launch vehicle) for every warhead that it destroys.

In practice, this would mean a force composed of single-warhead missiles (not multiple independently targetable re-entry vehicles, MIRVs). Some might be land-based, either in widely dispersed silos or on mobile vehicles. Others would be carried on either submarines or aircraft. Each "platform" (as the jargon calls it) should carry as few missiles as practicable. Submarines might have eight, ten or a dozen missile tubes—certainly not the 24 mounted in all the ballistic missile submarines launched in the last decade by the US Navy. Aircraft might carry only a handful of cruise or ballistic stand-off missiles, not the 20–30 carried by current generations of American bombers.

The purpose of such a force structure would be to diminish the probability that a state would either suffer a pre-emptive attack or feel compelled to launch one. A state with such a force would know that an adversary had few incentives to attack first, since it could not significantly degrade the victim's capability for retaliation. Similarly, the adversary would have minimal fears of itself being pre-emptively attacked. That would stem in part from the small size of the first state's nuclear force: its commanders would have fewer warheads available for nuclear war-fighting. It would also stem from the fact that the first state's force was deployed in such a manner as to minimise its commanders' fears that unless they were to launch pre-emptively, they would run a large risk of having their own force destroyed in its bases.

Here, it should be noted, one must speak of diminishing probabilities rather than eliminating opportunities. So long as nuclear weapons exist, the probability that a force will be the target of a pre-emptive attack or itself be used to attempt one can never be regarded as zero. That is inherent in the mutual hostage relationship that binds nuclear adversaries together. Indeed, it has been the effort to escape from this condition of hostage that has both driven the nuclear arms race between the superpowers and has led to the adoption by both of strategic doctrines that demand the early use of nuclear weapons once a war starts, lest they be destroyed first.

"Atmospheric" Arguments

A third set of arguments is altogether different. They relate to the atmosphere surrounding the relations between the two superpowers. During the last few years Washington and Moscow have made great progress towards reaching a far-reaching accommodation between them. Deep reductions in the size of their nuclear forces and the deployment of those forces in a posture emphasising their retaliatory roles would assist this process of easing tensions and make it easier for the two governments to address other divisive issues, such as the management of their current and potential conflicts in the Third World. Conversely, the knowledge that present nuclear force deployment patterns and use doctrines place a high premium on pre-emption and escalation would make it more difficult, if a crisis occurs, for the two political leaderships to preserve the gains they have made thus far. Tensions would ratchet upwards to match the hair-trigger readiness of the opposing nuclear forces.

Defining Minimum Deterrence: What Is to Be Minimised?

Numbers of Weapons Versus Types of Roles and Missions

There is no necessary relationship between the size of a nuclear force and the strategic concepts that govern its potential uses. The crucial distinction between a minimum deterrence force and the current US and Soviet forces is not size, but the fact that the latter are configured and deployed for fighting nuclear wars as well as for deterring them. Both governments have spent large sums developing forces sufficiently accurate to destroy "hardened" targets such as missile silos, command bunkers and communication nodes. At the outset of a war, even one in which battlefield nuclear weapons have not been used, both Washington and Moscow would feel strong pressures to use their "hard-target counterforce" capabilities before they were destroyed by the opponent. This is why many specialists feel that a nuclear war between superpowers would escalate uncontrollably.[6]

A large force can certainly fulfil minimum deterrence criteria, however. There is no logical reason why a large force cannot be guided by the principle of No First Use—a determination that it would not be used so long as another party to a conflict had not initiated the use of nuclear weapons. But possession of a large force provides a powerful incentive for developing the capability for using it pre-emptively for hard-target counterforce missions. Conversely, if it is accurate enough even a very small force can be used for such missions. But it would make little sense for a government to deploy only a few hundred warheads and yet adopt a nuclear war-fighting strategy, including spending large sums to develop high accuracy and the real-time intelligence necessary to take advantage of that accuracy. Such a force might be exhausted in an attempt to destroy its opponent's forces without having sufficient numbers of warheads in reserve to deter a retaliatory attack.

Warhead Numbers and Target Sets

While there is no necessary relationship between force size and types of target, it is likely that such a relationship will exist in practice. American force planners—and, no doubt, their Soviet opposite numbers—have found that in acquiring nuclear weapons, as with so many other activities, the appetite grows with the eating. The more weapons they have at their disposal the larger grow their lists of "essential" targets that "must" be hit in order to undermine their opponent's military capabilities. Conversely, they then call for larger numbers of warheads in order to cross-target the more important installations on their list to assure penetration of defences—to make sure, for example, that a given command bunker is targeted not only by two ballistic missile warheads but by a cruise-missile warhead (or two) as well.[7]

Few force planners would proclaim themselves in favour of overkill. Many might even say they are guided by the notion of minimum deterrence—that they are buying just enough forces to do a necessary job. Only the opponent's knowledge that its entire military/industrial structure is at risk would deter it from initiating war. Placing that structure at risk, however, means in practice placing at risk most of the Soviet (or American) population as well. Even a counterforce strike aimed strictly at destroying only a small part of that structure on the other side's intercontinental ballistic missile (ICBM) silos—would result in the deaths of 10–20 million people.

Such scenarios lie near one end of a spectrum. Near the other end lies the threatened destruction of either the adversary state's capital or a small number of its major cities. That is what British and French planners have in mind. They calculate that Soviet leaders would not value the devastation of Great Britain or France more highly than they would the preservation of Moscow or, say, of Leningrad, Kiev and Odessa.[8] Should American planners make a similar calculation? Some who have themselves wrestled with the problem say yes. McGeorge Bundy's often-cited 1969 formulation cannot be improved upon:

> Think-tank analysts ... can assume that the loss of dozens of great cities is somehow a real choice for sane men. They are in an unreal world. In the real world of real political leaders—whether here or in the Soviet Union—a decision that would bring even one hydrogen bomb on one city of one's own country would be recognised in advance as a catastrophic blunder; ten bombs on ten cities would be a disaster beyond history; and a hundred bombs on a hundred cities are unthinkable.[9]

Between these two ends of the spectrum lies a range of other possible target lists. They might include the following types:

All significant military/industrial targets *except* hardened silos and command bunkers. This would require fewer warheads and less accuracy.

Major *soft* military targets, such as airfields, naval bases, large barracks and other troop concentrations. This would require still fewer warheads, and also less precision.

Major industrial facilities. Most are located in population centres. Again, fewer warheads and less accuracy would be required.

Finally, an alternative to targeting major cities (and one that lies even further down the spectrum) might be the targeting of 20 or so nuclear power-generating plants with the intention of creating Chernobyl-type disasters. These would result in far fewer deaths than any of the other target sets, but would cause profound disruption to the functioning of society. This target set would require relatively few warheads, but large numbers of fatalities could be avoided only if low-yield weapons were used. Therefore it would require relatively high accuracy.[10]

No nuclear-weapon state has ever published its target lists. But with the exception of China, and with varying degrees of specificity, each has made known the general categories of targets that its planners intend to put at risk. It is usually assumed that deterrence depends on one's adversary knowing one's intentions. However, the question remains whether such efforts are necessary. It seems likely that deterrence is achieved primarily by potential adversaries knowing that a state has a relatively capable, relatively reliable, and relatively invulnerable nuclear force, and that specifying types of targets adds little to the deterrence already achieved by virtue of the targeted state's knowledge of these capabilities.

Both the United States and Great Britain have let it be known in recent years that their target planners now emphasise the key elements of Soviet power, especially the destruction of the Soviet leadership within even hardened command bunkers. The stated assumption behind such a strategy is that Soviet leaders value their own lives more than they do the lives of large numbers of their fellow citizens. But it seems at least as likely that the knowledge that they are prime targets, rather than adding to the strength of deterrence, would induce key leaders to take the desperate step of launching first in a crisis, in the hope of limiting damage.

Superpower Relations Versus Other Relationships

Our focus thus far has been on the superpower relationship. Only they are capable of overkill. The other three states with acknowledged nuclear forces—Great Britain, France and China—each possess a few hundred warheads. Theirs are minimum deterrence forces already. Even if the British and French go through with their present ambitious plans for modernising and expanding their forces so as to make more certain their ability to penetrate the kinds of defensive systems the Soviet Union might deploy over the next 30 years, each would possess considerably fewer than 1000 warheads. Whether they would agree to reductions in warhead numbers as part of a global arms-control regime would depend on how they evaluate both Soviet defences and the vulnerability of their own forces to a pre-emptive attack. The leaders of the three nuclear middle powers currently seem confident that even though their forces are orders of magnitude smaller than the Soviet force, enough of their weapons would survive an effort to destroy them (especially those deployed in submarines) so that the prospect of their retaliation deters Moscow. In the future, as technologies advance (anti-submarine warfare, for instance), they might feel less confident. Then they might perceive numerical disparities as mattering more.

We will discuss the further proliferation of nuclear weapons later in this chapter, but here it should be noted that none of the potential proliferants—not even Israel, which already almost certainly possesses a nuclear force formidable within its region—seems likely to deploy a force capable of posing a deterrent threat to either superpower. So far as those states are concerned, the relationship between them and the two superpowers will not change by virtue of the latter reducing their nuclear forces to a level of, say, 2000 or fewer warheads.

Implementing Minimum Deterrence

Multilateral Negotiations Versus Unilateral Actions

The touchstone of the process of negotiated arms control between Washington and Moscow has been numerical parity. When the 1972 Interim Agreement limiting strategic offensive weapons codified a Soviet lead in some systems, the United States Senate vowed not to ratify a subsequent agreement unless it included equal ceilings.[11] Yet there exists not a shred of evidence that deterrence depends on numerical equality. So long as a state possesses a sufficient number of weapons capable of surviving an adversary's attack and then penetrating its defences and destroying some targets of substantial value, what matters is the adversary's assessment of the probability that the threat will actually be carried out. That probability certainly need not be anywhere near 1—and in practice, of course, it will never be. Even if a state has suffered grievous harm from an adversary's nuclear attack but some substantial portion of its population still lives, why should its leaders risk those remaining lives by retaliating? Yet who can be confident they will not?

American (and, no doubt, Soviet) nuclear planners have tried to ease the burden of decision on their leaders by giving them nuclear "options"—mixes of weapons and targets that first destroy objects the adversary values less than their own lives, or population as such, in order to limit damage or demonstrate resolve but not to bring on massive retaliation.[12] Yet war game after war game has suggested that this strategy would result in escalation: rather than accept an end to the conflict on less than satisfactory terms, the adversary replies with a similar attack, once again placing the burden of escalation on the first state. That way lies the slippery slope towards nuclear war-fighting, and towards widespread destruction of civilian as well as military targets. It might be preferable instead to de-emphasise these nuclear options and to threaten highly valued objects from the start. That is the choice that London, Paris and Beijing all perforce have made.

These arguments support the contention that a state can adopt a minimum deterrence posture unilaterally. That would not even be an issue if such a posture were simply a change in strategy—in declaratory policy and in the orders given to force commanders. It becomes an issue when the state adopting the posture does so in the context of deep reductions in the number of nuclear weapons it deploys and when the state in question is either the United States or the Soviet Union. While there are no logical reasons for insisting upon rough parity between them, there are cogent political reasons. Unlike the other three acknowledged nuclear-weapon states, the superpowers are not (or have not allowed themselves to be) constrained by economic circumstances to buying only small, minimum deterrence forces. It seems certain that neither domestic political system could tolerate substantial numerical inferiority. For that reason, they should seek through negotiations to institutionalise a regime of minimum deterrence.

The Question of "No First Use"

The easiest way for a state to implement a minimum deterrence strategy is to declare that it will use nuclear weapons only in response to another power's use of such weapons against itself or its allies. China has had such a declaratory policy from the time (1964) she exploded her first nuclear device. So, in fact, has the Soviet Union. In the case of China, given the small size of her nuclear force, the posture is credible. First use by Beijing would invite massive retaliation in response. In the case of the Soviet Union, given not only her large force, but her reliance upon large, highly MIRVed ICBMs that would be lucrative targets for American missiles unless they were launched first, it is less credible. (The Soviets, indeed, solve this problem rhetorically by equating a launch to escape a "certain" attack with retaliation, rather than pre-emption).[13]

For "No First Use" to be fully credible, it should be accompanied by changes in deployment and in the ways in which military forces are trained. The former means phasing out weapons, like the Soviet SS-18 or the US MX missiles, which are pre-eminently first-strike weapons. The same applies to battlefield nuclear weapons that must be used early in a war lest they be destroyed by enemy nuclear—or conventional—strikes. Those of very short range (artillery shells or short-range missiles) should be phased out. Others should be pulled back from front-line locations. The latter means that field commanders should be told that nuclear weapons will not be released to their control until after the adversary has used one, and their exercises should assume second use only.

No First Use may be an element of a minimum deterrence posture, but it need not be. French policy is clearly one of minimum deterrence—given the size of the *force de frappe* it could be no other—but it certainly does not include No First Use. On the contrary, according to official declarations, France would use *préstratégique* nuclear weapons—short-range missiles and air-dropped bombs—against invading Soviet troops. Such weapons would be intended as a warning. If it were not heeded and the attack continued, French policy is rapidly to escalate to the use of strategic weapons against targets in the Soviet Union.

Minimum Deterrence and Extended Deterrence

For the United States, No First Use poses particularly difficult problems. The United States is committed to defend her allies of the North Atlantic Treaty Organization (NATO) against a Soviet attack. It would much prefer to deter that attack than defeat it. American policy-makers are by and large convinced that one reason why the Soviet Union has never come close to using military force against the European members of the alliance is the threat of a US nuclear response, and that without that threat NATO's ability to deter a Soviet attack would be diminished. It might be argued that so long as its NATO obligation continues, the United States should not formally commit itself to No First Use even if Moscow agrees to conventional force reductions that would eliminate the advantages the Warsaw Pact now enjoys. Yet it might also be argued that a declaration of No First Use necessarily expresses only an aspiration, not a certainty. Who can be certain that if war were to break out in Europe, and if conventional defences failed to hold, a desperate American president might not resort to nuclear weapons even if he were pledged to No First Use? So long as both the US nuclear arsenal and its commitment to NATO continue to exist, there will be extended deterrence.

To suggest such a possibility is not mere cynicism. Nor, especially, is it to imply that a stated commitment to No First Use has no value. Rather, so long as it is accompanied by the other elements that add to its credibility, it may have real value in a crisis. Leaders would know that so long as they do not initiate the use of nuclear weapons, there is at least some probability—perhaps a large one—that an adversary committed to No First Use would not do so. That would put a brake on pre-emption. It would slow the rush over the brink, if not stop it. And it might indeed stop it. Since there is good reason to think that the crucial threshold is the one between non-nuclear and nuclear war, and that once that threshold is crossed, further escalation would be formidably difficult to prevent, No First Use may indeed be an important source of stability in a crisis.

Concomitants and Preconditions

Conventional Balance. Many analysts have argued that achieving a conventional balance is a precondition for deep reductions of superpower nuclear arsenals. Here again, there is no logical reason for insisting upon such a linkage until nuclear reductions are very drastic indeed—down to the level of a few hundred rather than, say, 2000 warheads. Assuming that half the warheads of a 2000-warhead force survive a surprise attack (surely a conservative assumption given the crucial importance of survivability, especially for a small force), and that 500 of those surviving are earmarked for retaliatory use against urban-industrial targets, 500 are therefore available for use against the attacker's "soft" military targets. That is a far cry from the thousands of warheads now provided in American (or Soviet) war plans, but even they would be capable of inflicting almost unimaginable damage. Whether a military force could survive such an onslaught, or whether—given the collateral damage such strikes would inflict—the attacker's political leaders would still think the campaign worth pursuing, are questions not easily answered in the affirmative.

Nevertheless, while there may be no strong logical reasons for making balanced conventional reductions a precondition of movement towards a US–Soviet deep-reductions, minimum-deterrence regime, the political reasons for doing so seem overwhelming. The military relationship between West and East has been extraordinarily stable. This is of course the case now, when US–Soviet relations seem better than they have ever been, but it has also been the case in the most tense days of the Cold War, including the first four years of the Reagan Administration. Nuclear weapons have been a central ingredient of that stability, and while there are technological developments on the horizon that might unless checked make the nuclear relationship less stable, there is still no reason to move precipitously to alter the nuclear relationship until there are strong grounds for believing that the conventional balance, also, is stable.

Defensive Doctrines and Postures. Achieving a conventional arms balance at substantially lower levels may be a politically necessary precondition for very deep cuts in the US and Soviet nuclear arsenals, but it may not be a sufficient precondition. It can be argued that, just as the two sides' nuclear forces have been shaped both in their size and in their composition by offensive, war-fighting strategies, the same has been true of their conventional forces. Both sides—the Warsaw Pact, especially, but also NATO—would seek at the outbreak of war to carry the battle to the other's territory. Such strategic doctrines would make a severe crisis especially dangerous. Not only would escalation across the nuclear threshold be difficult to restrain, but so would the conventional escalation that preceded it.

Western analysts have drawn attention to this danger for a number of years, but Western governments have never acknowledged it. Quite the contrary: NATO military doctrines have been more admittedly offensive in their orientation during the last decade than ever before. Perhaps that is why Gorbachev has recently focused on the problem. Not only has he acknowledged that there are asymmetries in the conventional balance, many of them favouring the East, but he has called for especially deep reductions in those elements of each side's force posture that the other finds the most threatening. For the West that means Warsaw Pact tanks, artillery and other armoured vehicles. For the Warsaw Pact that means NATO strike aircraft and short-range nuclear weapons.

Negotiating conventional force reductions—especially asymmetrical ones—will not be easy. But the two sides can meanwhile continue to make progress in an area where they have recently achieved considerable success: negotiating confidence-building measures, such as mutual notification of large troops movements, exchanging observers for manoeuvres and the like, aimed at reducing fears of surprise attacks. Equally encouraging are the informal talks that the superpowers have recently begun between their highest uniformed officers, also aimed at reducing the uncertainties that prompt the worst-case analyses which in turn feed fears.

Chemical Weapons. Western governments have for years been concerned about the Warsaw Pact's preparations for fighting a war with chemical weapons. These include both offensive and defensive capabilities, and they far exceed the West's. Yet Western governments have never elected to spend large sums enhancing their own chemical warfare capabilities. One reason they have not is the view that nuclear weapons—the other weapons of mass destruction—would deter the use of chemical weapons as well. One Western government, that of France, has gone so far as to make the substantial reduction of Soviet chemical stockpiles a specific precondition of its participation in any nuclear arms reduction regime. There is some prospect that the on-going chemical warfare negotiations in Geneva might achieve that objective.

Warhead-Launcher Ratios. A concomitant to a minimum deterrence regime, if not a precondition, should be that the ratio of warheads to launchers should be as low as possible, ideally unitary. The Strategic Arms Reduction Treaty (START) agreement that the superpowers are now in the process of negotiating has justly been criticised because it violates this principle and actually makes the ratios larger. However, that agreement would leave each side with so many warheads—more than 7000 strategic warheads, let alone tactical ones, with which the proposed treaty does not deal—that neither could seriously expect to be able to destroy a substantial part of the other's force with a first-strike attack, and thus the problem is not serious. In the instance of, say, a 2000-warhead force, it would be.

Strategic Defences. A second concomitant should be that no power deploys strategic defences capable enough to allow its leaders to feel that, combined with a first strike, they could prevent a targeted state from effectively retaliating. Small offensive forces are particularly sensitive to defensive developments. That is why it is essential that the 1972 US–Soviet Anti-ballistic Missile (ABM) Treaty severely limiting defensive deployments, and research and development, should be extended and perhaps made even more stringent.

Short-Range Weapons. Whether a minimum deterrence force should include short-range, "battlefield" weapons is a question that raises issues that need not be addressed at length here. It can be argued that nearly all of the missions that can be performed by nuclear weapons based near the front lines of a conventional battle can be performed by "strategic" weapons launched from distant bases. But it can also be (and often is) argued that basing weapons in allied countries—in the US case, particularly West Germany—both helps to reassure allies who fear they may be abandoned in a crisis and makes it less likely that their use would trigger a "strategic" response. Nevertheless, it seems highly desirable that relatively few of the weapons in a minimum deterrence force should be short-range weapons. That is in order to avoid basing them in locations where they might offer an adversary an incentive for destroying them early in a conflict, and thus perhaps promoting nuclear use by the targeted side rather than allowing them to be lost.

Verification

The fewer the nuclear weapons on each side in an arms-control regime, the more important an issue verification will become. Verification—making sure that each party knows the number and types of weapons in the other's nuclear forces—raises three distinct types of sensitivities.

Technical Sensitivities: How Many of What Can Be Detected? The first type relates to the capabilities of monitoring systems. Inevitably, they will be able to locate and count some kinds of nuclear weapons more easily, and therefore with greater confidence in the monitors' accuracy, than other kinds. Nuclear weapons coupled to fixed, land-based, long-range, ballistic missiles surely raise the fewest problems. It would probably be impossible for a state to conceal even a handful of such vehicles. The smaller and more mobile delivery systems become, the more difficult they are to detect. For example, many specialists contend that even with intrusive on-site inspection, it would be virtually impossible to know *how many* sea-launched Cruise missiles (SLCMs) a state has. On the other hand, it would almost certainly be possible to detect *whether* a state had such weapons. For such systems, a total ban may be the only effective limitation.

Stability Sensitivities: How Many Make a Difference? Successful concealment might have destabilising consequences. How many nuclear weapons would a state need to be able to conceal before its leaders had the confidence that they had the requisite edge for successfully launching a disarming first strike? That would depend in good measure on their perception of the vulnerability of the other state's forces. If a sizable proportion of its nuclear weapons were mounted on reliable launch vehicles, such as ballistic missiles, based so as to make them invulnerable, such as on submarines, even the ability of the first state to "break out" of an arms-control regime with, say, twice its permitted number in one or another category of weapons would make little difference to the stability of deterrence.

Political Sensitivities: How Many Make an Impact? Very large numerical disparities might have little impact on stability in a crisis, but they would surely have a large impact on pre-crisis peacetime domestic politics, which in the United States, at least, cannot tolerate any significant Soviet quantitative advantages in strategic nuclear forces. Nor is it likely that Soviet internal politics could again tolerate large US numerical advantages, particularly if they came about through the violation of an arms-control regime. Political sensitivities are therefore the most critical sensitivities of all, even though the imbalances that rub domestic nerves raw may not be of the sort that could lead an adversary to think it could derive meaningful military advantage by striking first.

Protection Against Breakout

As these reflections on verification make clear, for an arms-control regime to be viable—that is, for it to provide no party with incentives either to use nuclear weapons pre-emptively in a crisis or to embark upon a peace-time arms race—it must be designed in a way to make it evident that any party has "broken out" of it before the dimensions of that "breakout" are so large that their consequences are assessed as fundamentally threatening. That means that if one side cheats by secretly stockpiling nuclear weapons the fact should be detectable with high reliability and in a timely manner.

That may indeed mean eschewing some types of weapons regardless of their supposed military utility—SLCMs or some (perhaps all) types of land-mobile missiles—if their presence or absence cannot be monitored. It also means not setting ceilings for numbers of weapons so low that the sudden detection of another party's evasion could bring on a crisis in the domestic politics of the detecting state.

That is one of the principal reasons why very deep reductions of superpower nuclear forces—down, say, to the 500 or so warheads that both Robert S. McNamara and a working group of Soviet specialists have suggested[14]—would seem to be politically unwise. With ceilings for permitted forces so low, the sudden discovery (or revelation by the evading state) of 200 or 300 previously hidden weapons would be likely to have tumultuous political effects, even if they probably would not give their possessor the confidence to launch a first strike. When force levels are high, such an occurrence would undoubtedly provide evidence of another state's lack of good faith, but it could not easily affect the basic balance of power. Even if superpower forces were reduced to 50 per cent of their present levels, the clandestine production and deployment by either side of several hundred additional missiles not only would provide no usable military advantage, but the imbalance would not be large enough to be useful for political intimidation or blackmail.

Force levels, verification, breakout—and stability—are thus intimately related. Really deep reductions in the size of superpower nuclear arsenals should proceed only at the pace that confidence in verification arrangements increases. That does not, of course, mean merely one state's capabilities for remote monitoring of events within other states—what the jargon of arms control refers to as National Technical Means (NTM). The Intermediate-Range Nuclear Forces (INF) Treaty has made it the norm that verification must now include a wide range of cooperative activities based upon very intrusive on-site inspection. The START treaty currently under negotiation, because it seeks to limit so many more weapons, will provide for even more intrusive measures. It will introduce foreign inspectors into military and industrial facilities where heretofore their presence has been simply unimaginable.

A deep-reductions regime would build upon these experiences with highly intrusive on-site means of verification, and would almost certainly be even more intrusive. There will be plenty of grumblings on the part of those intruded upon, and accusations galore that the monitoring personnel have behaved improperly, but the likelihood is that within a relatively short time the transparency of this kind of verification will be accepted by the major powers as merely part of the cost of doing business in the modern world. That continued acceptance, in turn, will be a signal that the probability of destabilising danger is low.

Enforcement

A deep-reductions regime would necessarily have no enforcement mechanisms other than the actions affected states might take to defend themselves. In that respect it would be like all other contemporary arms-control regimes, and unlike any viable abolition regime. The detection of a violation of a deep-reductions regime might conceivably give rise to nuclear blackmail—or to a nuclear war—if the state that had achieved a marked advantage attempted to exploit it. But that kind of extraordinary risk-taking would be most unlikely so long as other states retained a retaliatory capability. Much more likely an outcome would be an unconstrained arms race as other states sought to restore a balance acceptable within their own polities. By contrast, the successful violator of an abolition regime could use its illegally possessed nuclear weapons to blackmail states that did not possess them. Only a supranational authority, itself armed with nuclear weapons, could prevent that occurrence. It is, of course, extraordinarily difficult to construct a plausible scenario for the empowerment of such an authority. That is why it is so difficult to imagine a workable abolition regime.[15]

Getting There from Here

At their Reykjavik meeting in 1986 Reagan and Gorbachev found that they could easily reach agreement on the desirability of deep reductions, and even abolition, of their nuclear arsenals. Their rapid convergence undoubtedly owed much to the fact that they were dealing only with hypothetical goals, not with their nations' actual deployed nuclear forces. Shaping those existing forces into a reliable and effective deterrent at low numerical levels would be more difficult. That is because existing US and Soviet forces have been designed with effective deterrence as only one of several criteria.

Another criterion, minimising cost, has resulted in the current high ratio of relatively cheap nuclear warheads to relatively expensive delivery systems—large missiles that carry many MIRVs, large submarines that carry many missiles, and large bombers that carry many stand-off missiles or gravity bombs. In theory, placing many eggs in each basket was perhaps risky, as it increased the payoff for launching first in a crisis. But the risk of losing many eggs to a strike by only a few of the adversary's weapons has been judged acceptable because the number of baskets has been so high.

Economic incentives towards deploying what nuclear jargon calls "highly fractionated" forces are reinforced by another criterion—the need for compliance with previous arms-control agreements and with the START treaty now under negotiation. These agreements limit warhead numbers. But because they also limit launcher numbers, they actually produce disincentives towards creating a force in which the ratio of warheads to launch vehicles is as low as possible, which as we have seen is an important desideratum for effective deterrence when warhead numbers are drastically reduced.

To convert existing superpower forces into minimum deterrence forces, warhead numbers should be cut without proportionately reducing numbers of launchers and platforms. A first step would be converting missiles from MIRVed into single-warhead vehicles. Certainly all land-based missiles should be so treated. If sea-based missiles were included, it would be possible to retain something like existing submarine fleets with their full complements of missile tubes. Maximising numbers of submarines would provide a hedge against possible breakthroughs in Anti-Submarine Warfare (ASW) technology. Alternatively, if MIRVed missiles are retained at sea it would be necessary to block up a sizable proportion of the launch tubes aboard ballistic missile submarines, so as to make them permanently unusable. These are measures that present no technical obstacles, and which could almost certainly be confidently monitored through a combination of on-site and remote inspection and observation of missile flight tests.

It would be much more difficult to alter bombers so that they would be incapable of carrying more than a given number of bombs or missiles. Rather than attempt to do so, it would be better to try to limit stockpiles of aircraft-delivered nuclear weapons and to rely on the prudence of governments not to load too many eggs into too few baskets.

The political obstacles to measures like these, that discard capabilities acquired at great expense and convert highly complex and costly weapons into simpler, less capable ones, might at one time have seemed insurmountable. But military bureaucracies and tax-paying publics that have (so it seems) stood by and watched the breaking into scrap metal of an entire category of the most modern weapons that had cost their state large sums of money—Soviet SS-20s and US Pershing II and ground-launched cruise missiles—would be unlikely to lodge more strenuous objections if their heads of state agreed to the kinds of reductions proposed here.

An altogether different approach would be to scrap all, or nearly all, components of existing forces and to construct a new force especially designed to provide maximum deterrence with minimum size. That is what a working group of well-known Soviet physical and social scientists had in mind in suggesting that both superpowers should limit themselves to a force of 600 single-warhead land-based (some mobile, some in fixed silos) long-range ballistic missiles.[16] Given the small size of that notional force, there might be no real cost savings or other reasons to attempt to convert elements of the existing force.

Minimum Deterrence and the Proliferation of Nuclear Weapons

For much of the nuclear era, "horizontal" proliferation, the spread of nuclear weapons to additional states, has been linked rhetorically with "vertical" proliferation, the growing size of the two superpowers' arsenals. Many Third World statesmen (and some First World analysts) have tirelessly called attention to the double standard implied in an international order that permits the superpowers to deploy more than 12,000 nuclear warheads each while constraining all but a handful of other states (the acknowledged nuclear weapons powers) from deploying any at all. In the 1968 Treaty on the Non-Proliferation of Nuclear Weapons (NPT) the superpowers undertook "to pursue negotiations in good faith on effective measures relating to cessation of the nuclear arms race at an early date and to nuclear disarmament" (Article VI).

Concerning the initial chain of nuclear causation it is possible to be relatively clear. Once the United States possessed the bomb, it was inevitable that the Soviet Union would, too. The British, French, and Chinese bombs seem nearly as inevitable. They were, after all, the other states formally accorded great-power status by virtue of permanent seats on the United Nations Security Council. The bomb was another outward sign. More important, they each felt directly threatened by one of the nuclear-armed superpowers—in China's case, although at different times, by both. But in all these instances it seems clear that the crucial consideration was the *fact* of the US or Soviet nuclear arsenals, not their size.

After China, however, the trail becomes less distinct. India is the only other state that acknowledges having fabricated and exploded a nuclear "device." Its doing so almost certainly had nothing to do with the size of the superpower arsenals, but, rather, with her regional rivalry with nuclear China and (as yet) non-nuclear Pakistan, and her wish to be recognised as the preeminent power in South Asia.[17] Israel is another case. There is every reason, save the "smoking gun" of an actual explosion, to believe that Israel not only has a nuclear force but a fairly large (as many as 200 warheads) and technologically sophisticated one.[18] Here again, it is not the size of the superpower forces that has been influential but the very existence of those—and other—forces. If comparatively secure states like Great Britain and France can have nuclear weapons, so an Israeli might argue, it is certainly legitimate for a state so much more directly threatened to have them.

A decision by the superpowers drastically to reduce the size of their nuclear arsenals would be likely to have cross-cutting effects on the decisions of potential proliferating states. On the one hand, there would surely be some persons within the governments of such states themselves already opposed to acquiring nuclear weapons—who would use such a change of policy on the part of the superpowers to support their own position in internal debates. Washington and Moscow are at last disarming, they would say; now is not the time to be adding to the number of nuclear-weapons powers.

But other voices would undoubtedly argue the opposite. If Washington and Moscow had concluded that small nuclear forces provided, on balance, more stability and security than either a large force or no force at all, why should the same logic not apply to other states? Small nuclear forces—like the British, French, or Chinese force—would now look more respectable. It would of course remain true that to acquire a force anything like as sophisticated and survivable as any of these three would require enormous investments of resources and of human talent. But the path would look somewhat more inviting and the objective somewhat more attainable.

Yet it is surely the case that, for both the proponents and opponents of going nuclear, the superpower example would be of only marginal importance. It would be used by advocates to buttress positions previously held, in the hope of swaying colleagues whose mind had not yet been made up. The key element in any decision would probably remain a government's perception of its own security situation within its own region. There will undoubtedly be additional states like India and Israel—and Pakistan—that will decide that, on balance, given the present or potential threats they face from powers within their region, they should acquire nuclear weapons. But their number will probably not be large.

There remains the special case of a state that might choose to acquire nuclear weapons because it feels threatened by a superpower. Libya might be one example. It is difficult to think of others. A farfetched but not entirely inconceivable example might be Cuba. In very different circumstances from those that prevail today, so might West Germany, or Japan. Here again, it seems unlikely that the size of the relevant superpower's force would have much to do with such a decision, given the disparity that would exist between the force that the smaller state might develop and the one the superpower would continue to deploy. Yet deep reductions on the part of the superpower might in this instance, also, be a marginal but none the less real influence, making the disparity appear smaller and the decision to acquire a putative deterrent capability a bit easier to rationalise.

The situation would be sharply different if there were ever an international agreement on the abolition of nuclear weapons. Abolition would be a powerful leveller. In a purportedly nuclear-free world, all states would be on a roughly equal footing. The United States and the Soviet Union would perhaps find it somewhat easier, but probably not significantly so, clandestinely to produce and stockpile nuclear weapons than would any of a number of advanced, industrialised states, and the latter would in turn not have a significant advantage over many developing Third World states. The incentives that a threatened state might have to use a hidden reserve of nuclear weapons in a dire crisis would be enormous. They would surely be greater—because the probability of retaliation (or punishment) by a nuclear-armed state would be less—in the context of an international abolition regime.

Geopolitical Consequences of Minimum Deterrence

Critics of minimum deterrence as a principle for structuring nuclear forces, like critics of No First Use as an enunciated policy, argue that either measure would lead to a diminution of the caution, induced by fear of escalation across the nuclear threshold, which has caused the superpowers always to draw back from any clash of their conventional forces. The international system, so it is predicted, would thereby be made more safe for conventional war.[19]

Caution, however, exists in the minds of the fearful—or the prudent. It is impossible to say how much caution is enough to provide sufficient friction so that states will not tumble down the slippery slope to war. So long as nuclear weapons exist, however, they will induce caution. Especially are they likely to do so in the minds of military establishments and political leaderships all of whose members have spent their entire working lives under the shadow of the bomb. Such socialisation is not easily forgotten. So long as there exists a rough balance between their conventional capabilities, the superpowers are unlikely to risk war with one another.

Nor is there any reason to think that their willingness to use military force against other powers would be significantly affected by their transition to a minimum deterrence posture in the nuclear realm. Those hypothetical encounters would, after all, be fought with conventional arms. Yet it should also be observed that for the foreseeable future the slope of the curve that charts the superpowers' propensity to intervene is likely to be downward. In recent years, both superpowers have seemed to reach the realisation that their vital interests are very rarely affected by developments within the Third World. Even in the special instance of Eastern Europe, and for reasons that have little to do with the military balance, Moscow seems likely to be able to tolerate much more far-reaching change than in the past.[20]

The overall geopolitical effects of the adoption by Washington and Moscow of a posture of minimum deterrence are likely to be small. The international hierarchy would not be much affected. Nor would it be even in the very unlikely event that the two emerged from the arms-control negotiating chambers with nuclear forces not much larger than those of Great Britain, France and China. The United States and the Soviet Union are superpowers not because of their nuclear arsenals but because of their ability to project substantial conventional forces far from their own frontiers. Other states have relatively large conventional forces; the various armies in the Middle East are perhaps the most obvious examples. But they are capable of operating only in their own region.

No longer, in fact, can one speak of a single international hierarchy. There are at least two, one military, the other economic. In the economic hierarchy the Soviet Union scarcely figures, and the United States is only one important actor among several. That hierarchy is quite distinct from the military one. Indeed, military spending has long been a drag on the economies of both the superpowers. Strategic forces have not been the largest element in their military budgets, but they have been far from insignificant. As we have seen, adopting a posture of minimum deterrence would be unlikely to yield substantial short-run savings. But over the span of, say, 20 years, significant reductions in military spending might be achieved.

Such reductions would, of course, depend on how much Washington and Moscow chose to spend on conventional forces. If tensions between them were high, they might pour resources into strengthening their defences. But a strand of argument that has run through this chapter is that there is every reason to suppose that US–Soviet tensions will not be high and that mutually perceived success in reducing the nuclear threat each poses to the other could be an important means of lowering them.

In the final analysis, that is why maintaining deterrence at much lower levels of nuclear forces is so worth attempting. By putting a halt to the nuclear arms race and by eliminating the most worrisome sources of crisis instability, the risks of nuclear war occurring would be drastically reduced—a boon not simply for the American and Soviet populations but for all humankind. And the chances are that such measures would go far towards creating the kind of environment that would make much easier the resolution of the other disputes that divide the two nations.

Notes

1. For a discussion of the effects of 50 per cent reductions on the superpowers' ability to cover their likely target lists, see Michael M. May, George F. Bing, and John D. Steinbruner, "Strategic Arsenals after START: The Implications of Deep Cuts," *International Security* 13, no. 1 (1988–9): 90–133. For likely damage by various types of Soviet counterforce attacks on the United States, see William Daugherty, Barbara Levi, and Frank von Hippel, "The Consequences of 'Limited' Nuclear Attacks on the United States," *International Security* 10, no. 4 (1985–6): 3–45.
2. Robert S. McNamara, *Blundering into Disaster: Surviving the First Century of the Nuclear Age* (New York: Pantheon Books, 1986): 122–124.
3. Harold A. Feiveson, Richard H. Ullman, and Frank von Hippel, "Reducing U.S. and Soviet Nuclear Arsenals," *Bulletin of the Atomic Scientists* 47 (August 1985): 144–151.
4. "Statement by M.S. Gorbachev, General Secretary of the CPSU Central Committee," *Pravda,* 16 January 1986, 1–2; translated in Foreign Broadcast Information Service (FBIS), *Daily Report, Soviet Union,* 16 January 1986, AA-1 to AA-9. A Soviet translation (dated, however, 15 January) was published as a full-page advertisement in *The New York Times,* 5 February 1986, A13.
5. For a clearly stated summary of the physical issues, see Barbara G. Levi and Tony Rothman, "Nuclear Winter: A Matter of Degrees," *Physics Today* 38 (September 1985): 58–65.
6. For a lengthy elaboration of the arguments in this paragraph, see Morton H. Halperin, *Nuclear Fallacy: Dispelling the Myth of Nuclear Strategy* (Cambridge, MA: Ballinger, 1987).
7. On the history of US nuclear planning see David A. Rosenberg, "U.S. Nuclear War Planning, 1945–1960," in *Strategic Nuclear Targeting,* eds. Desmond Ball and Jeffrey Richelson (Ithaca, NY: Cornell University Press, 1986): 35–56; and Desmond Ball, "The Development of the SIOP, 1960–1983," in ibid., 57–83.
8. See Lawrence Freedman, "British Nuclear Targeting," in ibid., 109–26 and David S. Yost, "French Nuclear Targeting," in ibid., 127–156.
9. McGeorge Bundy, "To Cap the Volcano," *Foreign Affairs* 48, no. 1 (1969–70): 10.
10. See Bennet Ramberg, "Targeting Nuclear Energy," in *Strategic Nuclear Targeting,* 250–266. Ramberg predicts very high levels of casualties, but he posits a much larger attack than one limited to, say, 20 nuclear power-generating plants.
11. Stobe Talbott, *Endgame: The Inside Story of SALT II* (New York: Harper and Row, 1979): 24.

12. See Ball's discussion of nuclear options in "Development of the SIOP." For Soviet targeting, see William T. Lee, "Soviet Nuclear Targeting Strategy," in *Strategic Nuclear Targeting*, 84–108.
13. See Stephen M. Meyer, "Soviet Nuclear Operations," in *Managing Nuclear Operations*, eds. Aston B. Carter, John D. Steinbruner, and Charles A. Zraket (Washington, DC: The Brookings Institution, 1987): esp. 495–512.
14. For McNamara's suggestion, see his *Blundering into Disaster*. For the Soviet group's, see below, n. 16.
15. For a thorough discussion of these issues, see John H. Barton, "The Proscription of Nuclear Weapons: A Third Nuclear Regime," in *Nuclear Weapons and World Politics*, eds. Michael Mandelbaum et al. (New York: McGraw-Hill, 1977): 151–211.
16. Committee of Soviet Scientists for Peace, Against the Nuclear Threat, *Strategic Stability Under the Conditions of Radical Nuclear Arms Reductions: Report on a Study (Abridged)* (Moscow, April 1987). The co-chairmen of the working group were Academician Roald Sagdayev, Director of the Institute of Space Research, and Andrei Kokoshin, Deputy Director of the Institute of the USA and Canada Studies. A summary of the report appeared as Andrei Kokoshin, "A Soviet View on Radical Weapons Cuts," *Bulletin of the Atomic Scientists* 44 (March 1988): 14–17.
17. For a discussion of India's nuclear weapons programme, see Leonard S. Spector, *Going Nuclear* (Cambridge, MA: Ballinger, 1987): 73–100.
18. For the Israeli programme, see ibid., 130–145.
19. For an example of such an argument (in this instance, by four distinguished German commentators), see Karl Kaiser et al., "Nuclear Weapons and the Preservation of Peace," *Foreign Affairs* 60, no. 5 (1981–2): 1157–1170.
20. For an argument to this effect, see Richard H. Ullman, "Ending the Cold War," *Foreign Policy* 72 (Fall 1988): 135–139.

Richard H. Ullman (1933–2014) was Professor of International Affairs at Princeton University. He was a staff member of the US National Security Council, a member of the policy planning staff of the Office of the US Secretary for Defense, and the director of studies at the Council on Foreign Relations. He also was a member of the policy planning staff at the US Department of State.

PART III

After the Cold War

Introduction to Part III

The third part of the book offers a collection of reflections, analyses, and proposals presented and discussed in the courses of ISODARCO during the decade following the end of the Cold War. This period was marked by great relief for the end of the superpower rivalry that had kept humanity hostage. Drastic cuts were made in the US and Russian arsenals since the mid-eighties, together with changes in the nuclear postures of the two former archenemies.

Even if a nuclear catastrophe had been averted, numerous issues of great concern remained while new threats materialized. For example, an alarming problem became the security of about a million kilograms of weapon-grade plutonium and highly enriched uranium resulting from the dismantlement of the US and former Soviet Union nuclear weapons arsenals. This issue and the need for effective US–Russia cooperative efforts are addressed in great detail by Frank von Hippel and Oleg Bukharin. Another question involved the danger of the militarization of outer space; Richard Garwin argues that this needs to be prevented by starting with a ban on space and anti-satellite weapons that could have otherwise provided new means of initiating a nuclear war.

Nevertheless, the changed security environment was generally favourable for a thorough reconsideration of the role of nuclear weapons in international relations and some cautious optimism emerged that conditions were finally ripe for progress towards nuclear disarmament. Several observers questioned the utility of nuclear weapons in the new security environment. In his study on the German decision to pursue non-proliferation, Harald Müller argues that the value of nuclear weapons in the post-bipolar system of the twenty-first century is rather limited. For a "trading state" like Germany, which places welfare and influence by persuasion above traditional power politics, nuclear weapons would be more of a liability than an asset.

While hopes for nuclear abolition lost momentum during the second half of the nineties, a new wave of disarmament arose during the mid-noughties, driven partly by the growing concern for nuclear accidents, the risk of proliferation by "rogue states," and the threat of nuclear terrorism in a post-9/11 security context. Prompted by an appeal from the US statesmen George Schultz, William Perry, Henry Kissinger, and San Nunn that was published in the *Wall Street Journal* in 2007,[1] the call for a "world free of nuclear weapons" was followed by similar initiatives worldwide and gained the support of well-organized and highly motivated civil society movements. Most importantly, this appeal was endorsed by the United States under the leadership of President Barack Obama, who made nuclear abolition a guiding principle of his foreign and defence policy. Although no one ever doubted that the road to nuclear disarmament would be long and uncertain, "global zero" moved to the top of the international political agenda and created a favourable climate in which further de-legitimization of nuclear weapons could thrive.

The contribution of ISODARCO to this debate has been substantial. It is reflected in the study by Patricia Lewis of the new verification technologies, mechanisms, and procedures to ensure compliance with the obligations that sovereign states undertake in their non-proliferation and disarmament agreements. Alexei Arbatov provides a thorough analysis of how international relations would unfold in a scenario of a nuclear-free world. He argues that this would require a fundamentally different system of international security and governance. Finally, Matthew Evangelista discusses the complex question of military strategy in a world beyond nuclear deterrence, with special attention to the position of the United States and her perspective on international relations.

Clearly, things proved more complicated and "getting to zero" was quickly seen as an ambitious goal under existing circumstances. The topics of the following pages, however, continue to provide relevant insight for today's world, depicting the criticality of debates that serve as the foundations for understanding why and how to move, even if slowly, towards a world with fewer and fewer weapons capable of global annihilation.

Notes

1. George P. Shultz et al., "A World Free of Nuclear Weapons," *Wall Street Journal*, 4 January 2007.

CHAPTER 15

Weapons on Earth and in Space: Global Security in the New International Situation

Richard L. Garwin

It is certainly more accurate although less satisfying to recognize "the new international situation" instead of "the new world order," since order is most evident by its absence on the world scene. But the situation could become much worse, and I believe it is our purpose to understand the ways in which it could worsen and how we might prevent that.

Introduction

The concept of security involves more than the absence of violent death imposed by other nations, or the absence of domination by another nation, and it is just those additional features that further complicate the achievement of security. Evidently, the definition of security should involve access to at least a minimum supply of food and water, and help with health care, at least for those who contribute to society or are incapable of doing so.

Originally presented to the Third ISODARCO Beijing Seminar on Arms Control, *Global Security in New International Situation, Arms Control in Outer Space, Nuclear Disarmament, Nuclear Test and Verification Techniques*, Beijing, 21–26 October 1992.

R.L. Garwin (✉)
IBM Research Division, Yorktown Heights, NY, USA

P. Foradori et al. (eds.), *Arms Control and Disarmament*,
https://doi.org/10.1007/978-3-319-62259-0_15

The fact that these further conditions for security will not be much discussed during this symposium does not signify their lack of importance. That hundreds of thousands of people are killed or forcibly displaced in "ethnic cleansing" in Yugoslavia or die from warlord-induced famine in Somalia is no less serious for them than if they were destroyed in nuclear war. The desperate situation of individuals, whole communities, and even nations can lead to nuclear war, particularly by the proliferation of nuclear weapons.

In this sketch, I shall make brief comments on all of the topics in the outline of this symposium, knowing that my colleagues have prepared detailed analyses and presentations on most of them.

Arms Control on Earth

Proliferation of Nuclear Weapons

Only one year ago, the Soviet Union disintegrated into its constituent republics, all but the three Baltic republics remain loosely affiliated in the Commonwealth of Independent States (CIS).

Clearly it was necessary to do something about the many thousands of tactical and strategic weapons spread among many of the ex-Soviet republics, and President Bush and President Gorbachev, and later Presidents Bush and Yeltsin, announced a history-making agreement essentially to eliminate almost all of the tactical nuclear weapons. This removed tactical nuclear weapons from ships and submarines and brought most of the deployed American tactical nuclear weapons back to the United States.

Even under the current difficult political and economic circumstances, all of the former Soviet tactical nuclear weapons have been returned to the territory of the Russian Republic, leaving Belarus, Ukraine and Kazakhstan still housing some strategic warheads on silo-based missiles. Further agreements have been reached so that the number of US nuclear warheads is to decline to 3500, while the number of Russian warheads will fall to 3000. Furthermore, Russia, Belarus, Ukraine and Kazakhstan have all agreed to join the Non-Proliferation Treaty (NPT), Russia as a nuclear state and the other three nations as non-nuclear states.

That is the encouraging scene from the point of view of agreements already concluded. From the point of view of physical reality, the situation is not so reassuring. Although the tactical nuclear weapons are now all returned to Russia, there is no indication that any significant number have been destroyed. Even those that are "destroyed" could serve as a source of nuclear material in just the right amount to make an advanced nuclear weapon.

The multiple political authorities, the presence of armed conflict and economic hardship in some of these territories, all raise concerns that the extensive safeguards over the custody of nuclear weapons may be compromised by miscreants in the interest of anarchy, of profit, or of ethnic goals. I will discuss this aspect of Russian and US nuclear weaponry further under the nuclear security and nuclear disarmament titles.

More classically, the discussion of nuclear proliferation has involved the decision of nations or sub-national groups to acquire completed nuclear weapons by purchase or theft, the capability to build nuclear weapons outright, or a dual-purpose capability, in which nuclear reactors would be deployed for the production of electrical power, with the prospect that at some future time the plutonium produced would be separated and used to make nuclear weapons. The most explicit example of this in recent times became clear after the Gulf War, when the United Nation sanctions imposed because of the invasion of Kuwait by Iraq mandated the discovery and destruction of the Iraqi stocks of chemical weaponry and chemical warfare agent, and of the capability for developing nuclear and chemical weapons.

The lesson to be learned from this is that a wealthy country has many avenues towards nuclear weaponry, not only those that have proved most economical or traditional in building large numbers of nuclear weapons. Thus, the high-performance gas centrifuge is but one approach, together with the traditional gaseous diffusion plant, but the old electromagnetic separator (calutron) process would be much easier to use in the nineties than it was in the forties, and eventually laser isotope separation will be feasible.

Much more access than is usual would be required to verify that a nation was not undertaking a programme with any of these varied means of isotope separation of uranium-235. On the other hand, many nations have legal facilities for the separation of plutonium from spent reactor fuel, either for research purposes or on the basis of a supposedly economical recycle of the plutonium into the reactors. Thus, on the one hand, totally clandestine activities might go on in violation of the NPT, or with equal import if the nation is not an NPT signatory. On the other hand, the timely warning of nuclear weapons manufacture could be eroded if a nation openly created a stock of plutonium metal subject only to the accounting safeguards of the International Atomic Energy Agency (IAEA).

To my mind, a large part of the solution is so-called societal verification, long advocated in particular by Joseph Rotblat. I have also urged in many publications and congressional testimony that more emphasis be placed in the formulation of treaties to the requirement that the treaty be published widely in the nations adhering to it, and that individuals working on programmes related to the area of the treaty constraints should sign annually a certification that they have read the treaty and that their own work is in compliance with the treaty. I have indicated that this has long been the case in the United States for those involved in missile defence activities, which must not conflict with the US-Soviet Anti-ballistic Missile (ABM) Treaty of 1972.

But one should go further, in fact, to ensure in the constitution or other basic law that treaties are automatically the "law of the land," without the necessity for any further legislation to make them into domestic law. And one should state explicitly in the treaty that it is the duty of anyone having evidence of violation to report it not only to the government of the nation concerned but also or alternatively to the international body responsible for verification of the treaty.

With the nuclear inventories of the United States and Russia scheduled to be reduced by a factor 10 within the next decade or so (and which could with greater benefit be reduced by 97 per cent—to about 1000 warheads on each side—within the next five years), there will be little sympathy for additional nations to acquire nuclear weaponry. As will be discussed in the next topic, if the nuclear security needs of nations can reasonably be addressed without their acquiring nuclear weapons, there will surely be little tolerance for their acquisition of nuclear weapons in violation of the NPT, and probably not much patience for nations which do not adhere to the NPT.

Nuclear Security

In international discussions, "nuclear security" has two quite different meanings. The first is "nuclear safety," against accidental detonation of a nuclear weapon in storage or in transit, against fire that would disperse plutonium, or even against accidental launch or misuse of nuclear weaponry.

These questions have been quite publicly addressed over the last couple of years in the United States, especially as a consequence of the report of the "Drell Committee," composed of Sidney D. Drell, Charles H. Townes and John S. Foster, reporting to the US Congress on "Nuclear Weapons Safety."[1] This committee authoritatively stated that some weapons that had been calculated to be "one-point safe" were not in fact totally safe against nuclear yield from impact detonation of the explosive. Other weapons might give a full nuclear yield if involved in a crash under some very unlikely circumstances, because they did not have the most modern electrical systems that would prevent the accidental application of a firing signal to the nuclear weapon. Still other weapons did not have a so-called fire-resistant pit (FRP) and so might disseminate plutonium smoke if they were consumed by fire.

With the scheduled reductions in nuclear weaponry, there is the opportunity for eliminating those weapons that do not meet the most modern safety standards, and the substitution of warheads that already exist and do meet the standards for some of the other sub-standard warheads. Ray E. Kidder has provided Congress with an independent study on these points. In a recent article in *Arms Control Today* Kidder discusses the measures that have been taken following the Drell report, such as storing bomber weapons in secure warehouses rather than on aircraft, and concludes that if it is not required to modify bombs and cruise missiles to incorporate FRPs, the improvements to ballistic missile warheads could be accomplished "by substituting existing safer warheads."[2] The only nuclear testing required would be if it were decided to substitute for the W-89 warhead carried by the Trident II missile—the W-89 that was developed for the now-cancelled Short-Range Attack Missile SRAM II. According to Kidder, no more than four nuclear tests would be needed to adapt the W-89 for use in the W-88 mark 5 re-entry vehicle used in the Trident II.

My own personal view is that the weapons are safe enough, given the change in Navy practice to load the warheads onto the missile after the missiles have been lowered into the launch tubes of the submarines.

In addition to nuclear safety, the Drell panel considered nuclear *surety*, which deals with security and "use control." I personally was involved in the final considerations that lead to the deployment of the Permissive Action Link (PAL) on all US land-based and bomber-carried nuclear weapons. It has long been my conviction that the security of the United States and the world would be enhanced if *all* nuclear weapons employed such mechanisms and procedures, including those based on US strategic submarines. At present, US submarine-deployed weapons are protected against unjustified use by procedural methods, rather than by a foolproof mechanism on each warhead that requires the insertion of a unique code in order to obtain a nuclear explosion.

Nuclear Disarmament, Test Ban and Verification

The incidence of nuclear testing has diminished greatly with the moratorium on Soviet testing over the last year (following environmental and local objection to nuclear tests at Semipalatinsk and even at Novaya Zemlya). France has also voluntarily adopted a one-year moratorium on nuclear testing, and the US Congress has just passed legislation imposing a nine-month moratorium and setting an end to testing by a date certain—1996. Under this legislation, no more than 15 US tests would be allowed all together during this period. President Bush has issued an order restricting testing to the enhancement of safety, but the Department of Defense has included in its own statement following the President's order additional reasons for testing, so the situation is unclear.

In the context of the consensus that "the principal objective of US nuclear policy should be to deploy nuclear weapons solely as a deterrent to their use by other and to use them only in response to nuclear attack"[3] tests to explore the "effects" of nuclear explosions are no longer contemplated by the United States. I believe that a total ban on tests is desirable as soon as it can be realized, and I applaud the congressional legislation to set a date certain of 1996 for the end to nuclear testing by the United States. A recent study sponsored by the government of Norway describes the state of the art of monitoring techniques for verifying a comprehensive test ban.[4] Cooperative measures can also be used, such as the pre-announcement of any large chemical explosions and, of course, societal verification.

As for nuclear disarmament, I do not at present see the way to eliminate nuclear weapons, in view of the power that would give to a nation that had a small number while no other nation or organization had any. I do see great merit, however, in extending negative and positive nuclear guarantees by the nations still possessing nuclear weapons. The negative guarantee is that they will under no circumstances use nuclear weapons except in response to nuclear attack (or to biological warfare) while the positive security guarantee (more difficult to implement) would be actually to use nuclear weapons on behalf of a non-nuclear state attacked by nuclear weapons. I have discussed this in a 1977 publication[5] and believe that multiple, independent such guarantees have now become feasible.

More generally, nations, which possess nuclear weapons should strive to reduce their numbers and to put them explicitly at the service of the United Nations (UN). This would encourage non-nuclear states to solicit positive nuclear guarantees from the UN, which could serve as a kind of broker for the period during which nuclear weapons were actually held by individual nations.

An inhibition to more rapid reductions and to a lower level of remaining weapons than the 3000 or 3500 to which Russia and the United States are committed is the uncertain future of the nuclear stockpiles held by the United Kingdom, by France and by China. I believe that in return for a commitment by Russia and the United States to reduce rapidly to 1000 nuclear warheads, and also to abstain from ballistic missile defences, these three nations could accept a limit of 300 warheads each. Surely, also, continued underground testing by China without any commitment to limiting numbers or duration would imperil the comprehensive test ban to which the United States is legislated to adhere in 1996. Of course, that legislation could be modified in the future, and that is exactly the peril.

Arms Control in Outer Space

Uses of Space and Potential for Conflict

The limitation of space weapons is usually discussed in conjunction with the limitation of anti-satellite weapon (ASAT) capabilities, because one of the potential uses of a space weapon is to defend valuable satellites against attack. But some potential uses for space weapons have nothing to do with ASAT or defence against ASAT (DSAT)—for instance, weapons for use against warheads or vehicles passing *through* space (ABM capabilities) or even for use against targets on the ground, on the sea or in the atmosphere. These later could be “bombs in orbit,” “long-rod penetrators” or other mechanical or explosive systems, or directed-energy weapons—mainly lasers for use against aircraft or ground targets.

In general, the space-to-ground weapons are less effective and more costly than weapons that are kept on the ground until they are needed, and in addition they are vulnerable to ASAT. This is surely the case for nuclear weapons in orbit, most of which would take much longer to reach their targets than if they were based more survivably on the ground, in silos or in the oceans.

Of course, the Gulf War showed clearly the value of satellite systems in support of conventional warfare, not only communications but also the benefit of precision navigation. These military uses do not constitute weapons in space, but they enhance the capabilities of those able to use them.

A few international agreements control some aspects of space activities, in particular the Outer Space Treaty of 1967, which bans weapons of mass destruction and also nuclear weapons of any kind in space. Also the US-Soviet ABM Treaty (now becoming the US-Russian ABM Treaty) of 1972 bans space-based elements that can substitute for the launchers, interceptors or radars of the normal ABM system of 1972.

Among space assets are weather satellites, satellites for navigation observation, communications, geodesy and technological experiment. There are also satellites that provide early warning of the launch of ballistic missiles anywhere in the world (the so-called DSP satellites). Among the observation satellites, there are not only Landsat, SPOT and other imaging and earth monitoring satellites, but also satellites that provide images and other intelligence information to governments. In the context of the US-Soviet arms control agreements, these so-called national technical means (NTM) are protected against interference in verifying compliance with the treaty. It is generally accepted that world security is enhanced by the openness contributed by satellite monitoring. Space could be the home of further beneficial systems, in principle, for instance satellites that transform solar power to microwave beams, transmitted to earth to feed the electric power grid. Although there is no agreement that such systems are economically viable, they should not be precluded arbitrarily.

Weapons have indeed been proposed for space, both nuclear and non-nuclear; defensive weapons to counter elements of ballistic missile systems as proposed in the Strategic Defense Initiative (SDI) and even as early as 1960 in the ballistic anti-missile boost interceptor (BAMBI) system for non-nuclear boost-phase intercept. Among the strictly SDI-proposed weapons are not only the Brilliant Pebbles (small individual space-based interceptors [SBI]) but also larger SBI housed multiply in "space garages," as well as directed-energy weapons—massive lasers, neutral particle beam generators, microwave generators, some powered by nuclear explosives, while others would be electrically powered.

With the vast reduction in threat from the weapons complex of the former Soviet Union, SDI is currently involved with responding to the Missile Defense Act of 1991 (MDA-91), which has focused SDI responsibility on countering accidental launch of a couple of Russian nuclear-armed ballistic missiles, providing a defence against a third-world intercontinental ballistic missile (ICBM) armed with nuclear warhead, and defending US troops and allies abroad from a theatre ballistic missile (TBM) threat armed with chemical or high-explosive munitions.

What is the problem with space weapons? Why should we propose to control or ban those things, which don't even exist? By implication, space weapons have the capability to destroy something important to a potential adversary. As we will see, they are neither very effective in general nor unique in this capability, and their vulnerability (to get to the end of the story) means that they are not in general a serious contributor to national security. But they can be dangerous for the same reason. There could be a competition to install such weapons and to counter them, since countering is fairly feasible and the space weapons are, by definition, residing in a region which is not national territory. Proposals to dominate space or to divide it up would raise enormous problems and, in my opinion, should not even be acceptable to the United States; in addition, they would violate the Outer Space Treaty. But space weapons and ASAT are a technological challenge, and even Japan and Germany, barred from developing nuclear weapons, have no bar to responding to that challenge.

Space weapons would be countered because it is relatively easy to do so, and in general easier than modifying the system against which the weapons would otherwise be used. Of course, if one side maintained an enormous, self-protecting weapon system in space, it would not be so readily countered, but that nation would then control access to space by everyone else, which is presumably unacceptable to other states.

But if space weapons are deployed, and countermeasures adopted, there will be counter-countermeasures and great crisis instability, especially because space is the home of valuable observation systems that do a lot to provide stabilizing information. These systems, like anything deployed in space, are fragile and could readily be destroyed in a battle between individual space weapons and their counters. They could be further hardened at considerable expense and penalty to performance, or otherwise protected, but they would not be invulnerable.

Because space belongs to no nation but instead is common property, there is great potential for competition and interaction. Furthermore, we are having a serious problem with space debris in low earth orbit (LEO), with the damage rate to satellites now dominated by human-created debris rather than by natural meteors. In the last couple of years it has been generally agreed that it is irresponsible to create space debris in LEO, because of the potentially damaging impact on later activities in LEO.

As has been indicated, space weapons are not particularly effective, nor are they necessary. They are out of the way of their targets, they are vulnerable and they are costly. In LEO they are moving at 8 km/s, which can cause serious difficulties for an intercept in the upper reaches of the atmosphere—excessive heating, and the like. As for "necessary," consider the capability of space weapons against the threats defined in MDA-91. Against the accidental launch of Russian ballistic missiles, better protection can be achieved at less cost and sooner by the implementation of additional launch control measures, especially a separate in-silo mechanism to disable a normal self-destruct after launch.

For protection against a third-world nuclear-armed ICBM, it is more effective and certainly far less expensive to use ground-based interceptors, even from a single site, launched on information from DSP. For protection against TBM, space-based weapons have little or no effectiveness because the apogee of the TBM can be held below any effective intercept altitude of the Brilliant Pebble, and that threat also can be handled better by taking advantage of the launch information from DSP or other launch detection satellite.

Even for ASAT purposes, ground-based ASATs are superior, except for the specific case of space mines, which can be effective and are not readily countered if their response to a countermeasure is to destroy their quarry. Space mines would be banned by any agreement banning space weapons and ASAT.

A Treaty Limiting Anti-satellite Weapons

Although there have been proposals for "rules of the road" for behaviour in space, it seems that such "rules" should be deduced from a guiding principle, which I take to be the banning of space weapons and anti-satellite tests. I favour the formulation, which I helped introduce on 18 May 1983 to the United States Senate Committee on Foreign Relations, in cooperation with the Union of Concerned Scientists[6]; the first three articles of "A Treaty Limiting Anti-satellite Weapons" follow. Because the

Treaty is drafted as one to which all nations could adhere, the demise of the Soviet Union does not force major changes in the text.

ARTICLE I Each Party undertakes not to destroy, damage, render inoperable or change the flight trajectory of space objects of other States.

ARTICLE II (1) Each Party undertakes not to place in orbit around the earth weapons for destroying, damaging, rendering inoperable, or changing the flight trajectory of space objects, or for damaging object in the atmosphere or on the ground. (2) Each Party undertakes not to install such weapons on celestial bodies, or station such weapons in outer space in any other manner. (3) Each Party undertakes not to test such weapons in space or against space objects.

ARTICLE III (1) For the purpose of providing assurance of compliance with the provisions of this treaty, each Party shall use national technical means of verification at its disposal in a manner consistent with generally recognized principles of international law. (2) Verification by national technical means shall be supplemented, as appropriate, by such cooperative measures for contributing to the effectiveness of verification by national technical means as the Parties shall agree upon in the Standing Consultative Commission. (3) Each Party undertakes not to interfere with the national technical means of verification of the other Party operating in accordance with paragraph 1 of this Article. (4) Each Party undertakes not to use deliberate concealment measures, which impede verification by national technical means of compliance with this treaty.

Questions of Definition and Verification

Alternative formulations could define precisely what is banned, define precisely what is permitted, or provide a set of thresholds for discussion. The first option clearly will not work in a field in which the technology and its implementation change rapidly, so that items, which are clearly harmful would not be listed now and would therefore be permitted. Similarly, the second approach is likely to miss the permission of important and peaceful activities in space. As an implementation of the proposed ban, one might consider the "dual threshold" approach, according to which one can do anything which is compliant with the treaty without discussion up to a certain threshold of laser power, brightness, speed or whatever, and do anything up to a second threshold so long as it is revealed and discussed. One may even be able to deploy or test lasers or particle beams or other items beyond the second threshold (although there is no presumption that one can do so) *if* one can show that the item is actually compliant with the treaty.

The dual threshold scheme is particularly useful for avoiding the presence of directed energy weapons in space while not inhibiting scientific research. I have previously published a discussion of this approach.[7]

By banning ASAT and space weapons, in almost all cases the quantitative limits could be replaced by *de minimus* thresholds, below which brightness, radar power, et cetera, activities are clearly of no interest in ABM or ASAT, and above which there must be discussion among the parties, in the context of a total ban on ASAT and space weapons, and in the context of continuing adherence to the ABM Treaty.

In the case of the space-based laser (SBL), for instance, it is easier to monitor that no significant laser energy is generated and projected in space, than to distinguish the 100 J/cm^2 that would damage even fairly robust satellites from the 500 J/cm^2 that might damage current boosters unequipped with even small amounts of hardening against laser energy.

I pass from that one example to a list of functions (and objects) that would be limited and monitored. To buttress the outright prohibition on test of space weapons and on activities in contravention to the ABM Treaty, the regime would mandate presentation for discussion certain activities that exceed "discussion thresholds." Such presentation in the Standing Consultative Commission would allow a Party to explain the nature of the activity, why it is not a prohibited activity and to provide cooperative self-verification means (on-board cameras, tests performed within view of competent NTM, etc.).

Discussion Thresholds for Functions and Objects

Space-Based Interceptors. Space tests of SBI are prohibited. In support of this ban, tests in space or objects placed in orbit, which contain devices designed to give a mass equal or exceeding 100 grams an incremental velocity exceeding 0.5 km/s, are subject to discussion and agreement. Rocket propulsion with acceleration below 10 g is exempt from this limitation.

Fast Passage. The passage within 10 km of another space object at a relative velocity exceeding 10 m/s is subject to discussion and agreement.

Lasers. Lasers of potential brightness exceeding 10^{15} W/sr or pulses exceeding 10^{15} J/sr may not be directed from space or to space without prior discussion and agreement as indicated above. "Potential brightness" is defined as the maximum power output of which the laser is capable, multiplied by the area of the exit aperture (mirror) of the laser platform, and divided by the square of the laser wavelength.[8] This limit can be attained through various combinations of fluxes of electromagnetic radiation, wavelength and diameter of relay mirrors or emitting apertures.[9]

Particle Beam Accelerators. Not to exceed 10 MeV particles or 10 kW of beam power without discussion and agreement.

Sensors. Passive sensors are not to be limited. So long as a satellite or probe does not radiate electromagnetic power beyond that reasonably necessary for communication, it is not limited in size or aperture of optics. (Satellites are not limited in their radiation of thermal infrared power, nor of reflected sunlight.)

Fissile Materials. Fissile material (uranium-235, plutonium-239, or uranium-233) in excess of 1 kg may not be launched into space, except in the form of a nuclear reactor intended for power in deep space and which has been subject to non-invasive on-pad inspection.

Discussion

Just about any technology could be developed and tested under the combined permissions of ASAT mission and (discrimination tool in aid of) fixed-site ABM. On the other hand, the protection of the ABM treaty could be extended to effectively bar space-deployed ABM systems if the unmodified treaty were supplemented by a ban on space weapons. This would avoid the competitive evolution of weapons in the guise of "discrimination aids" and would ease the problems of verification.

Of course, scientific experiments with all results and details to be available to all nations might well warrant an exception from limitations that are, after all, intended to bar the evolution of *weapons*, and which may unintentionally limit scientific or commercial experimentation that might benefit all.

The verification of a commitment to ban all weapons from space, as well as test of weapons in space and from earth to space, is far simpler than the verification of particular limits on space-weapon activities. A primary tool in this verification programme is pre-announcement of almost all launches and space activities.

On-Pad Inspection in Support of a Ban on Fissile Material in Space

Together with pre-announcement comes the possibility of on-pad inspection for certain banned activities or satellite elements. For instance, in support of a ban on the launching of fissile materials, although the verification limit might be no more sensitive than to detect 1 kg of fissile material, the ban would extend to launching *any* fissile material, in order to aid verification by eliminating the requirement to determine accurately the quantity if it is present in near-threshold amounts. Very briefly, plutonium-239 and uranium-238 can be measured by passive detection of the nuclear gamma rays, if the fissile material is not heavily shielded. Thus, a combination of passive detectors weighing 100 kg and operating at a metre or two from the surface of the launch vehicle for a period of 10 min would suffice to verify the absence of such materials, in the absence of shielding.

In order to ensure that the fissile material was not shielded and thus hidden, the vehicle would be radiographed according to strict rules and with agreed equipment, while still on the pad. To preserve secrecy (as might be warranted), the radiographic detector would provide pixel-by-pixel reports on attenuation of the radiographic beam of neutrons or gamma rays to a computer memory, and only those pixels with attenuation greater than 100, for instance, would be reported to the two sides. Since the side in control of the launch knows perfectly well the distribution of mass and shielding within the launcher and payload,[10] they could avoid excessive attenuation or could provide explanations, or could ask for skew radiography, so that innocent elements of space launches would not appear like shields. Large fuel tanks for on-orbit propulsion or upper-stage activities could be fuelled through normal plumbing on the pad.

Finally, serious attention should be given in the overall verification regime by the human and legal contribution to verification. It would not break new ground to require that a specific treaty be published widely in the nations concerned, and distributed widely to scientific workers and military officers in the fields concerned. As part of the verification regime, these individuals should be required periodically to certify to a responsible body in their own nation that their activities are in compliance with the treaty obligations.

Strictly standard satellites would not normally need inspection, but others might be secret in nature, subject to non-revelatory on-pad inspection and perhaps formal testimony from a number of intermediaries trusted by both sides.

There is a question of the venue and the mechanism for an international treaty. This framework was frankly devised for US-Soviet initiative, to support and to extend the ABM Treaty and the long suspended ASAT negotiations. Presumably only space-faring nations would join the international treaty, and that would not be an overwhelming number.

Epilogue
At a time when it finally has become accepted to control and to reduce nuclear threat with a pen instead of with the sword, there is still time to prevent the emergence of a new means of initiation of nuclear war.

In reality, the United States would be losing little of value in limiting itself in this way and on balance it would gain very much by closing an interaction that could be destabilizing. To those who imply that there is no reason to institute control mechanisms when there is no apparent hazard, I point out the worth of safety rods in a nuclear reactor, which are all the more essential when the reactor is shut down and essentially no neutrons are present. But remove them suddenly, and it is too late to control the reactor when the "threat" is apparent.

It is naive to imagine that war-supporting satellites would survive and continue to operate during large-scale nuclear war, if only because it is so easy to provide a nuclear-armed ASAT system such as the United States deployed in the sixties with nuclear warheads atop intermediate-range ballistic missiles (IRBMs). Banning ASAT tests and space weapons will not change this fact, but it will eliminate a wasteful and destabilizing interaction in peacetime.

Stabilizing and Destabilizing Aspects of Space Utilization

In our model of the strategic nuclear confrontation, with missiles assumed vulnerable in principle, but not vulnerable in practice because of equality of numbers, realistic accuracy and reliability, space observation provides several stabilizing influences. First, it can help to verify agreements to deploy equal numbers of warheads and may help to verify commitments not to produce additional warheads. The assurance of detection of launch (confidence that no missile has been launched) is of further value, although not critical as it would be in the case of vulnerable MIRVed missiles. And space observation could verify that threats to the stability of this strategic nuclear regime were not arising, such as widespread deployment of ABM, threatening ABM test, and the like.

On the other hand, space assets such as the Global Positioning System (GPS) could play a role that would be destabilizing under some circumstances. For instance, modifications of re-entry vehicles or guidance systems might provide the offensive force with much more accurately placed nuclear explosions, and surveillance could determine accurately the location of fixed targets such as silos.

To a considerable extent, the accuracy improvement of strategic missiles is irrelevant to the US offensive capability, since several generations of improvement of inertial navigation systems (INS) have provided US weapons with essentially perfect accuracy against all but the most durably hardened targets. The idea contained in the US Government-funded study "Discriminate Deterrence" (1988)[11] that in the coming decade accuracy could be so good that non-nuclear warheads could destroy strategic targets, including silos, on the other side deserves some scepticism, not because that accuracy could not be obtained, but because such vulnerability could be eliminated by other means, such as passive or active defence which need extend only a few metres from the target in the case of non-nuclear warheads, to the hiding of the precise location of the target under a protective cover of one kind or another, and so on.

The introduction of such weapons systems which could in principle have very good effectiveness against an unmodified adversary posture would not have such an effect at all, in view of the feasible countermeasures and modifications that could be made. But the adversary could do something else to counter the system, and that would be to have the ability to destroy in a timely fashion the basis of that capability—in this case, the real-time GPS navigation complex of 24 satellites in 12-hour orbit. This could be done by the development and deployment of an ASAT capability adequate for the task, or (perhaps more reliably) by the deployment of space mines, each located within a lethal distance (perhaps a few kilometres) of its target GPS satellite, ready to explode on command or if blinded.

Evidently, a new element of instability has been introduced; while attempting to restore stability to the strategic nuclear confrontation, one of the feedback elements has been converted from a passive system to a system which itself is sufficiently complex and interactive that it is unstable either from the point of view of crisis or arms race.

Of course, focusing on strategic stability misses aspects of human and national interaction that are essential to understanding and security. For instance, one could well believe that the British and the French nuclear weapon delivery systems are totally vulnerable to attack by US weapons, but that aspect of US-United Kingdom or US-French interaction does not affect behaviour, independence or relations between the peoples and the nations.

For the many nations, which have chosen not to acquire nuclear weapons "strategic stability of the nuclear confrontation" has no first-party relevance. Nevertheless, it is important for them that there is no *instability* and it is clear that for most nations the acquisition of a nuclear force in order to have a stable confrontation with the United States would have quite the opposite effect.

Space and Non-strategic, Non-nuclear Military Activities

For the case of specificity, we focus on a dispute between two or more nations in a limited theatre of operations, perhaps 1000 km on a side, or 1/500 of the surface area of the globe. Space resources of relevance to non-nuclear conflict include weather satellites, communication satellites (ComSats), navigation satellites (GPS), and imaging satellites, including LANDSAT and the military imaging satellites of much better resolution.

The Soviet Union, for example, has long emphasized concealment and deception, and radio-electronic countermeasures (RECM) in her tactical operations. Every military obviously pays attention to such matters, and they thus clearly degrade the capabilities of the satellite systems thus far mentioned.

Not much can be done to make weather unobservable, and it is difficult to jam the receiving ground station so as to deny the receipt of the weather data (pictures). The jamming task is all the more difficult because of the possibility of multiple ground stations.

ComSats can be used for communications within the theatre (intra-theatre communications) or for communications to or from the theatre. Ground stations are likely to have an antenna with some directivity for ComSats in geostationary orbit (GEO), but not necessarily for the store-and-forward satellites that are occasionally used in LEO. In the future, there may be entire networks of LEO satellites as in the commercial Iridium system under development by Motorola, but I concentrate here on the GEO ComSats. Even commercial ComSats have a great deal of capability, as evidenced by the fact that any one of nearly a million satellites "dishes" in the United States, 3-m diameter, receives a high-quality TV signal of 4-MHz bandwidth and has a choice of 100 or more such signals. This is typically achieved with a transmitting antenna on the satellite giving continental coverage with a transmitting power of about 20 W for a single one of 12–20 transmitters, each transmitter carrying on the order of six commercial TV signals.

For these microwave signals, it is very difficult to jam the downlink, both because of the substantial gain of the individual ground antennas, and because a jammer does not typically have line of sight coverage to many of the receivers in the theatre. The satellite *receiver* is a different matter, especially if the satellite is used for intra-theatre relay. In this case, a jammer in the theatre would be within the beam-width of the receiving antenna on the satellite, and it is quite easy to reduce very greatly the communications capability of the satellite. In the expectation of such tactical jamming, military satellites often employ transmitter signal power far in excess of what would be needed against a noise background alone, and implement other anti-jam techniques such as frequency hopping, spread spectrum, and the like.

The difficulty of preserving communications against jamming is indicated by the lengths to which the United States has gone in developing the MILSTAR satellite, with heavily degraded capability in the interests of providing reliable, minimal communications in a jamming environment. Other options exist for theatre communications, including rapidly deployed microwave links, fibre optic communications, and local balloons or high-flying unmanned air vehicles (UAV, "drone") for theatre communication communications relay. Other options exist for reliable satellite communication as well, including especially insensitive receivers for the uplink, redundant site laser communications uplinks, and the like. The point is that a responsible military will not depend solely on the peacetime ComSat capability for the conduct of a theatre war.

Similar considerations hold for GPS. All 24 GPS satellites operate on the same frequencies of L1 = 1575.42 MHz and L2 = 1227.6 MHz. Without going into unnecessary detail, we note that the transmitted power of about 5 W is spread over the hemisphere of the earth, resulting in a signal output from a typical GPS antenna of 10^{-16} W. This is equal to the thermal noise power in a bandwidth of 20 kHz.

Various levels of jamming deny various aspects of the GPS signal. With a data bandwidth of about 100 Hz on the clear and acquisition (C/A) signal, this threshold comes for a ratio of jamming to signal power (J/S) of the ratio of the bandwidths—20,000/100 or about 200. This would be achieved by a 5 W noise jammer at about 200 km distance. But if the GPS receiver is linked to an inertial measurement unit (IMU) to allow an integration time constant of about one second or a noise bandwidth of perhaps 0.3 Hz, an increase by a factor 300 in jamming power can be tolerated for maintaining signal track, so a 10 W jammer would be effective only out to about 10 km.

Nevertheless, given the considerably higher radiated power capabilities available to jammers, and the benefit to the jammer of using a directional antenna, it is unreasonable to expect that GPS can be used on munitions to attack defended targets or those within a heavy jamming perimeter. Remedies to the GPS user include using that same IMU to carry the vehicle or the weapon from outside the jamming perimeter at a range of some 200 km, for instance right into the target, without further GPS reception. That this is feasible is clear from the use of an IMU to guide the Tomahawk missile from its last TERCOM measurement to its target.

If IMU of the requisite performance, cost, and size are not available, the GPS user can deploy mock GPS transmitters based on balloons or aircraft in the theatre, which can provide much higher signal strengths and thus raise substantially the threshold for jamming. If one goes to a transmitter power of 100 W and needs to illuminate a region only 100 km in diameter rather than 12,000 km, the received signal is increased by a factor 3000, as is the necessary jammer power. Such GPS augmentation or replacement is necessary for a robust capability in the face of jamming, and this has been obvious for a long time.[12]

Imaging Satellites for Tactical Observation

Satellites in LEO are typically over the assumed theatre of operations less than one per cent of the time, and a single satellite in LEO suffers gaps in its viewing capability of the theatre which are 20 hours or greater in extent. At a nominal satellite altitude of 500 km, even a satellite with a mirror the size of the Hubble Space Telescope (some 2.8 m diameter) would have resolution and light gathering power equivalent to a lens or mirror some 25 times smaller in diameter (11 cm) at an altitude of 20 km. For this reason and because of the greater flexibility and capability for repeated coverage by aircraft or UAV carrying such modest cameras, I have long emphasized the desirability of employing such vehicles for theatre observation. Operating at supersonic speed of some 1 km/s (3600 km/h), and with a swath width of 40 km, a single UAV could film every (unobscured) point in the theatre every eight hours. The capability to capture on film these high-resolution pictures is far in access of what could be done from space over any available communication channel.

In any case, aerial reconnaissance both by aircraft and by UAV is an existing capability and a historic fact. Imaging from satellite has many uses over the entire world, and it can, of course, be used also within a localized theatre, but it would be foolish to depend only on satellites to provide the timely information about disposition of the opposing military force in the theatre or for identifying targets for conventional weapons—shells, bombs or the like.

The Policy Question

The free use of space by individual nations is enshrined in international law, and the benefits from satellite communication, navigation, observation, and the like are enormous. National satellite systems contribute also to national economic goals, in some cases providing economic advantage in return for the investment. It is no different for the military applications, in the absence of outright conflict.

In time of limited theatre war, however, either of the participants does what it can to protect its forces and to overcome, destroy, or injure the forces on the other side. War is dangerous and destructive, to the military forces engaged, to the civilians in the theatre and, potentially, not only to the military forces not engaged but to the national homelands and populations and to the other nations of the world.

Obviously, every effort should be made to extend the rule of law and of justice so that armed conflict does not take place, and it is the path of wisdom as well to constrain armed conflict by international treaties that proscribe certain weapons or actions, as with the recent successful conclusion of the negotiations banning chemical weapons. Beyond that, it is important, in my opinion to strengthen the UN and to attempt to move towards the norm of coalitions in support of the UN, when armed conflict is unavoidable.

According to the IAEA, the United States has provided intelligence information in support of the IAEA mission under various United Nations resolutions to discover and to destroy installations in Iraq related to nuclear weaponry. Other nations might also have supplied such information, which could be obtained not only from satellite imagery but also from aircraft observation. In fact, the Open Skies Treaty provides the right of access for aerial photography by one nation over another, which in my opinion ought to be extended to lower altitudes and to the use of other sensors such as synthetic aperture radar (SAR). This will provide the benefits of such observation to other nations in peacetime.

As indicated, it would be perfectly feasible to use ASAT weapons in wartime to destroy the satellite capabilities of the other side, especially those in LEO. The side operating the satellites has the option of hardening them against such threats, using manoeuvre, and the like, but as I have emphasized in many publications, the advantage is heavily on the side of destruction rather than preservation.

It has been proposed that the United States needs ASAT in order to counter ASAT activities by others, but this argument is clearly not valid, in view of the much larger cost of the US satellite systems compared with any adversary's investment in space. To a large extent, the preparation to respond in kind legitimizes the initial use of force, just as the Geneva Convention of 1925 which bans the first use of chemical weapons permitted the development, possession and planning for the (second) use of chemical weapons by nations adhering to the Convention.

In principle, satellites could also be defended actively, either by a system of space weapons that could counter ASAT rockets as they were launched, or by more local defence to destroy ASAT interceptors as they approached their quarry in LEO. No doubt such systems could be built, and with some effectiveness. In fact, it would be somewhat easier to destroy ASATs in boost phase or soon after than to do the close-in defence, but I know that the United States would not be happy if some other nation had the capability to deny us access to space, and my own view is that other nations should not be content if the United States deployed such a capability as well.

I believe that large LEO satellites can be preserved and should be preserved against destruction by a commitment to retaliate not in kind but against targets of very great value and uniqueness, probably on the home territory of those employing the ASAT. Of course, not everybody has this same view, because it clearly implies the spreading of conflict from theatre war to a much wider and significant conflict. In my view, the spreading would have occurred by the destruction of the satellite, which has the primary mission on world surveillance or providing a world navigation capability, and is only peripherally involved in the theatre conflict.

This is a problem for arms control and world stability. In my opinion, it should be solved by a ban on both space weapons and ASAT, which systems, if allowed to run free, can lead to a feedback loop exhibiting first arms race instability and then crisis instability.

Some might propose instead to break this feedback loop by banning military use of satellites, or by banning the use of GPS by weapons or by military aircraft, or by banning satellites from imaging in a theatre of conflict. There was a good deal of such discussion in the early days of the involvement of the UN Conference on Disarmament (CD) with space arms control, but the consensus that emerged allows such activities so long as there are no weapons in space.

A robust space arms control regime, however, could not be based solely on a ban on *use* of space weapons or of ASAT, but would need to extend to a ban on test and also a ban on possession of dedicated ASAT systems. Given the small size of potential ASAT interceptors, a ban on possession could probably not be verified by NTM, but such a treaty could nonetheless be effective by the use of other means of verification, including the involvement of technical personnel working on such programmes in each nation.

It seems to me both essential and feasible at this time to begin again a serious effort to achieve soon a ban on actual space weapons and on anti-satellite weapons.

Notes

1. Panel on Nuclear Weapons Safety of the US House Committee on Armed Services, *Nuclear Weapons Safety* (Washington, DC: US Government Printing Office, 1990).
2. Ray E. Kidder, "How Much More Nuclear Testing Do We Need?," *Arms Control Today* 22 (September 1992): 11–14.
3. Committee on International Security and Arms Control, National Academy of Sciences, *The Future of the U.S.-Soviet Nuclear Relationship* (Washington, DC: National Academy of Sciences, 1991).
4. Steve Fetter et al., "Expert Study on Questions Related to a Comprehensive Test Ban Treaty," in *Towards a Comprehensive Test Ban Treaty* (Oslo: Royal Norwegian Ministry of Foreign Affairs, May 1992).
5. Richard L. Garwin, "Reducing Dependence on Nuclear Weapons: A Second Nuclear Regime," in *Nuclear Weapons and World Politics: Alternatives for the Future (1980s project/Council on Foreign Relations)*, by David C. Gompert et al. (New York: McGraw-Hill, 1977).
6. Richard L. Garwin, John Tike, and Yevgeny P. Velikhov, "Space Weapons," *Bulletin of the Atomic Scientists* 40, No. 5 (1984): 1S–15S.
7. Richard L. Garwin and Roal'd Z. Sagdeev, "Verification of Compliance with the ABM Treaty and with Limits on Space Weapons," in *Verification: Monitoring Disarmament*, eds. Francesco Calogero, Marvin L. Goldberger, and Sergei P. Kapitza (Boulder, San Francisco, Oxford: Westview Press, 1991): 23–44.
8. See Richard L. Garwin, "How Many Orbiting Lasers for Boost-Phase Intercept?," *Nature* 315 (23 May 1985): 286–290.

9. This limitation on potential brightness would correspond to illumination of about 1 W/cm^2 at 300 km distance. It would be reached with 1 kW of laser light at 1 micron wavelength, into a mirror of 1 m^2 area.
10. In fact, they could do preliminary radiography on their own.
11. Commission on Integrated Long-Term Strategy, *Discriminate Deterrence* (Washington, DC: US Government Printing Office, 1988).
12. Richard L. Garwin, "Effective Military Technology for the 1980s," *International Security* 1 (Fall, 1976): 50–77, esp. 66–71.

Richard L. Garwin is IBM Fellow Emeritus at the Thomas J. Watson Research Center in Yorktown Heights, New York. As a consultant to Los Alamos National Laboratory working on nuclear weapons, Garwin was the author of the actual design used in the first hydrogen bomb (code-named Mike). During his career he divided his time among applied research, basic science and consulting to the US Government on national security matters. Garwin served on the US President's Science Advisory Committee under Presidents Kennedy, Johnson and Nixon.

CHAPTER 16

The Non-proliferation Treaty and the German Choice Not to Proliferate

Harald Müller

Treaties, Norms and Power

Why should any treaty be of importance for the political life of states and people? There are those, scholars who call themselves "realists" (a misnomer, I believe), hard-nosed politicians, journalists specializing in security matters and so on, who maintain that it is only national interest, power and security that drive states' behaviour; treaties are merely temporary fixations of constellations of powers which will lose their validity as power relations change and states' interests evolve.[1]

This school of thought ignores the influence of international law and research results emerging from the focus on international regimes, which the "institutionalist" school of international relations has produced over the last decade. International lawyers have never tired of pointing to one eminent fact of international life: that most states abide most of the time by most international treaties and agreements, just as

Originally published in David Carlton et al. eds., *Controlling the International Transfer of Weaponry and Related Technology* (Aldershot: Dartmouth, 1995): 173–186.

H. Müller (✉)
Peace Research Institute Frankfurt, Frankfurt, Germany

P. Foradori et al. (eds.), *Arms Control and Disarmament*,
https://doi.org/10.1007/978-3-319-62259-0_16

in domestic affairs, where most citizens faithfully observe most laws most of the time, but each might breach a stipulation here and there—think of crossing a street when the lights are red—without thereby invalidating the law as such.[2]

Regime analysts have developed this point. First, international regimes containing principles, norms, rules and procedures for a given area of foreign, economic, environmental or security policy lead to an overwhelming record of state compliance. But this is by no means all. As states comply with the rules they agreed to—maybe initially with some reluctance and concern—compliance becomes part of their habit and custom. Compliant behaviour does not result from ever-repeated recalculation of pros and cons. The fact that a treaty is there, and that it was complied with for the last five or ten years, leads states to think it normal and proper to comply for the next five years as well.

Second, with changing habits, prevailing attitudes are influenced. Protectionism in international trade was something normal and good a hundred years ago for most states, with the exception of Great Britain, the world market leader. Today protectionism is not uncommon, but each step in this direction requires lengthy justification, claims of being treated unfairly by one's trading partners and promises to revoke the measure if the grievances are met by partners' actions. In other words, the international trade regime, built on liberal principles and norms, has changed the framework of attitudes towards foreign economic behaviour wherein states' policies are framed.

Last, and most importantly, even states' interests are affected. This is nowhere more true than in the field of security. In the classical system of states, self-help was the rule and the security dilemma—the fear of being attacked in an inferior position when under-armed, and the certain prospect of propelling one's neighbours into reciprocal behaviour when overarming—was the natural state of international affairs. In a world of interlocking security regimes, as currently emerging in Europe, the security dilemma is mitigated if not completely overcome for a great part of the European family of states. The regimes provide instruments to isolate, and sometimes deal with, the few hot spots where conflict breaks out. The meaning of "national interest" has shifted drastically in comparison with previous periods. It has moved towards demonstrating to neighbours that one's own intentions are peaceful by reliable compliance with agreements, and towards receiving reciprocal reassurances from them in turn, in as many respects as is warranted by the political situation and the nature of modem military forces and technology.[3]

The importance of the Non-Proliferation Treaty (NPT) to Europe as a whole lies in its being part and parcel of an overall security order, in its interconnection with other treaties and agreements covering various aspects of armament, foreign and security policy. The emergence of this regulated security order puts European politics on a new plane, present turmoil notwithstanding. This kind of order is of pre-eminent importance at a time when turmoil and turbulence are inevitable, given the hasty transition to a new socio-economic and political system which will take time to gain roots. In times of turbulence and social upheaval, primordial political themes, based on ethnic or religious cohesion, necessarily gain higher importance in terms of social and political mobilization. Their characteristic of providing identity by sharply differentiating between "us" and "them" involves a high propensity to degenerate into armed conflict. The presence of "rules of the road" between states then serves to contain the violence in isolated pockets rather than having them spread out like bushfires. It is no coincidence that armed violence has been confined to new sovereignties emerging from decaying multi-ethnic empires. Friction between the more established states belonging to the European system has emerged, but without a shot being fired; think, for example, of Hungary and Romania.

Let us briefly pause and consider what we have learned from some 20 years of research into the motivations of proliferating states. The overarching motive for programmes designed to acquire nuclear weapons has been related to matters of security or, in a few cases such as Iraq and perhaps Libya, unhealthy ambition. Most of the time, governments were afraid of possible conventional superiority of hostile neighbours or their coalitions, or sinister intentions held by a neighbouring nuclear weapon state. If we want to examine the propensity of a region to become a proliferation hot spot, we have to look into the security situation of the region's states. Prestige and status, misgivings over discrimination and the bad example of the superpowers, bureaucratic momentum and vested interests, and attempts at diverting public opinion from blatant failures of domestic policy—all these factors played a role, but they were effective only within the framework of regional power and security considerations.

In terms of these proliferation motivations, we can distinguish between five regions in post-Cold War Europe. The first region, and the one of the least concern, is ironically one with fairly well developed capabilities. The members of the European Community (EC) and European Free Trade Area (EFTA) in Western and Northern Europe are all technologically well-advanced states. Some of them, such as Sweden, Belgium, Switzerland,

Germany and the Netherlands, have a sophisticated nuclear sector, including the mastery of sensitive technologies or even weapon-usable material on their territories. Germany, in fact, has been widely discussed as a ready candidate for proliferation after reunification, and many remember that Germany was one of the primary targets when the NPT was negotiated in the late sixties.[4] Yet in terms of motivation and observable policy the country is unambiguously not ambitious in the nuclear area, along with other Western European non-nuclear weapon states.

The second group is the former allies of the Soviet Union. Their whole identity has undergone radical change. Their main problem is to establish a position where threats that might emerge from sudden changes in the former Soviet Union, especially Russia, are properly balanced, and where ethnic and territorial disputes with their neighbours that had been subdued by 40 years of inter-bloc confrontation can be properly dealt with without resort to force. Another security issue, the power of the reunified Germany, should be laid to rest through the treaties between Germany, Poland and the former Czechoslovakia. Though the negotiations were difficult and not without friction, the results have created a solid basis for good neighbourly relations, which should go a long way to eliminate residual fears.

The possession of nuclear weapons might be seen by Eastern European states as a guarantee of survival against an expansionist Russia (or Ukraine). However, it is only a remote possibility, as none possesses the capability rapidly to acquire nuclear weapons. A nuclear weapon programme would be protracted and impossible to hide. It would open a dangerous transition time, full of tension and resentment, even from formerly well-disposed neighbours. It would also divert precious and scarce resources, which these countries need badly to fire the engine of their market economies.

The alternative is fairly obvious, and it is the one sought by these countries themselves: to achieve a step-by-step integration into the Western sphere, via association with the North Atlantic Treaty Organisation (NATO) and the EC with the option of later membership that would afford them a security equal to their Western neighbours. This need has been understood in the West and most strongly promoted by Germany. NATO, however, is somewhat shy about opening the prospect of full membership, partly owing to the concern not to alienate or isolate the states of the former Soviet Union, on whose political future so much of the European security situation hinges. On the other hand, there is little doubt that, if the security situation in East-central Europe should worsen dramatically, NATO would find ways to react quickly that would satisfy the security needs of central eastern Europe.[5] These more palatable and less costly alternatives for Poland, Hungary, the Czech and Slovak Republics, Romania and Bulgaria to provide for their security makes it extremely unlikely that serious motivations might emerge to "go nuclear."

The third group of countries is the former member republics of the Soviet Union. For them, the security problem is far more acute. The case of Bosnia provides a warning. Within the republics there is a wave of nationalism that might provide a fruitful ground domestically for demagogues who would propose brandishing nuclear weapons. The lack of ambition for nuclear weapon status of most of the former republics is contingent on the survival of present leaderships and the absence of serious disputes over territory and people among them.[6] Some of the republics (wherein at least a solid base for nuclear research exists) are candidates for pariah status. Armenia, with her traditional quarrels with Turkey and Azerbaijan, is one example[7]; Georgia is a second. The influence of Islamic fundamentalism, with all kinds of consequences, on the Central Asian republics cannot yet be adequately assessed, but it must be watched at any rate. Finally, Ukraine, Kazakhstan and Belarus have agreed, in the Lisbon Protocol to the Strategic Arms Reduction Treaty (START), to accede to the NPT as soon as possible as non-nuclear weapon states, though it remains to be seen whether all, and Ukraine in particular, will actually comply in full. There are other signs of hope. The Baltic states, which must certainly have serious concerns for their future security, have so far chosen to seek close relations with the West and adhere to the NPT. This course of events points to the most promising avenue for managing the problems of the republics: to approach them with the very firm request that they preserve non-nuclear status, and with the promise of good relations and economic cooperation if they do so; with the corollary of advice to Russia not to create security fears for these republics.

Among all countries that have emerged in a quite changed security environment since 1989, one stands out as a real problem: Serbia, which might become a classical pariah, isolated and despised for her backward Stalinist system and her violent conduct in Yugoslavian affairs. The international community appears not to be ready to confront Serbia with armed force, but it will not lend cooperation, political sympathies and support either. Serbia seems likely to have quite a number of disputes with her neighbours, Hungary, Albania, Austria, Bulgaria and Slovenia, whether or not the dream of a Greater Serbia comes true. Serbia has at her disposal a small but sound nuclear research community. It would take the country a long time to develop her present capacities into threshold status, but, in view of the constellation of her motivations, Serbia is likely to be a target for proliferation control.[8]

Lastly, the southern rim of Europe, from Spain to areas east of the Caucasus, is the region most directly threatened by proliferation elsewhere, notably the Middle East. The South-western part of the European Mediterranean looks southwards to the Maghreb, where the ferment of social change, political ambition and religious fundamentalism creates uncertainty about future security threats. Fear is focused on attempts by Muammar Ghadaffi to bypass his ailing nuclear technological base by directly acquiring a nuclear weapon; and the strange ambiguities that accompanied the acquisition of a research reactor from China by Algeria, which may presently have a "fundamentalist" government and which has some able nuclear scientists. Despite these dangers, acquiring nuclear arms has not been discussed by Italy or Spain. On the contrary, Spain joined the NPT in 1987, and Italy has cosponsored a Group of Seven (G-7) resolution on its indefinite extension.

If we pull this brief regional analysis together, it is clear that security problems, which could lead to proliferation, are not absent in Europe, to say the least. Yet the prognosis is rather for the continuation of the *status quo*. The existence of the NPT plays a pivotal role, giving rival European states the assurance that their neighbours will abstain from a nuclear race as long as they do so themselves. It serves as a standard of behaviour for newcomers who may otherwise be tempted to consider a nuclear option to provide for their security. It establishes in international relations the necessary compromise to national sovereignty demanded by safeguards and verification procedures.

Thus the NPT is an important building block within an interlocking European security order. It provides a frame of reference for the shaping of security principles and legal commitments for the newly independent states. It works together with the Conventional Forces in Europe (CFE) talks and the Stockholm and Vienna confidence building agreements regulating holdings of offensive conventional weapons, troops and their movements in peacetime; with the Intermediate-Range Nuclear Force (INF) and START Treaties; with the Open Sky Agreement, which is not without importance for the nuclear proliferation issue; with the Biological Weapon Convention; and, in particular, with the Chemical Weapon Convention signed in January 1993. All these agreements bring order and predictability to the European security scene. Without them it would have been difficult, for example, to establish agreed figures on troop strength and armour holdings in the Commonwealth of Independent States (CIS) and Eastern Europe, or to get the agreement by Belarus, Kazakhstan and Ukraine to consider non-nuclear status.

The NPT and Proliferation Threats to Europe

Proliferation chains represent a time-honoured model in the analysis of the spread of nuclear weapons.[9] This model recognizes that the motivation of a government to embark on a nuclear programme is largely affected by sudden changes in its security environment, and that the sudden emergence of a new nuclear weapon state nearby is one of the more dramatic strategic shifts possible. Acquiring nuclear weapons in such a situation might even be seen as an act of prudence under such circumstances if no alliance with a reliable protector is available as an alternative option.

The most obvious proliferation chain runs from the Middle East through the Balkans and the Caucasus, with Turkey playing a key role. Turkey would be directly affected if any of her neighbours—Iraq, Syria, or Iran—were to successfully develop nuclear weapons. She can live, and has been living, with the nuclear weapon status of Israel as it presents no direct threat. But, on the other hand, Israel's nuclear weapons provide a permanent stimulus for Arab (and Iranian) leaders to reconsider their security position. The Iraqi example is a telling one.[10]

Turkey possesses the necessary infrastructure to respond with a sizeable emergency programme if needed. She possesses significant research facilities, an industrial basis that is better than Iraq's, and a sufficient number of scientists and engineers. Also Turkey's military organization is fairly efficient by Middle Eastern standards. The infrastructure is likely even to improve as experience is gained through the joint development programme for the KAREM reactor with Argentina.[11] And one should not forget that it took Turkey a long while and much discussion to become a party to the NPT.

A Turkish nuclear weapons programme would send shock waves in three directions: to Greece, to Bulgaria and to Armenia, which, as already indicated, is already in a precarious security situation. These countries may be hard put to respond in kind, given their lack of facilities and capabilities. Yet a change of heart in these countries would spark concerns in Albania, Serbia and Romania. At this stage, the chain would have reached central Europe. Armenia would feel immediately threatened, for historical and geopolitical reasons, by a nuclearized Turkish state. An Armenian reaction could in turn destabilize the whole Caucasian region. To be sure, we may be talking about a process that could take up to two or three decades. Yet Europe, with her many uncertainties and instabilities in this precarious transition period, could not afford and sustain the additional strain stemming from this enormous politico-strategic momentum.

Again the NPT serves as a useful brake on this momentum, particularly since the safeguards system has been strengthened through recent decisions in the International Atomic Energy Agency (IAEA) Board of Governors. All Middle Eastern, southern European and Caucasian countries are parties to the NPT or can be expected to accede to it in the foreseeable future, with the exception of Israel. Distrust reigns with regard to Iranian activities, but it would be very difficult indeed for Iran to mount an effort comparable to Iraq's without detection under prevailing circumstances. Iraq, for her part, is still firmly controlled by the United Nations Security Council, the United Nations Special Commission on Iraq (UNSCOM) and the IAEA. NPT full-scope safeguards, along with special inspections, should help provide enough confidence to uphold the objective of non-proliferation.

Moreover, the picture becomes brighter if we look at the political conditions. As the key to the chain is Turkey, Turkish security must be guaranteed in the first place. This points to the crucial purpose of NATO, not as a deterrence instrument, but as a non-proliferation tool. This function derives from NATO's basic mission to guarantee the security of its member states through mutual alliance, notably with the United States. To preserve this function for Turkey is one of the keys to upholding the European non-proliferation system. In addition it is of critical importance to find a procedure to deal with security issues in the Balkans. The Yugoslav experience does not bode well for the present capability of the community of European states to calm down fears and to contain violence that, in combination with potential capabilities growing in the Middle East, might seriously affect security and strategic thinking in the Balkan countries. A more flexible and reliable system, containing elements of collective security enforcement, is needed.[12]

This corresponds to the persistent request of non-nuclear weapon states within the NPT to receive reliable security guarantees in return for their renouncing nuclear weapons. In other words, a three-layered non-proliferation regime must encompass the NPT on the first layer, comprehensive arms control on the second and a credible system of collective security on the third. If such a multi-layered system emerges, we can lay all our proliferation fears concerning the "old continent" to rest.[13]

Germany and Nuclear Proliferation

There is general agreement that the most important renunciation is that of Germany. Contrary to predictions, or even recommendations, that after unification and the end of the East-West conflict Germany would eagerly hasten to acquire great power status, there are no signs of interest in nuclear weapons.[14] On the contrary, if anything *is* significant in this regard it is a leap in the German commitment to an active and determined non-proliferation policy. In 1968, a reluctant Germany was forcefully persuaded to come to terms with the superpower agreement over the NPT. It was just ten years after West Germany, in a secret agreement with France and Italy, had tried in vain to start an atomic weapons programme. Germany had fought tooth and nail to minimize the impact of safeguards on her nuclear industry, to adopt liberal language concerning constraints on nuclear exports and to insist on an earlier termination date for the NPT than the 25 years finally written into Article X.[15] But in 1992, both Germany and Italy joined the G-7 consensus that the NPT (and their "inferior" nuclear status) be made permanent. Germany was also a driving force behind the sharpening of IAEA safeguards, and she worked hard to achieve nuclear suppliers' agreement on full-scope safeguards as a condition of supply, as agreed in Warsaw in April 1992. After three major changes in law and more than 30 substantial amendments to regulations, Germany, a primary source of sensitive technology so far, possesses one of the most radical and effective export control systems in the world.

The change is enormous, but it is explainable. For 20 years the nuclear issue in Germany was connected with the deprivation of power and status suffered as a consequence of the Second World War. To gain nuclear sovereignty and equality, in both the military and civilian field, was part of the political rehabilitation programme during that period. When military equality was ruled out by the strength of opposition, the peaceful atom gained all the more prominence as a symbol of technological, economic and political achievement. In the seventies and eighties, this symbolism was eroding. Nuclear power lost its attraction because of high cost, public opposition, environmental problems and the two big accidents of Harrisburg and Chernobyl. Germany gained self-confidence as an economic and financial power, a leader within the EC and a mover of European politics through her detente policy and her successful commitment to arms control. Slowly political considerations such as world security and the need to abide by NPT obligations entered German considerations in nuclear matters; concomitantly, the Foreign Ministry became more and more an advocate of non-proliferation proper.[16] Two events at the end of the eighties then propelled an incremental adaptation of policy into a sudden and visible shift. First, the export scandals, whose revelation started in 1988 and reached high points in 1989 and 1991, damaged Germany's reputation and attracted political attention. Second, German unification created a political obligation to adopt "global responsibility," and non-proliferation of weapons of mass destruction was chosen by the government as an area where such responsibility could be demonstrated. The negotiations on the unification treaty forced Germany to reconsider the basic choice of non-nuclearism. It was obvious that not only the Soviet Union, but Germany's allies as well would prefer a reaffirmation of Germany's non-nuclear status. There was little resistance; Hans-Dietrich Genscher, as Foreign Minister, solemnly declared the continuing adherence to the NPT by the united Germany when he addressed the Fourth NPT Review Conference in August 1990, and a little later this obligation became part of the unification treaty. Genscher's speech, which also publicized the recent decision by the cabinet to require full-scope safeguards as a condition of a new nuclear supply agreements, marked the beginning of a series of initiatives, some of which have already been mentioned.[17]

One should add German diplomacy towards South Africa and France to encourage accession to the NPT; persuasion of Argentina and Brazil to adopt a full-scope safeguards arrangement analogous to that of the European Atomic Energy Community (EURATOM); requests to North Korea for strict adherence to NPT obligations; and various approaches to Third World countries within and outside the treaty. Germany pushed for a Security Council statement on non-proliferation and pursues an interest in establishing a system of sanctions against proliferators. Since 1991, her development aid policy takes the recipient's proliferation status into account. Germany has conveyed to Ukraine, Belarus and Kazakhstan her desire for them to accede to the NPT as non-nuclear weapon states and she has initiated the setting-up of a programme to discourage CIS nuclear weapon specialists from emigrating.

All these moves were made in full consciousness that Germany is, and will remain, a non-nuclear weapon state. As such, she has no interest in other states joining the nuclear club and feels her national and security interests would be threatened if countries took this path. The NPT is properly seen as a pillar of both European and global security that serves immediate German interests. Rather than a reluctant target of the system, as during the sixties and seventies, Germany sees herself as a leading advocate, as one of the two most important non-nuclear parties of the NPT, and as a custodian of the non-proliferation regime. There is a widespread desire to extinguish the stain on German reputation caused by careless exporting of sensitive technology.

The Rationale for Non-nuclearism

The plain fact is that Germany has none of the motivations necessary to reconsider her non-nuclear status. It would be far too abstract an approach only to consider her geopolitical situation, to recognize her *Mittellage*, being sandwiched between powers against each of which Germany might be stronger, but against a combination of which she is bound to lose; that would be to focus mistakenly on legends that allegedly formed the history of the first half of this century. Security is not in practice primarily a matter of geopolitics; geopolitics are shaped and transformed through politics and political institutions. Seen from this perspective, Germany's security situation is better than at any time in her history. She is surrounded by friends and lacks enemies. Rather than being a pariah (as after the First World War), Germany is well integrated in international organizations that afford her security and material wellbeing, most notably the EC and NATO. She is allied to three nuclear weapon states, and the fourth, Russia, is seeking close ties for the sake of her own economic interests. Germany has no revanchist-aggressive schemes for which nuclear weapons could be put to use. Her prestige and status are well established, based on a strong and well-organized economy that could only suffer if resources were diverted to an ill-advised nuclear adventure. And German public opinion is fairly anti-nuclear, in the military field even more than in the civilian one. Germany has realized all her primary goals of the early fifties—she has become rich, influential, secure and now united. All this was achieved without nuclear weapons and, it is submitted, only because she did not have them. Any politician trying to promote a nuclearized Germany would be laughed out of parliament.[18]

Finally, we must ask why it is that a major country like Germany is positively uninterested in nuclear weapons. The answer is simple and straightforward: because, for a "trading state" such as Germany,[19] nuclear weapons are of no use, a burden rather than a blessing. Above all, the difficulties facing nuclear weapon states in and around Germany have not gone unnoticed. This has served first to reduce, then to eliminate, claims about "discrimination" which Germany shared with many countries when the NPT was first negotiated, and which many Third World governments still espouse.

Various other factors have served to reinforce this German approach. First, the danger of superpower nuclear war that had fuelled the resentment of hapless and helpless non-nuclear bystanders has disappeared. Second, the inefficiency of nuclear weapons as a means of projecting power has been shown in Vietnam and Afghanistan. Third, the Gulf War demonstrated that power inequality is based on the asymmetry in conventional power projection, not in nuclear weapons. Fourth, the steady rise of Japan and Germany as compared to France and Great Britain has proved that status and prestige are not dependent upon possession of nuclear weapons. Third World nuclear threshold countries have also gained little from their positions of ambiguity. India has lost her primary position in the non-aligned movement in the last 20 years. Pakistan is overshadowed in the Islamic world by Egypt, Indonesia, Saudi Arabia and Iran. Israel is isolated, and if the isolation is breached it is certainly not because of Israel's nuclear arsenal. Argentina, Brazil and South Africa have gained favour and influence from their renunciation of the nuclear option. Fifth, the need to maintain substantial nuclear weapons industrial sectors has imposed a drain on funding for high-technology research and development and on scarce skilled manpower resources, harming the competitiveness of the civilian sector of nuclear weapon states, while hoped-for spill-overs have been rare. Japan, Germany, Sweden, the Netherlands and others have economies that have performed better and more competitively than all nuclear weapon states. Sixth, the nuclear establishments in nuclear weapon states have been major polluters because they were not subject to adequate civilian controls. The cleaning up of the nuclear weapon complexes will cost hundreds of billions of dollars, depleting resources needed for urgent civilian investments.

In short, nuclear weapons are widely recognized as more a curse than a privilege. Anger over discrimination may be replaced by pity. In the world of the twenty-first century, military means will play a relatively minor role compared to economic, financial and environmental skills. Germany is well placed in these fields. She is a trading state whose population and political elite place welfare, influence by persuasion and, increasingly, environmental issues above conventional and traditional power politics.[20] The enormous reluctance shown on all sides in the present debate on out-of-area deployments is just more proof of this basic attitude, rather than of a decisive shift. For a trading state so well placed to take on the difficult tasks of the twenty-first century, what would nuclear weapons do, other than consume badly needed resources? Germany clearly opposes the further spread of nuclear weapons as a source of potential turmoil and disruption not conducive to her national interest, but nurtures no ambitions to have such weapons for herself. This assessment is valid *rebus sic stantibus,* that is, as long as Germany's present solidly democratic institutions remain unchanged, as long as the welfare state upholds the dominant combination of materialist and post-materialist demands placed upon the political system, and as long as the security environment does not become overwhelmingly threatening and one of self-help rather than of interlocking security regimes and interdependent alliances. If these conditions prevail, the main characteristics of Germany's policy will be to stay within the boundaries of the NPT, to try her best to universalize this instrument, to punish outsiders and to help the extension of the treaties. This policy serves Europe as well as it serves Germany.

Notes

1. Kenneth N. Waltz, *Theory of International Relations* (New York: McGraw-Hill, 1979).
2. Thomas M. Franck, *The Power of Legitimacy among Nations* (New York, Oxford: Oxford University Press, 1990).
3. Robert O. Keohane, *International Institutions and State Power* (Boulder, CO: Sage Publications, 1989); Volker Rittberger, ed., *International Regimes in East-West Politics* (London: Pinter, 1990); Oran R. Young, *International Cooperation: Building Regimes for Natural Resources and the Environment* (New York: Cornell University Press, 1989); Joseph S. Nye, "Nuclear Learning and the Evolution of U.S.-Soviet Security Competition," in *Windows of Opportunity: From Cold War to Peaceful Competition in U.S.-Soviet Relations,* eds. Graham Allison and William L. Ury (Cambridge, MA: Ballinger, 1989): 131–162.
4. Matthias Küntzel, "Die Bundesrepublik Deutschland zwischen Nuklearambition und Atomwaffenverzicht: Eine Untersuchung der Kontroverse um den Beitritt zum Atomwaffen-Sperrvertrag," (PhD diss., University of Hamburg, 1991); Dieter Mahnke, *Nukleare Mitwirkung:*

Die Bundesrepublik Deutschland in der Atlantischen Allianz 1954–1970 (Berlin: De Gruyter, 1972); Uwe Nerlich, *Der NV-Vertrag in der Politik der Bundesrepublik Deutschland: Zur Struktur eines aussenpolitischen Prioritätenkonfliktes* (Ebenhausen: Stiftung Wissenschaft und Politik, 1973).

5. See, for example, Curt Gasteyger, "The Remaking of Eastern Europe's Security," *Survival* 33 (March/April 1991): 111–124; NATO, "North Atlantic Council Statement on Partnership with the Countries of Central and Eastern Europe," *NATO Press Service*, 6 June 1991; *Atlantic News* 25, no. 2363 (25 October 1991): 1–2; ibid., no. 2368 (9 November 1991): 1–2; ibid., no. 2369 (13 November 1991): 1–2.
6. Igor Malashenko, "Behind the Arms Cuts: a Jousting of Republics," *International Herald Tribune*, 22 October 1991; Karl-Heinz Kamp, *Die Sicherheit der Sowjetischen Atomwaffen* (St. Augustin: Konrad-Adenauer-Stiftung, 1991).
7. William C. Potter, "What if Armenia Joins Nuclear Club?," *Los Angeles Times*, 25 October 1991.
8. It should be noted that Yugoslavia indicated the possibility of withdrawing from the NPT in response to insufficient implementation of the nuclear weapon states' disarmament obligation during the NPT Review Conference in 1975.
9. Lewis A. Dunn and Herman Kahn, *Trends in Nuclear Proliferation, 1975–1995: Projections, Problems, and Policy Options* (Croton-on-Hudson, NY: The Hudson Institute, 1976).
10. See, for example, Geoffrey Kemp, *The Control of the Middle East Arms Race* (Washington, DC: Brookings Institution Press, 1991): 71–74.
11. *Eye on Supply* 1 (Spring 1990): 9–10.
12. Harald Müller, "A United Nations of Europe and North America," *Arms Control Today* 21 (January–February 1991): 3–8; Dieter Senghaas, "Friedliche Streitbeilegung und kollektive Sicherheit im neuen Europa," *Europa-Archiv* 10 (1991): 311–317; Richard H. Ullman, *Securing Europe* (Princeton, NJ: Princeton University Press, 1991).
13. See, for example, Ullman, *Securing Europe*.
14. John Mearsheimer, "Back to the Future: Instability in Europe after the Cold War," *International Security* 15 (Summer 1990): 5–56.
15. Matthias Küntzel, "Die Bundesrepublik Deutschland".
16. Harald Müller, "The Politics of Technology Transfer," in *Germany and the Middle East; Patterns and Prospects*, ed. Shahram Chubin (London: Pinter, 1992): 154–176.
17. Wolfgang Kötter and Harald Müller, *Germany and the Bomb: Nuclear Policies in the Two German States and the United Germany's Nonproliferation Commitments*, PRIF Report no. 14 (Frankfurt: PRIF, 1990);

David Fischer and Harald Müller, *A Treaty in Trouble: Europe and the NPT after the Fourth Review Conference*, PRIF Report no. 17 (Frankfurt: PRIF, 1991): 44.

18. For the attitudes of Germans towards nuclear weapons and on general foreign policy orientations of the present Germany, see, for example, Thomas Risse-Kappen, *The Zero Option: INF, West Germany and Arms Control* (Boulder, CO: Westview Press, 1988); Barry M. Blechman and Cathleen Fisher, eds., *The Silent Partner: West Germany and Arms Control* (Cambridge, MA: Ballinger, 1988); Harald Müller, "No Need to Fear United Germany," *Bulletin of the Atomic Scientists* 46 (April 1990): 14–15. The recipe of "managed proliferation" of nuclear weapons to Germany meets deaf ears in the target country. See, for example, Mearsheimer, "Back to the Future".
19. Richard Rosecrance, *Der neue Handelsstaat: Herausforderungen für Politik und Wirtschaft* (Frankfurt: Campus Verlag, 1987).
20. Hans W. Maull, "Germany and Japan: The New Civilian Powers," *Foreign Affairs* 69 (Winter 1990/91): 91–106.

Harald Müller is Executive Director (ret.) of the Peace Research Institute in Frankfurt, Professor of International Relations at Goethe University and an adjunct professor at the Paul H. Nitze School of Advanced International Studies at Johns Hopkins University. He has been a member of the German delegation to the Nuclear Non-proliferation Treaty Review Conferences since 1995 and has served on the United Nations secretary-general's Advisory Board on Disarmament Matters (1999–2005) and as the vice president of the European Union Consortium for Non-proliferation (2011–2016).

CHAPTER 17

US-Russian Cooperation on Fissile Material Security and Disposition

Frank von Hippel and Oleg Bukharin

Introduction

Reductions in the nuclear weapons arsenals of the United States and the former Soviet Union (FSU) have created the challenge of securely storing and disposing of about a million kilograms of weapons plutonium and highly enriched uranium (HEU). Reprocessing in Western Europe, Russia, Japan and India has added to this total a stockpile of almost 200,000 kilograms of separated "civilian" but weapons-usable plutonium.

There is particular concern about the short-term security of fissile materials in Russia because of the enormous economic and political dislocations in that country. Cooperative international efforts, to assist Russia in strengthening the security of her fissile materials and in arranging for their disposal, are described below and their adequacy is assessed. However, international cooperative efforts to increase the security of weapons-usable fissile materials and eliminate surplus stocks must focus on the long-term global problem as well as the special short-term problem in Russia.

Originally published in Alessandro Pascolini and Dieter Schroeer, eds., *The Weapons Legacy of the Cold War* (Aldershot: Ashgate-Dartmouth, 1997): 51–66.

F. von Hippel (✉)
Princeton University, Princeton, NJ, USA

O. Bukharin
US Nuclear Regulatory Commission, Rockville, MD, USA

P. Foradori et al. (eds.), *Arms Control and Disarmament*,
https://doi.org/10.1007/978-3-319-62259-0_17

Fissile Materials

According to current usage in the arms control community, "fissile" isotopes are fissionable isotopes which can sustain an explosive chain reaction if assembled in a "critical mass."

Uranium-235

Uranium-235 is the only relatively abundant naturally occurring fissile isotope. It is diluted in natural uranium with 140 times as much uranium-238, which cannot by itself sustain an explosive chain reaction. Uranium has to be isotopically "enriched" to at least 20 per cent uranium-235 before it becomes "highly enriched" uranium and weapons-usable according to IAEA (International Atomic Energy Agency) convention. "Weapons-grade" uranium, used in fission nuclear explosives, is enriched to more than 90 per cent in uranium-235. The advantage of higher enrichment is that the critical mass is smaller. The IAEA uses 25 kg of uranium-235 in highly enriched weapons-grade uranium as its standard for a "significant quantity" required to make a first-generation implosion fission explosive.

Many processes have been developed to enrich uranium. Gaseous diffusion has been the primary enrichment process in the past. The state-of-the-art method today uses gas centrifuges. Using such enrichment processes, the FSU and the United States each produced several hundred thousand kilograms (several hundred tonnes) of weapons-grade uranium. The United Kingdom, France and China each produced in the order of 10,000 kg, and South Africa produced and Pakistan may have produced several hundred kilograms. Israel and India are both believed to have enrichment programmes but are not known to have produced significant quantities of weapons-grade uranium. The United States, Russia, France, United Kingdom, Germany, the Netherlands and Japan have also enriched uranium to low enrichment levels (2–5 per cent uranium-235) to fuel light-water reactors and, in some cases (notably the United States and Russia), have enriched uranium to weapons-grade to fuel their naval power, research and weapons materials production reactors.

Plutonium-239

A number of fissile isotopes have been created in nuclear reactors by neutron capture on heavy non-fissile isotopes. The artificial fissile isotope that has been created in largest abundance is plutonium-239 (half-life of 24,400 years). It is created by neutron capture on uranium-238, making uranium-239, which converts by two subsequent radioactive decays to neptunium-239 and then to plutonium-239.

The United States produced about 90,000 kg of weapons-grade plutonium[1] and the FSU somewhat more. The IAEA uses 8 kg as its estimate for the amount of plutonium required to make a first-generation implosion weapon, including production losses. The Trinity and Nagasaki bombs each contained 6.1 kg. For the plutonium coming out of excess US nuclear warheads, a "planning figure" is 4 kg.

The rate at which a nuclear reactor fissions uranium determines the rate at which fission heat is produced. A useful number to remember is that the fission of 1 kg of uranium-235 per day releases heat at the rate of about one gigawatt (10^9 W). Power reactors convert about one-third of the released thermal energy into electrical energy. Therefore a typical one-gigawatt (electric, GWe) power plant will have a thermal rating of about three gigawatts (thermal, GWt). A 3-GWt light-water power reactor, operating at a typical average of 75 per cent capacity, would fission about

$$3\,\text{GWt} \times 0.75 \times 365\,\text{days} \times 1\,\text{kg/GWt} \approx 800\,\text{kg} = 0.8\,\text{t of uranium-235}$$

and produce almost 0.6 t of plutonium-239 per year. However, almost two-thirds of the produced plutonium-239 is fissioned while the fuel stays in the reactor core, reducing the amount of plutonium in approximately 20 t of discharged spent fuel to about 0.2 t. The combined capacity of the world's nuclear reactors is about 350 GWe, about 90 per cent of it in light-water reactors. The approximately 7000 t of spent fuel discharged annually from these reactors contains about 65 t of plutonium (65,000 kg).

When a neutron in a nuclear reactor is absorbed by a plutonium-239 nucleus, the probability of fission is about two-thirds. The other third of the time a plutonium-240 nucleus is produced. Plutonium-240 can also absorb a slow neutron to become plutonium-241, which in turn can absorb a slow neutron to fission or become plutonium-242. All these plutonium isotopes are weapons-usable. However, weapons designers prefer relatively pure plutonium-239, and "weapons-grade" plutonium is defined in the United States as containing less than 6 per cent plutonium-240. The production of weapons-grade plutonium requires that the uranium fuel in which plutonium is produced not be exposed too long to neutrons.

Otherwise, an increasing fraction of the plutonium-239 will capture neutrons and fission or be converted to plutonium-240. The production of weapons-grade plutonium therefore requires much more frequent refuelling than is economically optimal.

The fresh fuel used in light-water reactors today typically contains 4 to 5 per cent uranium-235. When the "spent" fuel is discharged, it contains about the same percentage of fission products, about 1 per cent un-fissioned uranium-235, and about 1 per cent plutonium formed as a result of neutron capture in the uranium-238 in the fresh fuel. Currently, about 2300 t of spent light-water fuel is being reprocessed and over 20,000 kg of plutonium is being recovered annually in Britain and France. Russia has a smaller-scale commercial reprocessing plant, where she reprocesses some East European (Bulgarian and Hungarian) and FSU spent fuel. India and Japan also have small reprocessing plants for spent power reactor fuel.

However, virtually all nuclear utilities in Europe and Japan have found reprocessing a convenient interim measure to deal with spent fuel in the absence of politically acceptable solutions involving interim surface or permanent underground storage. Reprocessing contracts require that foreign customers take back their separated plutonium and "high-level waste" (concentrated fission and transmutation products immobilized in glass). Thus the owning country still faces the problem of storing or disposing of its radioactive waste and must have its separated plutonium fabricated into fuel for other nuclear reactors, such as "mixed-oxide" (MOX) uranium-plutonium fuel. Plutonium recycling is both controversial and expensive. In order to keep their reprocessing customers, France and Britain have been forced to store "temporarily" increasing stockpiles of both high-level waste and plutonium. Russia is similarly storing the radioactive waste and plutonium of her customers. At this point, France and Britain have each accumulated at their reprocessing plants over 50 t of stored separated power reactor plutonium (mostly foreign in France, mostly domestic in Britain) and Russia has accumulated about 30 t.

Disposal of Excess Weapons Materials

Russia's serious internal economic and political instabilities have created great concern in the West about the possibility that Russian plutonium or HEU might appear on the black market. This has led to various initiatives to help Russia dispose of her excess stocks of fissile materials.

From Weapons Components to Fissile Materials

In the United States, the dismantling of nuclear weapons takes place at the Pantex plant near Amarillo, Texas. Plutonium "pits" from the primary fission triggers of the dismantled weapons are stored in bunkers at Pantex. HEU containing secondary bomb components are disassembled at the Y-12 site in Oak Ridge, Tennessee. After removal from the secondary, the HEU is melted into ingots. The United States has put about 10 t of weapons-grade uranium under IAEA safeguards in a storage vault at the Y-12 facility.

In Russia, weapons components containing nuclear materials ("physics packages") are dismantled at Arzamas-16 and Yekaterinburg-45. Plutonium pits and uranium secondaries are placed in containers. Some are then sent for long-term storage to Chelyabinsk-65 and Tomsk-7. Much of the weapons-grade uranium is sent to Tomsk-7 to be oxidized for transport to a uranium-enrichment facility, where it is blended down to low-enriched uranium (LEU).[2] Neither the United States nor Russia has yet undertaken large-scale conversion of plutonium "pits" of primaries to metal ingots or plutonium-oxide powder.

Uranium-235 Disposal

Under the February 1993 agreement, the United States has agreed, in principle, to buy from Russia over a period of 20 years approximately 500 t of excess 90 per cent enriched weapons-grade uranium. The rate of purchase would be at least 10 t per year during the first five years and 30 t per year for 15 years thereafter. Russia has agreed to clean the HEU of chemical contaminants and dilute it with 1.5 per cent enriched uranium to an enrichment level of about 4.4 per cent. At the prices prevailing at the time of the deal, the LEU would be worth about $12 billion. Deliveries of HEU-derived LEU to the United States began in 1995.

In order for the United States to verify that the LEU is derived from weapons-grade uranium, the United States is allowed to sample the material coming into the blending point, to witness shredded HEU metal from Russian warheads being converted into an oxide powder in furnaces, and to seal containers of oxide before they are shipped to the blending facilities. In return, Russia is permitted to monitor the further blending of the LEU to enrichments specified by fuel fabricators and then its shipment to the fuel fabrication facilities.

If deliveries of Russian LEU reach the equivalent of 30 t of weapons-grade uranium per year, they could satisfy approximately one-half of the LEU requirements of US power reactors. US and Canadian uranium miners and the US uranium-enrichment complex have therefore raised various objections to this deal. As a result, the US Department of Commerce has required that an equivalent amount of natural uranium, bought from non-Russian sources, will have to be given back to Russia. Russia will therefore be paid only for the difference between the value of this natural uranium and the LEU, that is for the enrichment work. As a compromise with the uranium miners, recently passed US legislation has finally created a gradually increasing quota for the sale on the US market of the stockpile of "Russian" natural uranium that will be created in connection with the importation of the Russian low-enrichment uranium.[3]

There have also been difficulties in getting political acceptance of the reduction of United States uranium-enrichment activity that will result from the imports of Russian LEU. In July 1993, a government-owned corporation was created to manage the US enrichment complex as a first step towards privatization. To compensate this emerging US Enrichment Corporation (USEC) for the loss of business that will result from the purchase of the Russian LEU, USEC has been made the sole agent for the contract. However, the management of USEC has complained about the price negotiated by the US government for the embedded separative work (SWU) in the LEU. It has attempted to establish SWU prices[4] that would result in profits comparable to those, which it would receive if the SWUs were produced in its own American enrichment plants.[5]

However, Russia's Ministry of Atomic Energy (MinAtom) rebuffed USEC's proposals to drop its price, so the Clinton administration has provided a short-term subsidy by giving USEC some surplus government-owned enriched and natural uranium. For the longer term, the administration expects that the USEC can be motivated to negotiate mutually acceptable prices by the threat that the government will otherwise allow other organizations to compete to be the agents for sale of the Russian LEU.[6]

It might be hoped that the European Atomic Energy Community (EURATOM) and Japan would also purchase blended down weapons uranium from Russia. The British fuel-cycle company, British Nuclear Fuel Services, and MinAtom have been exploring the possible purchase over

20 years of LEU produced by blending down, in Russia, 100 t of weapons-grade uranium.[7] Trade between Russia and Japan is complicated because of their political dispute over the Kurile Islands; Japanese fuel fabricators have thus far refused to purchase LEU derived from weapons uranium.

The United States also plans to dispose of some of her excess weapons uranium. In March 1995, President Clinton announced that he was declaring as surplus an additional 200 t of excess weapons-usable fissile material, of which 38 t were plutonium.[8] The part of this excess HEU which is suitable for use as nuclear fuel will be blended down to LEU.[9] An initial decision has been made to transfer 50 t for blending.[10] Some of the over 90 per cent enriched HEU which was already in the form of UF_6 is being blended down at the Portsmouth enrichment facility.[11] The IAEA has been invited to subject this down-blending process to international safeguards.[12] The quantity of uranium-235 that the United States is transferring to civilian use or disposal is smaller than that being bought from Russia because the United States originally produced less HEU and because the United States has decided to retain a very large stockpile of weapons-grade uranium for future use as naval reactor fuel.

Plutonium Disposal

One problem, which has slowed progress in planning for plutonium disposition is that, in contrast to the situation for surplus HEU, it would cost more to make reactor fuel with plutonium than it would cost to buy LEU fuel—even if the plutonium is free. Nevertheless, a large-scale industry is being established in Western Europe to fabricate separated power reactor plutonium into MOX fuel, in which several per cent of plutonium is mixed with natural or depleted uranium. France and Belgium have in operation—and Britain soon will have—industrial-scale facilities, each capable of fabricating annually several tonnes of plutonium into MOX fuel.[13] These facilities, however, do not currently have the capacity to keep up even with the rate of plutonium separation in West Europe. Nor, it appears, do either Western Europe or Japan have any extra power reactor capacity to absorb MOX fuel fabricated elsewhere.

To the countries that are involved in MOX production and their major customers (Germany and Japan) the obvious answer to this problem is to build a MOX plant in Russia.[14] Indeed, MinAtom would like to use its plutonium-disposition problem as a justification for building a new generation of 0.8 GWe "BN-800" fast-neutron reactors, originally proposed as plutonium breeder reactors.[15] However, Russia does not have the funds to build the new reactors. Another MOX proposal would have Russian and/or US weapons plutonium fabricated into fuel for use in Canadian heavy-water reactors. However, MinAtom opposes the idea of shipping weapons plutonium out of the country. Furthermore, it would presumably want to be paid for the MOX fuel if another country benefited from its fuel value.

The United States, which opposes reprocessing, fears that any increased production of MOX could undermine her anti-reprocessing position. The US government is therefore investigating alternatives to MOX for plutonium disposition, which would meet what a National Academy of Sciences study dubbed the "spent-fuel standard."[16] This standard requires that the plutonium be made as inaccessible as in spent fuel, where the plutonium is protected by a mixture of fission products which create a lethal gamma-ray field around the spent fuel. The leading alternative to MOX being considered in the United States is to embed small cans of plutonium-containing glass in large canisters of fission product-containing glass as the fission products, from which the plutonium was originally separated, are solidified in a waste form suitable for underground disposal.

In Russia, "vitrification" of the fission product waste from reprocessing has been carried out since 1986 at the Chelyabinsk-65 commercial reprocessing plant. This infrastructure could also be used to implement a "can-in-canister" plutonium-disposition approach. However, MinAtom rejects the idea of disposing of plutonium as waste.[17]

At the G-7 "nuclear summit" with Russia, which was held in Moscow in April 1996, it was agreed to convene an international meeting of experts in France by the end of 1996 "to identify possible development of international cooperation" in the disposition of weapons plutonium.[18] Whatever is ultimately done, it is important that it does not make the risk of diversion worse. This is a real danger because it is more difficult to safeguard nuclear weapons materials in process than in storage.

Securing Existing Stockpiles

Even if the disposition of excess weapons uranium and plutonium were to proceed at the maximum planned pace, much of this material will have to be stored for 20 years, and significant quantities of weapons-usable material will remain in military and civilian use indefinitely thereafter. It is therefore critical that the security of all weapons-usable fissile material be upgraded as rapidly as possible. This is of particular concern for Russian stockpiles because of the instability associated with the current very difficult political and economic transition in that country.

Amounts of weapons-usable fissile materials ranging from significant to huge are stored, processed or used at approximately 100 sites in Russia. The security of these materials reportedly varies from relatively stringent at warhead-production facilities to inadequate at large fuel cycle facilities and civil research institutes. Published accounts of the small diversions of fissile materials that have occurred thus far from naval fuel storage facilities, research institutes and fuel cycle facilities indicate that the principal threat of diversion of nuclear materials is posed by insiders.

The collapse of the Soviet Union has weakened many of the arrangements for monitoring and controlling the movements of personnel, greatly increased the permeability of Russia's national borders and made nuclear workers much more accessible to criminals interested in corrupting or intimidating them. In order to maintain nuclear material security, Russia must introduce a Western-style integrated material protection, control and accounting system designed to closely monitor fissile materials. Lack of Russian funds for such a system has led to a number of Western initiatives. The US "Soviet Threat Reduction Act" of October 1991 (now better known as the "Nunn-Lugar Program") has authorized the US Department of Defense (US DoD) to spend approximately $400 million a year in cooperative programmes to assist the FSU in "the transportation, storage, safeguarding and destruction of nuclear and other weapons [and] the prevention of weapons proliferation."

The Government-to-Government Programme

In September 1993, the US DoD and MinAtom signed an "Agreement on Cooperation" in the area of materials protection, control and accounting, with the initial project being the development of a model safeguards system at a LEU fuel fabrication plant in Electrostal, 50 km east of Moscow.

MinAtom's reluctance to offer its more "sensitive" facilities for cooperative efforts apparently stemmed from concerns about the audit and inspection rights upon which the US DoD insisted. The US DoD programme was also made less attractive to Russia's economically stressed nuclear facilities by the fact that it used US goods and services almost exclusively.

In 1994–1995, the United States offered reciprocal access to US facilities, starting with an invitation to visit a US plutonium storage facility at Hanford, Washington in July 1994. This led to an invitation to visit the plutonium storage facility at the reprocessing plant in Chelyabinsk-65 in October 1994 and then to an agreement to start upgrading security at this and a number of other major civilian research and fuel cycle facilities with large inventories of weapons-usable fissile materials.

The "Lab-to-Lab" Programme

The government-to-government programme did not really develop momentum, however, until after direct collaborations organized between US and Russian technical experts from their national nuclear laboratories focused on rapid upgrades of nuclear material security. In contrast to the government-to-government programme, about half of the lab-to-lab programme funds are spent in Russia on salaries and equipment.

The lab-to-lab programme began in 1994 with two pilot projects. At the Kurchatov Institute of Atomic Energy in Moscow (the FSU's first nuclear reactor development laboratory), US and Russian experts designed and built a comprehensive physical security system at a critical assembly building containing about 70 kg of weapons-grade uranium for a zero power mock-up of a space reactor. At the Institute of Experimental Physics in Arzamas-16 (the FSU's first nuclear weapons design laboratory), the lab-to-lab programme helped to fund a project to develop a modular computerized detection and tracking system for HEU and plutonium items. These initial successes established the bona fides of the lab-to-lab programme and encouraged Russian and US advocates of international cooperation on nuclear security issues.

Participation in the lab-to-lab programme has increased and, as of June 1996, included six US national laboratories (Los Alamos, Brookhaven, Livermore, Sandia, Oak Ridge and Argonne) and eight institutions in Russia (Arzamas-16, Chelyabinsk-70, Institute of Automation, Institute of Inorganic Materials, Institute of Physics and Power Engineering, Kurchatov Institute, Production Association "Eleron" and Tomsk-7). The funding level has climbed to about $100 million a year.

At this point the government-to-government programme covers virtually all MinAtom facilities where large inventories of weapons-usable materials are used for non-military purposes. The lab-to-lab programme overlaps with the government-to-government programme and, in addition, covers MinAtom's weapons design institutes. A US Department of Energy (US DoE)-MinAtom cooperative effort has also been launched on upgrading safeguards and security at Russia's warhead-production and disassembly facilities, as well as improving the security of fissile materials in transit.

In May 1996, a team of US experts visited a nuclear navy base near Murmansk as a first step in a US DoE initiative to improve the security of the storage facilities for fresh fuel for Russia's nuclear submarines. And the US DoE and *Gosatomnadzor* (Russia's civilian nuclear regulatory agency) are working to improve nuclear safeguards at research reactors and facilities not affiliated with either MinAtom or Russia's Ministry of Defence. Thus far, however, there has been no systematic effort to enhance the capabilities of the guard forces at Russian nuclear facilities. In this case, an organizational obstacle stems from the fact that the guard forces are provided, not by MinAtom, but by the Ministry of Internal Affairs.

Secure Storage

As Russia is drawing down her nuclear forces, large amounts of HEU and all plutonium are going into storage in the form of warhead components.[19] Initially, the material is stored at the warhead dismantling sites and the existing storage facilities of the plutonium complexes of Chelyabinsk-65 and Tomsk-7.[20] However, it is planned later to put the plutonium—and perhaps some HEU—into long-term secure storage.[21] The first of these high-security storage facilities is being built near the town of Chelyabinsk-65 with assistance from the US Nunn-Lugar programme. However, almost five years after this project was first proposed, only a concrete floor for a first unit has been laid. This facility, initially with a capacity for 25,000 containers, is currently scheduled for operation in 1999.

Progress in building this storage facility was delayed first by the decision to change the location of the first storage facility from Tomsk-7 to Chelyabinsk-65 and then by proposed changes in the design of the facility. Another significant delay was due to an impasse over MinAtom's request for US assistance in the funding of the actual construction of the facility, to pay for Russian salaries and materials. The Department of Defense insisted on limiting assistance to US goods and services for the design ($15 million) and construction ($75 million) phases. However, in the summer of 1995, the United States finally did commit herself to contributing to the cost of purchasing Russian construction materials. The Clinton administration has therefore requested and received permission from Congress to use an additional $29 million for this purpose and may ultimately provide an additional $46 million to bring its total support for the construction phase up to a total of $150 million—half of MinAtom's estimate of the total cost of the facility. In addition, $50 million in Nunn-Lugar funds have been budgeted for the manufacture in the United States of containers for the fissile material components.

In exchange for this assistance, MinAtom has agreed in principle to make the storage arrangements "transparent," so that the United States will be able to verify that the safeguards at the facility are adequate and that the materials are not removed for reuse in weapons. However, in negotiations over the implementation of these security arrangements, Russia has made it clear that she expects them to be made in the context of broader reciprocal transparency arrangements, which would cover excess US fissile material as well.

Thus far, only the storage of weapons plutonium has been discussed. However, in Chelyabinsk-65, Russia also has about 30,000 kg of plutonium separated from East European, Finnish and FSU spent fuel, contained in over 10,000 containers stored in a 50-year-old warehouse. Although the monitoring of the interior of this facility and its access controls are being upgraded under the government-to-government programme, the inadequacy of its design is being ignored while a modern, high-security storage facility is being built nearby for weapons plutonium. Thus far, the Nunn-Lugar programme has apparently not seriously concerned itself with the upgrading of the storage of such weapons-usable fissile materials in Russia that are not actually derived from dismantled weapons. Weapons-grade plutonium is also accumulating in storage facilities at Tomsk-7 and Krasnoyarsk-26, at a combined rate of about 1.5 t a year. This plutonium too should be securely stored.

Stopping Further Production of Fissile Material

Given the huge surplus of HEU and separated plutonium, it would make sense to have a moratorium on further production. In fact, the final four of the US plutonium-production reactors were shut down in 1987 and

1988.[22] Production of HEU for weapons in the Soviet Union ceased in 1988 and a gradual phase-out of the production of plutonium for weapons began in 1987.[23] By 1993, 10 out of Russia's 13 plutonium-production reactors had been shut down.

Russia, however, continues to operate three plutonium-production reactors (two in Tomsk-7 and one in Krasnoyarsk-26) because they produce by-product heat and electricity for the populations of the neighbouring cities. The aluminium-clad uranium-metal fuel used in these reactors cannot be stored long-term in water. It is therefore reprocessed, producing approximately 1.5 t of plutonium per year. The Russian government has reported that, starting in October 1994, new plutonium from these reactors is being converted to stable oxide and placed in storage.

In June 1994, Russia reconfirmed that it would shut down these three reactors by the year 2000; the United States agreed to work with Russia to replace them as regional sources of heat and electricity; and Russia and the United States agreed to develop a system of bilateral monitoring of the plutonium produced in the interim. Subsequently, the United States sponsored feasibility studies of energy replacement options for the Tomsk and Krasnoyarsk reactors. These studies found that replacement power plants would cost billions of dollars and, in the case of replacement nuclear reactors, would take up to 10 years to complete.[24]

Thus it has been agreed that the most feasible approach in the short term is conversion of the cores of the existing reactors to a fuel cycle that does not generate weapons-grade plutonium and improves the safety of the reactors. According to the estimates in the joint study, core conversion for the three reactors could be implemented in 32 months at a cost of $80 million to the United States and a similar cost to Russia.[25] In the meantime, the arrangements for bilateral monitoring of the plutonium being produced by the reactors are being delayed until Russia is satisfied with the level of US assistance.

Transparency and Irreversibility

Since 1995, efforts are being made to take advantage of Russia's new openness to on-site inspections to negotiate a comprehensive set of agreements between the United States and Russia to exchange information on stocks of nuclear warheads and fissile materials, and to arrange for the verification of their reduction. The US and Russian governments have repeatedly expressed their interests in transparency. Indeed, one of the statements issued at the May 1995 summit was titled "Joint Statement on the Transparency and Irreversibility of the Process of Reducing Nuclear Weapons."

This statement asked for reciprocal exchanges of detailed information on aggregate stockpiles of nuclear warheads, on stocks of fissile materials and on their safety and security. It has also asked for a cooperative arrangement for reciprocal monitoring at storage facilities of fissile materials removed from nuclear warheads. There has been considerable US-Russian discussion at the technical level, and diplomatic activity to work out the implementation of these agreements. However, all of these initiatives have become bogged down before achieving concrete results.

Transparency Arrangements for Dismantled Nuclear Warheads

In March 1994, US Secretary of Energy O'Leary and Russian Minister of Atomic Energy Mikhailov agreed to "reciprocal inspections by the end of 1994 of facilities containing plutonium removed from nuclear weapons [as a step towards concluding] an agreement on the means of confirming the plutonium and highly enriched uranium inventories from nuclear disarmament." Defining the procedures for such inspections has been difficult. Progress has been impeded by the fact that some marginally classified weapons design information would be revealed in the process. It was therefore decided that the inspections could not go forward until an "Agreement for Cooperation" had been negotiated.

The US DoD has proposed to the Russian Ministry of Defence specifics on the proposed bilateral stockpile declarations, including declarations of the historical stockpiles of warheads up to the present. The United States has also proposed declarations of stored surplus warheads together with verification measures such as "unique identifiers" (such as seals, tags and/or radiation "fingerprints"). In the same context, the United States has proposed declarations of total stocks of HEU and plutonium, by enrichment or isotopic composition, by form and by location, as well as historical production by site, back to 1970.

None of the proposals so far officially proposed or discussed includes verification of active stocks of warheads beyond those warheads on strategic missiles subject to being counted under the Strategic Arms Reduction Treaty (START) agreements. Nor, apparently, has declaration of the quantities of fissile material in individual warhead types so far been proposed, although they could be used to verify declarations of fissile material stocks in warheads.

Placing Civilian and Surplus Military Fissile Material Under IAEA Safeguards

In September 1996, Russia joined the United States in discussions about placing civilian and surplus military fissile material under IAEA safeguards. A major obstacle to such a placement appears to be financial: both the costs of preparing Russian and US facilities for the application of international safeguards and the costs to the IAEA of extending inspections to so many facilities.

An additional obstacle is concern about the possible leakage of weapon design information through the IAEA to potential proliferant states. The most important concern expressed within the US nuclear weapons establishment has been about revealing to the IAEA the amount of plutonium in modern warhead "pits." Either the governments will have to decide that knowing the amount of plutonium in a modern pit would not be that useful to a proliferant, or they should move ahead with the process of turning the pits into non-weapons form.

Defence Conversion

MinAtom's ten closed nuclear cities, with a combined population of about 700,000, are today in a critical situation. The cities receive only minimal governmental funding, they are isolated from Russia's developing markets by guarded double fences and their underemployed workers cannot relocate because of Russia's underdeveloped housing market. These cities cannot be left to economic and social collapse: such a collapse could result in a haemorrhage of nuclear materials into the black market. Russia's nuclear arsenal is shrinking, and its nuclear power capacity is likely to remain at its current level well into the next century. The collapse of the closed cities can be avoided only if useful commercial work is created for workers to replace weapons work that is no longer needed.

This problem has long been recognized by Moscow, and planning for the conversion of the nuclear complex began in 1989. However, the Russian conversion programme has largely failed because of a lack of market skills, secrecy, centralized control and, above all, insufficient investment. Some of the West's nuclear security initiatives discussed above are helping by creating jobs and helping the nuclear cities to develop new long-term missions. There is a need, however, for a dedicated defence conversion programme, which would focus on helping proliferation-sensitive nuclear facilities to establish commercially viable, non-nuclear enterprises.

The Defense Enterprise Fund established by the Nunn-Lugar programme spent a total of $152.7 million in 1994–1995 to build housing for retired missile officers and to start non-military production at four non-nuclear defence facilities. The International Science and Technology Center (ISTC) was established in 1992 in Moscow with funding from the United States, the European Union and Japan in order to give key weapons scientists an alternative to emigration. The Industrial Partnering Program (IPP), established in 1994 in the US DoE, has also focused most of its attention on research and development institutes.

The IPP appears to have the most potential for the nuclear cities as well, since it starts with direct technical contacts between US and Russian experts, and has the direct involvement of US industry. Ultimately, however, the success of this and other conversion efforts will depend upon the willingness of Western companies to invest significant funds in projects in the closed cities.

Conclusion

The Cold War nuclear arms race is over. However, the thousands of nuclear weapons and hundreds of tonnes of weapons-grade fissile material will not disappear overnight. Prevention of their diffusion to rogue states and terrorist groups is of paramount importance for international peace and security. In particular, there remains a serious problem of security of the huge stock of HEU and plutonium in Russia. The Russian government and the international community are working to upgrade the security of fissile materials.

Much, however, remains to be done. Dismantling of nuclear warheads, safe and secure management and disposition of fissile materials, and environmental restoration must become the principal tasks of the US and Russian nuclear complexes. Fissile material operations must emphasize international cooperation and transparency. The complexes themselves must be reduced in size and reconfigured to fulfil these post-Cold War tasks. Workers and facilities must be redirected to commercial non-nuclear activities.

Notes

1. US Department of Energy, *Plutonium: The First* 50 *Years: United States plutonium production, acquisition and utilization from 1944 through 1994*, DOE/DP-0137 (Washington, DC: US Department of Energy, 1996).

2. The cities are a part of MinAtom's network of ten "closed" cities. They now have names but they are still most widely known by these box numbers.
3. This legislation is described in *Nuclear Fuel*, 25 September 1995, 3.
4. Short for kilogram Separative Work Units, a measure of the amount of enrichment work provided by the separation facility. For example, 6.7 SWUs are required to produce one kilogram of 4.4 per cent LEU, with 0.3 per cent uranium-235 remaining in the depleted uranium "tails."
5. In fact, there is considerable uncertainty about the size of USEC's loss or profit—and even about its very existence. See "Outlook on USEC," *Nuclear Fuel*, 11 October 1993.
6. "USEC Privatization Preparations Mesh Opinions of Several Agencies," *Nuclear Fuel*, 20 May 1996, 3.
7. Private communication with a representative of British Nuclear Fuel Services, June 1995.
8. *Plutonium: The First* 50 *Years*, Table 15.
9. Of the excess HEU, 65 per cent has commercial value, 15 per cent does not, and the status of 20 per cent remains to be established. See "DOE Issues Draft EIS on HEU Disposition," *Nuclear Fuel*, 20 November 1995, 19.
10. William J. Broad, "Quietly, U.S. Converts Uranium into Fuel for Civilian Reactors," *New York Times*, 19 June 1995.
11. "USEC Says It Has Begun Blending US HEU," *Nuclear Fuel*, 19 June 1995, 19.
12. "IAEA Considers Applying Safeguards to HEU Down-Blending at Portsmouth GDP," *Nuclear Fuel*, 20 May 1996, 5.
13. The MELOX plant at the Marcoule site in France has a designed capacity of 120 t MOX per year and could be expanded to 160 t per year in the future. The PO plant in Dessel in Belgium has a capacity of 35 t per year. The Sellafield MOX plant in the United Kingdom has a designed capacity of 120 t MOX per year and is expected to come on line in 1997/8. See "COGEMA Preparing Addition of MOX Capacity at Melox, La Hague," *Nuclear Fuel*, 25 September 1995, 10–12; "BNFL Wants to Be a Leading Player in Disposal of Weapons Plutonium," *Nuclear Fuel*, 8 April 1996, 12–13.
14. "Germans, French Propose Competing Technology for Russian MOX Project," *Nuclear Fuel*, 3 June 1996, 11–13.
15. Victor Mikhailov et al., "Utilization of Plutonium in Russia's Nuclear Power Industry," *Post-Soviet Nuclear Complex Monitor*, 18 March 1994, 9–17.
16. Committee on International Security and Arms Control, National Academy of Sciences, *Management and Disposition of Excess Weapons Plutonium* (Washington, DC: National Academy Press, 1994);

Management and Disposition of Excess Weapons Plutonium: Reactor-Related Options (Washington, DC: National Academy Press, 1995).
17. Russian Information Agency Novosti, "Russia's Nuclear Fuel Cycle," *Moscow Summit on Nuclear Safety*, 19–20 April 1996, 8.
18. Declaration of the G-7 nations plus Russian leaders, "Moscow Nuclear Safety and Security Summit Declaration, 20 April, 1996," https://www.iaea.org/sites/default/files/infcirc509.pdf.
19. Assuming that Russia dismantles 1500–2000 warheads a year and an average warhead contains 15–30 kg of HEU, then 20–60 t of HEU a year is recovered from dismantled weapons.
20. Valery Menshikov, "On the Situation with the Storage of Plutonium and Enriched Uranium in Tomsk-7," *Yaderny Control* (February 1995): 3–4 (in Russian).
21. At least originally, the storage facility was intended for storing equal numbers of HEU and plutonium containers for a period of 80–100 years. See Vladimir Golozubov, "Main Design Principles for Russia's Weapons Fissile Materials Storage Facility," presented at the International Workshop *Reprocessing of Spent Fuel, Storage and Use of Reactor and Weapons Plutonium*, Moscow, 14–16 December 1992.
22. *Plutonium: The First* 50 *Years*, 25, 30.
23. Vladimir Petrovsky, *Statement at the 44th UN General Assembly*, 25 October 1989.
24. US-Russian Commission on Economic and Technological Cooperation, *Report on the Nuclear Energy Committee* (Washington, DC: Library of Congress, Congressional Research Service 1996): 5.
25. US Department of Energy, *Stopping Weapons-Grade Plutonium Production in Russia* (Washington, DC: US Department of Energy, 1996).

Frank von Hippel is a senior research physicist and Professor of Public and International Affairs Emeritus at Princeton University's Program on Science and Global Security. He is a founding co-chair of the International Panel on Fissile Materials. A physicist by training, his areas of policy research include nuclear arms control and non-proliferation, energy, and checks and balances in policymaking for technology. He served as assistant director for national security in the White House Office of Science and Technology.

Oleg Bukharin is a senior staff member with the US Nuclear Regulatory Commission, where he works on the issues of nuclear security and non-proliferation. Prior to joining the government, Oleg Bukharin conducted research and wrote on the topics of nuclear security and arms control at Princeton University's Program on Science and Global Security.

CHAPTER 18

The New Verification Game and Technologies at Our Disposal

Patricia Lewis

Introduction

Since the end of the Cold War, multilateral arms control negotiations have taken a more central role. Whereas, in the past, the two nuclear superpowers had well-defined attitudes to verification and compliance—attitudes that they sometimes changed in reaction to each other's proposals—today other states are also defining the approach to verification. Examples include the Chemical Weapons Convention (CWC), the "93+2" programme of the International Atomic Energy Agency (IAEA) to strengthen safeguards and the debate in Geneva on verifying a comprehensive test ban treaty (CTBT).

Technology has also changed. While some states may wish to turn the clock back and prevent, for example, satellite images being used in the verification regime of a treaty, non-governmental organizations (NGOs) and the news media—anyone with a few thousand dollars at their disposal—are able to commission commercial satellites or buy "off the shelf"

Originally published in Alessandro Pascolini and Dieter Schroeer, eds., *The Weapons Legacy of the Cold War* (Aldershot: Ashgate-Dartmouth, 1997): 147–162.

P. Lewis (✉)
Chatham House, London, UK

P. Foradori et al. (eds.), *Arms Control and Disarmament*,
https://doi.org/10.1007/978-3-319-62259-0_18

data. Although the spatial resolution of these data is not yet as good as that of superpower military satellites, much of it tells experienced analysts exactly what they want to know.

This chapter looks at the development of negotiations and of technology, and examines the "democratization" of verification, particularly the role for academia, NGOs and the media.

Verification

Verification is a process which establishes whether all parties are complying with their obligations under an agreement.[1] The process of verification consists of multiple steps that can be either unilateral or cooperative in nature, or both. It includes monitoring, collection of relevant information, analysis of the information and judging compliance.

Verification cannot provide total certainty that all parties are complying with a treaty. Each agreement will have its own standard of verification. For example, the requirements for the verification regime of a treaty banning all nuclear weapons will be far more stringent than, say, those for the Conventional Forces in Europe (CFE) Treaty. There is a general understanding that verification regimes should be designed so that violations are detected in time for appropriate action to be taken. The important role of verification is to ensure that a party contemplating cheating on a treaty realizes that it cannot do so without running a substantial risk of being found out. The design of verification regimes determines whether the likelihood of catching significant cheating is very high (say, 80 to 100 per cent) or is low (say, below 50 per cent). Generally, the more effort, money and resources put into verification, the higher the probability of detecting cheating.

There is a synergy between verification of arms limitation agreements and intelligence gathering for national security. Both processes include collecting information, collating information from a number of sources, analysing the information and distributing the information or analysis to interested parties.[2] Both verification and intelligence activities lead eventually to decisions on national and international security. The key difference between verification and intelligence gathering is that the former is carried out entirely in the open with the consent of all participating states, whereas the latter is a highly secretive operation. Intelligence agencies, however, play a role in verification, often by providing background information or by making suggestions for on-site inspection targets. The verification process also feeds information into the intelligence agencies, such as "ground-truthing" (establishing if the information on the ground supports the information gleaned from satellites).

Confidence-building measures are activities which are designed to build trust between parties. They are self-contained in the sense that they constitute an agreement between parties to carry them out. Verification, on the other hand, relates to an undertaking between the parties, which could stand without verification per se. Verification can often play a confidence-building role, for example in the process of election monitoring, verification of arms reductions and so on.

Verification in the Past

US-Soviet Union (Russia) Bilateral Treaties

From the end of World War II and the beginning of nuclear weapons until the rise of Mikhail Gorbachev in the Soviet Union, the role of verification in US-Soviet Union arms control treaties was greatly dependent on the technology available to carry out monitoring at a distance. Throughout the bilateral negotiations, until the 1987 Intermediate-Range Nuclear Forces (INF) Treaty, the issue of intrusive on-site inspections for verification purposes was guaranteed to stall or even halt negotiations. The United States pursued the concept of "anytime, anywhere" inspections,[3] whilst the Soviet Union viewed such proposals with intense suspicion, believing inspections to be a cover for espionage. The stand-off was so well established that it was said that "verification is becoming a shield for those not interested in arms control to hide behind."[4]

The Strategic Arms Limitation Treaties (START) and the Anti-Ballistic Missile (ABM) Treaty all relied for their verification on "national technical means," which, in the arms control context, meant monitoring by intelligence satellites. The Threshold Test Ban Treaty and the Peaceful Nuclear Explosions Treaty were not ratified for many years, partly because of the issue of verification. In 1991, when agreement was reached on intrusive verification, the treaties finally entered into force.

The breakthrough in intrusive verification between the two superpowers came when Soviet Union General Secretary Gorbachev introduced the policy of *Glasnost* (openness) and offered to open up sensitive military sites for inspections. The first bilateral agreement that took advantage of this change in policy was the 1987 INF Treaty, which not only included on-site inspections of INF bases, but also allowed monitoring of production facilities and of missile reductions. Since then the United States has back-tracked,[5] and the military and commercial agencies have raised concerns over the intrusiveness of on-site inspections and the cost of the verification regimes.

Multilateral Treaties

The main multilateral treaties in the field of arms control and disarmament are the Geneva Protocol (1925, entry into force—eif 1928); Antarctic Treaty (1959, eif 1961); Partial Test Ban Treaty (PTBT, 1963); Outer Space Treaty (1967); Tlatelolco Treaty (1967, eif 1968); Non-Proliferation Treaty (NPT, 1968, eif 1970); Seabed Treaty (1971, eif 1972); Biological Weapons Convention (BWC, 1972, eif 1975); Environmental Modification (Enmod) Convention (1977, eif 1978); Inhumane Weapons Convention (1981, eif 1983); Rarotonga Treaty (1985, eif 1986); Stockholm Accord (1986); Conventional Forces in Europe (CFE, 1990, eif 1992); Vienna Document (1990); Open Skies (1992, eif outstanding); Chemical Weapons Convention (CWC, 1992, eif outstanding).

Although East-West tensions were played out in multilateral negotiations (for example, in the CWC), their effects were often mitigated by states not participating in the Cold War. As a result, the arguments over intrusive verification in multilateral negotiations were of a different calibre from those in bilateral negotiations.

Treaties such as the 1968 NPT and the 1959 Antarctic Treaty have provisions for on-site inspection, although the 1963 PTBT and the 1972 BWC have no verification provisions at all. Adherence to the NPT is monitored by the IAEA in Vienna through bilateral safeguards agreements between the Agency and each member state. However, IAEA membership is not the same as NPT membership.

Before the end of the Cold War, but during the Gorbachev thaw, the highly significant Stockholm Accord was agreed between the participating states of the Conference on Security and Cooperation in Europe (CSCE), now called OSCE (Organization for Security and Cooperation in Europe). This was a series of confidence-building measures designed to increase transparency over military exercises in Europe. From the beginning of the Accord, states carried out challenge inspections of military exercises and exercise calendars and data on the exercises were exchanged between the states.

The execution of the Accord was very successful. The trust that built up between the CSCE countries as a result of the Stockholm Accord had a number of effects, including: (1) the formation of friendly relationships between East and West inspectors; (2) a reduction in the perception of "the other side" as an "enemy," so that there developed a sense of common purpose; (3) a pride in the inspection process itself, which led to friendly rivalry in, for example, seeing which team could offer the best food and wine; and (4) a reduction in the numbers and scales of the military exercises (partly as a result of lessening tension and partly as a result of attempting to reduce the cost of observation and inspection).

Thanks to the success of the Stockholm Accord, further treaties on conventional forces in Europe were negotiated (the Vienna Accord, the CFE Treaty and the Open Skies Treaty). All of these treaties have met with success, although the CFE Treaty has inherent structural problems due to the break-up of the Warsaw Pact.

At the Conference on Disarmament, the CWC was successfully negotiated, but with less stringent on-site inspection requirements than first postulated. As it is yet to come into force, it is not possible to say how the verification provisions will be viewed in practice. The BWC is undergoing a process whereby confidence-building measures and verification provisions are being worked out, and they will be integrated into the Treaty in the next few years.

Verification in the Present

Strengthening IAEA Safeguards: "93+2"

The NPT was severely undermined by the discovery of (1) Iraq's nuclear weapon programme, (2) the suspicion over the capabilities and intentions of the Democratic People's Republic of Korea (DPRK) and the long refusal of North Korea to fulfil her safeguard obligations, and (3) the protracted dispute over the ownership of ex-Soviet nuclear weapons on the territory of the Ukraine.

If the IAEA is to detect undeclared illegal activities in the future, safeguards need to be strengthened and reinforced. The IAEA has embarked on a programme (called "93+2") to evaluate the technical, financial and legal aspects of a wide set of safeguard measures. Proposals for new measures are extensive. They include new techniques, new types of on-site inspection and much more information to be provided by states parties.

There is clear emphasis on obtaining more information. This includes early provision of design information about the construction of and modification to nuclear facilities, and information on nuclear material and nuclear equipment transfers and on relevant non-nuclear material and equipment. A key component to obtaining information is an "Expanded Declaration" in which a state would provide information on all nuclear-related processes (including past and future facilities), production, research and development, and training. The thinking behind the Expanded Declaration is to make the total nuclear programmes of states more transparent.

A significant part of the new techniques aimed at increasing access to information is the procedure called "environmental monitoring." This is a technique, which takes samples from the air, soil, water, vegetation or exposed buildings' surfaces in the vicinity of the facility under investigation. The IAEA's studies thus far indicate that environmental monitoring is indeed a powerful tool. Isotopic signatures can be extremely specific and, the nearer to the facility the sample is taken, the easier it is to identify signatures of certain activities. Even past activities can be detected through this technique. Environmental monitoring thus has a significant deterrent effect. Even if inspectors are not allowed access to a building and even if an illegal activity was in the past, there is still a significant chance of a clandestine nuclear weapons programme being discovered.

Coupled with a greater degree of access to information and with environmental monitoring is a greater degree of physical access to nuclear and nuclear-related facilities. Increased physical access falls into two categories. The first is the concept of broad access, which includes managed access. The second is the underutilized provision for no-notice inspections. Broad access means: (1) access to any location on a site containing a safeguarded facility, (2) access to nuclear-related locations listed within an Expanded Declaration which are specifically declared as not containing nuclear material, and (3) access to other locations of interest to the IAEA on a voluntary basis. Managed access—a concept stemming from the negotiations for the CWC—is a process by which transparency is negotiated on site so that inspectors can have necessary access while valid, sensitive information can be protected, for example, by shrouding. Managed access is applicable to all three types of broad-access inspections and has been shown to be a practical approach to on-site inspections in commercially sensitive facilities.

In the case of no-notice inspections, states receive no advance warning of the inspection. The state and the facility know of an inspection only when the inspectors arrive at the facility gate. No-notice inspections are intended, in the main, to be for use at locations with declared nuclear material. In the future, however, it may be that a large number of states will opt for "anytime, anywhere," no-notice inspections within an Expanded Declaration in exchange for a reduction in routine inspections.

Fundamental to many of the proposed new measures is the intention to increase the IAEA's cooperation with states and state systems of accounting and control (SSACs). Although there is a variation in the technical capabilities of the SSACs, there is nonetheless a common core of capabilities, which would allow the IAEA to make use of the SSACs' activities and thereby reduce effort and costs. In the case of certain SSACs, cooperation with the IAEA is already established. In other cases—owing to history, capabilities and transparency concerns—the relationships between SSACs and the IAEA are underdeveloped. However, the IAEA is clearly making increased cooperation with SSACs a priority in its drive to reduce costs for routine inspections and thus make money available to carry out environmental monitoring and no-notice inspections.

Because there is going to be a large increase in the amount of information flowing into the IAEA, and this information is going to come from a variety of sources, it is vital that the procedures for analysis be improved. In addition to the already extensive computerized system in existence, the IAEA is developing a model to describe the known paths for nuclear weapons proliferation. The model will employ all types of information, including the potential uses of dual-use technology, non-nuclear materials and information from no-notice inspections and from environmental monitoring. The model is helping to determine the type of information required and how the information should be collected. It is thus an interactive part of the whole 93+2 process.

There are problems, however, with the implementation of the 93+2 process. Because some of the measures require a new legal agreement between the IAEA and each state, the implementation of 93+2 has been divided into two parts. The implementation of the Part 1 measures, which do not require any new agreement, began in 1995. The Part 2 measures, which do require new agreements, have yet to be implemented. The Part 2 measures include information on nuclear-related research and development not involving nuclear materials at research centres and universities; information on all buildings at nuclear plants and at other relevant locations; lists of domestic manufacturers of nuclear equipment; development plans; inspections at locations other than the "strategic points" at which inspections are already carried out; inspection access to other locations of interest; and no-notice inspections beyond the "strategic points" and other locations.

Comprehensive Test Ban Treaty

The mainstay of the verification regime for the CTBT will be an International Monitoring System (IMS) comprised of four basic technologies: seismic, radionuclide, infrasound and hydro-acoustic detector networks. This monitoring system will comprise 50 primary and 120 auxiliary seismic stations, a network of 11 hydro-acoustic monitors, 60 infrasound stations and 80 stations for measuring atmospheric radionuclides. It is likely that 40 of the 80 stations for increasing radioactive particles will also monitor the presence of noble gases such as xenon and argon. There are provisions for the "improvement of the verification regime," which allow electromagnetic pulse (EMP) detection, satellites or other technology to be incorporated in the IMS, subject to the consensus of the Executive Council, without requiring the full process of an Amendment Conference. The International Data Centre under the technical secretariat will process raw data from the IMS stations and send it to states parties.

On-site inspections and how to decide whether to carry one out had been a constant source of tensions within the negotiations. The final text allows an on-site inspection to be triggered by any relevant kind of information "consistent with generally recognized principles of international law," including national technical means but excluding espionage. The Executive Council must decide to carry out an inspection by a "majority of all members." There was an extensive argument between the United States and China on this issue: China insisted that the majority be two-thirds and the United States wanted a simple majority of those present and voting.

A decision on the on-site inspection has to be taken by the Executive Council within 96 hours of receiving a request and an inspection team has to arrive within six days of the receipt of the request. The timeframe for an inspection is 60 days, with the possibility of extending it by up to 70 days, subject to a majority decision of the Executive Council. Also included in the on-site inspection provisions are over-flights and managed access. States are allowed to protect sensitive facilities and information unrelated to compliance with the treaty. The inspection should move from less intrusive to more intrusive procedures. For a specific country, inspectors and access points have to be identified to the CTBT office within 30 days of the treaty's eif, and updated as appropriate.

The treaty also includes penalties if the Executive Council deems a request to have been "frivolous or abusive." Failure to comply with treaty obligations or abuse of the treaty's provisions can result in penalties ranging from suspension of membership rights to collective measures in conformity with international law, and the taking of urgent cases to the United Nations (UN).

Verification in the Future

Going to Zero

The destruction of South Africa's nuclear weapons demonstrated that, while it may not be possible to "dis-invent" nuclear weapons, it is possible to dismantle a nuclear weapon arsenal and to verify the dismantling of the whole programme.

There are two approaches to reductions in existing nuclear weapons arsenals, and both are needed, operating in parallel, if reductions are to be successful. The traditional approach is that of "top-down" reductions, that is to let the United States and Russia decrease their arsenals first, eventually bring in China, France and the United Kingdom, and hope for cooperation from Israel, India and Pakistan. The regional approach seeks to establish regional nuclear weapons-free zones through treaties such as the Treaty of Tlatelolco and the Treaty of Rarotonga.

Both of these approaches require stringent verification. The first requires the type of verification regime set in place by the INF and START Treaties with the addition of verifying warhead dismantling and cessation of fissile material production, then bringing other nuclear weapon states into the structure as and when necessary. The second approach needs a confidence building approach as the states in the region build trust in each other through a sequence of agreements and confidence-building measures.

The regional approach could follow on to the two nuclear weapon-free zones in existence (South Pacific and Latin America) and build from there. For example, the next step could be the establishment of a sub-Saharan nuclear free zone in Africa. The regions of South Asia, the North Pacific/

East Asia, the Middle East, North America, the Commonwealth of Independent States and Europe could all be involved in a process of confidence building, transparency and, where appropriate, nuclear weapons reductions in parallel with the "top-down" reduction process.

Much of the technology for verifying global nuclear weapon elimination is well known and has been used in the context of treaties to reduce nuclear-tipped missiles. However, the technologies required for verifying warhead dismantling are unproven within that context and there will have to be a period of preparation and practice inspections to ascertain and correct the problems and pitfalls associated with that verification process. Achieving a world free of nuclear weapons will also require the successful implementation of the comprehensive nuclear test ban and a ban on the production of fissile material for weapons purposes.

Getting to zero nuclear weapons will not be the whole story. Sustaining a nuclear weapons-free world indefinitely will be challenging. The verification regime will be intrusive and expensive. For how long will a high degree of intrusion be required or accepted? For how long will states be willing to fund such a verification regime and for how long can the world maintain its enthusiasm for such a process?

Regional Confidence Building and Sub-state Conflicts

A mechanism for reducing tension between groups within states before it reaches the point of conflict is needed within the international security system. In the first place, a mechanism is required for alerting the international community to tension that may escalate to violent conflict. Second, we should have a process of mediating between the hostile parties. Third, a set of tried and tested verification and confidence-building measures, appropriately chosen for the individual situation, should be put into operation.

The application of verification and confidence-building measures to sub-state conflicts has received very little attention until now, and it is this new task that provides the biggest challenge for the future. The main difficulty in applying confidence-building measures to sub-state and trans-border conflicts is that the situation is not one of state-to-state, but one of groups within states. Often, one of the groupings will be the government of the state, or one of the groupings may inhabit a region crossing state boundaries. There may be many vested interests in not allowing a mediation and confidence-building process to begin. These include an unwillingness to share power, a fear of exposure, deeply held prejudices and so on. Therefore there has to be a procedure whereby a group, or groups, who feel under threat, can approach the international community directly and be accorded some status so that they may be recognized and heard internationally.

During the process of mediation and negotiation, there are steps that can be taken to increase confidence in the intentions of the parties and to increase the likelihood of success for subsequent agreements. These range from building trust between local communities to providing data on levels of military equipment held by the state and by paramilitary organizations. Building trust through structured and agreed procedures between local communities or between the state and a minority group is called "civilian confidence building."

The application of verification and confidence building to sub-state and trans-border conflicts, such as ethnic or religious conflicts, is a new idea.[6] Such measures could include: (1) the setting up of youth organizations that include representation from all sections of the population; (2) establishing an independent newspaper that is mandated to take the concerns and aspirations of all sections of the population into account, and to help build bridges between minorities and majorities, and which is monitored by an independent agency; (3) establishing locally based committees, on which United Nations (UN) representatives also sit, to act as a forum for low-level complaints to lessen the risk of escalation into violent conflict; and (4) setting up, if appropriate, visits from communities in neighbouring states that have overlapping ethnic or religious communities to facilitate the exchange of ideas, information and solutions.

In the case of sub-state violent conflicts where there are military, paramilitary or militia (as in the case of Northern Ireland, for example), the military capabilities of all the groupings need to be known and monitored, and that information must be made available to all parties. Independent observers could be allowed to observe the military capabilities of each party so that each side has more confidence in the numbers they are given. During a negotiation, it is unlikely that parties will wish to give highly detailed data on the location, configuration and command and control of their military capabilities. However, once agreement has been reached, a detailed data exchange and verification regime could be established through an independent organization, such as the UN or the OSCE, and reductions, withdrawals, repositionings and reconfigurations could then be verified to everyone's satisfaction.

As in the case of state-to-state trust building, such measures would not solve major problems by themselves and they are no panacea. However, they could help to reduce tensions and improve the climate for negotiations and long-term agreements. For example, in a long-running dispute, there can be agreement that neither side wishes to enter into a violent conflict with the other and they may be able to identify a number of confidence-building measures to relieve tension. Such measures can be reinforced if there is a degree of verification built in, for example through on-site inspections and aerial over-flights. Tensions within negotiations are then reduced and parties may find that they reach agreement much more quickly—or they find that there is still disagreement, but that it is no longer so critical.

This process has been dubbed "agreeing when we can—negotiating when we can't." It requires the realization that, although confidence-building measures cannot solve a problem, they do help to reduce tension and increase understanding, and thereby facilitate creative discussion. This principle could be the foundation for experimenting with a range of new civilian and military confidence-building measures, which could set the scene for a more peaceful and prosperous twenty-first century.

Environmental Agreements

The role of verification in environmental agreements is now an established field of study and it is recognized as an important activity for future security. The effect that environmental degradation has on international security, in terms of poverty, migration, conflict over resources and so on, is becoming increasingly apparent. It is in the interests of all states that environmental agreements are complied with, just for the sake of the survival of the planet as we know it. For this reason, the implementation review processes and verification mechanisms are crucial in determining the success of environmental agreements. For example, it is in the interests of all states that competing industries in other countries are not gaining monetary advantage by ignoring their obligations under environmental agreements.

The countries with most to gain from checking that states are complying with their environmental obligations are often those which are most severely affected by the environmental degradation; these are frequently the poorer countries. It is very much in their interests that polluting states—in many cases the wealthier industrially developed states—are truly reducing their polluting emissions. This is particularly the case for climate changes and trans-boundary pollution.

However, some states, particularly those that are less industrialized, are generally suspicious of verification. One of the most important tasks facing the international community is to promote the values of implementation and verification and to point out the very real advantages that verified, meaningful treaties hold for disadvantaged states. This is particularly true for the ways in which appropriate technologies and methodologies could be used to implement the treaties. Industrially developing nations could make a significant contribution to the techniques being developed to monitor environmental agreements and, in doing so, they could shape the process more to their liking. Grand declarations and statements of intent on environmental issues are no longer enough. Environmental agreements must be backed up by strong implementation and checks on that implementation—the environment is an issue that concerns all.

The verification of environmental agreements has some overlap in the methodologies and technologies with the verification of arms control treaties, but there are also a number of differences. First, although some key environmental agreements have hard targets and timetables (such as the Montreal Protocol) and some ban explicit activities (such as the Whaling Convention), some agreements are much less quantitative than those in the field of disarmament.

Second, the measurable environmental data are very different from the data counting specific pieces of military hardware. There are often large margins of error and so there is much reliance on comparisons and consistencies of self-reported data with other variables. In addition, many of the baseline data are unknown. For example, with the Biodiversity Convention, it is impossible to know how many species there are in the world.

Third, even when there are individual items that can be measured, such as endangered animal species or species of plant life, they are subject to other forces in addition to malignant human activities that can destroy them. Trying to separate the activities that can be controlled and monitored from those that cannot is difficult, particularly when they are entangled. On the other hand, by far the greatest threat to animals and plants is the destruction of their natural habitat by humans. In addition, such destruction is relatively easy to measure accurately, and there is a large body of reliable data to support the monitoring of forest, marshlands and desert areas and so on.

The security picture is becoming more complex. In many cases environmental degradation, such as water shortages or water pollution, are forming part of the backdrop to violent conflict. Building confidence in environmental agreements via effective implementation will become an increasingly important part of global and regional security regimes.

Environmental agreements, particularly those which aim to protect plants and animals, often cover complex issues, which states may have neither the resources nor the inclination to address properly. To a large extent, the implementation and monitoring of such agreements is provided by NGOs and intergovernmental organizations (IGOs).[7] This is an increasing trend in all international agreements with large and small NGOs and IGOs playing crucial and sometimes leading roles.

The Role of NGOs

In the fields of arms control/disarmament, conflicts, human rights, the environment and so on, NGOs are playing an increasingly central role. The increasing importance of NGOs is due to a number of factors: (1) the deliberate move towards professionalism within NGOs, (2) the quality of research that NGOs carry out, (3) the quality of information that NGOs give to the media and general public, and (4) the reduction in expenditure by government ministries in industrial states which has led to a reduction in the quality of the research and information and of the decisions made within them. Cost cutting has also led to a reordering of priorities within governments so that there are a number of issues about which a government may have very little knowledge and there are many places in the world where a government may have no representation—there NGOs may be running significant operations.

There are both advantages and disadvantages to this situation. NGOs can be more flexible than governments and can therefore act quickly if necessary. NGOs are often more in touch with widespread public opinion (since they are largely funded by voluntary subscriptions) and they tend to see human rights as their main priority. Because they are not government people often trust NGOs more readily, and NGOs are therefore often the best mediators and reconciliators.

However, not all NGOs are worthy of trust. Some may set out deliberately to worsen a situation, although more likely such an outcome will be the result of incompetence, lack of discipline and overstretching of resources. There is widespread concern over the accountability of NGOs.

Democratic governments are generally accountable to their parliaments and to the public. Censure can be shown through both avenues. NGOs, on the other hand, are accountable to their funders (membership or foundations, or both). In the end, to whom are the funders accountable? If the membership is very large and spread throughout the population and there are good processes of accountability to the membership, NGOs are subject to censure similar to that of democratic governments: members can vote with their feet and their wallets. However, NGOs, which have limited membership, such as religious groups or small specialized organizations, are particularly vulnerable to the lack of checks and balances in their policy formation and action. Those organizations that entirely depend on foundations to fund them are aware that many of those funding bodies are highly unaccountable for their decisions and yet may be quite influential in their ability to affect policy.

There are no easy answers to these problems. NGOs are playing a larger part in international negotiations, in conflict resolution and in verification. In this respect the world has changed rapidly over the last 30 years and new communications technology has been one of the major factors in that change. It is likely, therefore, that NGOs will continue to increase their roles in international politics, and accommodation in the international system must be made for them.

Appendix: Treaties and Technologies

Non-Proliferation Treaty

IAEA safeguards, based on IAEA *Information Circular 153* (Vienna, IAEA, 1972), designed to detect the loss of a "significant quantity" of nuclear material within a "conversion time."

Significant quantities are plutonium: 8 kg; HEU: 25 kg; LEU: 75 kg; uranium-233: 8 kg.

Conversion times are: plutonium: 7–10 days; HEU: 7–10 days; oxides: 1–3 weeks; nitrates: 1–3 weeks; spent fuel: 1–3 months; LEU: 12 months; natural uranium: 12 months.

Nuclear weapons states do not have to accept safeguards.

Verification is carried out by on-site inspections, data verification and locks, seals and recording equipment.

IAEA membership is not the same as NPT membership.

"93+2" Strengthened Safeguards

Environmental sampling is a powerful tool for detecting traces of material. It is done in two ways: bulk sampling which allows the presence of illegal quantities to be detected but does not always provide a "smoking gun;" and particle sampling which allows tiny traces of illegal material to be detected unambiguously and thus does provide a "smoking gun."

Export Controls

Zangger Committee (nuclear technologies); London Suppliers Club (nuclear technologies); Australia Group (chemical technologies); Wassenaar Agreement (conventional technologies); Missile Technology Control Regime (missile technologies).

Trigger lists; common export controls; reliance on sharing intelligence.

Nuclear Test Monitoring

Seismic detection; radioactive debris monitoring; hydro-acoustic detection; infrasound detection; on-site inspections; aerial over-flight; satellite images; data transmission; international data centres; remote sensing satellites.

Nuclear Weapon Reductions (INF and START)

On-site inspections; national technical means (particularly satellites); radiation detectors; imaging techniques; seals; linear measuring devices; portal perimeter monitoring; infra-red profiler; x-ray cargo scanner; closed-circuit television.

Chemical Weapons Convention

Sampling equipment—leak-proof; portable analytic equipment; x-ray equipment; ultrasonic equipment; mobile mass spectroscopy; real-time x-ray fluorescence; protective clothing/masks; tags, seals, locks; data transmission; laboratory analysis; gas-liquid chromatography; high-performance liquid chromatography; mass-selective detectors; infra-red spectroscopy; nuclear magnetic resonance; mass spectroscopy; neutron activation analysis.

Open Skies

Cameras, 30 cm resolution; video recorders, 30 cm resolution; sideways-looking synthetic aperture radar (SAR), 3 m resolution; infra-red line-scanning devices, 50 cm resolution.

Conventional Forces in Europe Treaty and Stockholm Accord

Training: language, and so on; binoculars; tape recorders; communication equipment;

cameras—still and video; helicopters for overflying exercises and CFE sites.

Notes

1. United Nations, *Verification in All its Aspects: Study on the Role of the UN in the Field of Verification* (New York: UN General Assembly A/45/372, 28 August 1990).
2. Michael Herman, "Intelligence and Arms Control Verification," in *Verification Report 1991*, ed. John B. Poole (New York: The Apex Press for VERTIC, 1991): 187.
3. Article 10 of the United States Draft of the CWC, Conference on Disarmament Document CD/500, Geneva, 1984.
4. Sidney N. Graybeal, United States Commissioner to the United States-Soviet Union Standing Consultative Commission, 1973–7, quoted in Richard A. Scribner, Theodore J. Ralston, and William D. Metz, *The Verification Challenge: Promise and Problems of Strategic Nuclear Arms Control Verification* (Boston, MA: Birkhauser, 1985): 21.
5. See, for example, Gordon M. Burck, "The Chemical Weapons Convention Negotiations," in *Verification Report 1992*, eds. John B. Poole and Richard Guthrie (London: VERTIC, 1992): 126–128.
6. Dennis Sammut, "The CSCE and Russian Peacekeeping," in *Verification 1995: Arms Control, Peacekeeping and the Environment*, eds. John B. Poole and Richard Guthrie, (Boulder, CO: Westview Press for VERTIC, 1995): 291.
7. John Lanchbery, "Reviewing the Implementation of Biodiversity Agreements," in *Verification* 1995, 330.

Patricia Lewis is the Research Director, International Security, at Chatham House. A nuclear physicist by training, her former posts include Deputy Director and Scientist-in-Residence, Center for Non-proliferation Studies at the Monterey Institute of International Studies, Director of United Nations Institute for Disarmament Research (UNIDIR), and Director of the Verification Research, Training and Information Centre (VERTIC) in London. Dr. Lewis served on the 2004–2006 Blix WMD Commission and the 2010–2011 Ekeus Advisory Panel of the Organisation for the Prohibition of Chemical Weapons (OPCW); and was an adviser to the 2008–2010 Evans-Kawaguchi Commission on Nuclear Non-proliferation and Disarmament.

CHAPTER 19

Nuclear Deterrence, Disarmament and Non-proliferation

Alexei G. Arbatov

Since the end of the Cold War, nuclear deterrence between Russia and the United States has receded into the background in terms of day-to-day foreign policy and official public relations.[1] Although both countries retain thousands of nuclear warheads, they have ceased to be global rivals, and the chances of a deliberate war between them have fallen close to zero. There are serious differences on some issues, such as Yugoslavia (1999), Iraq (2003), Russian domestic politics and their effect on elections in Ukraine (2004), the eastward expansion of the North Atlantic Treaty Organization (NATO) (1999, 2003, and 2007), the war in Georgia (2008), a growing United States presence in several of the former Soviet republics, and the planned deployment of United States ballistic missile defence (BMD) sites in Europe.

Originally published in Catherine McArdle Kelleher and Judith Reppy, eds., *Getting to Zero: The Path to Nuclear Disarmament* (Stanford CA: Stanford University Press, 2011): 90–101.

A.G. Arbatov (✉)
Institute of World Economy and International Relations,
Russian Academy of Sciences, Moscow, Russian Federation

P. Foradori et al. (eds.), *Arms Control and Disarmament*,
https://doi.org/10.1007/978-3-319-62259-0_19

Nonetheless, Moscow and Washington are no longer the leaders of two coalitions of states and political-ideological movements that had made bipolarity and severe rivalry the global norm in international relations for almost five decades. Their relations—despite continuous ups and downs, friction, disagreements, and mutual recrimination—include numerous and important areas of cooperation.

This cooperation has embraced various economic and political spheres: peacekeeping operations, resolution of regional conflicts, non-proliferation of weapons of mass destruction (WMD), the struggle against terrorism, joint ground and naval exercises, programmes to secure and eliminate stockpiles of nuclear and chemical weapons, safe disposal of nuclear materials and decommissioned nuclear submarines, salvage operations at sea, and joint human space systems.

The legacy of the Cold War—mutual nuclear deterrence—is becoming less and less relevant as an instrument for dealing with post-Cold War international realities, threats, and risks. Eventually either nuclear deterrence will be abandoned, or the deterrence will lead to the collapse of international security through nuclear proliferation and actual use of nuclear weapons, either in combat, through an accident or as a terrorist act.

Post-Cold War Paradoxes

Since the early nineties, the United States and Russia (and in a more limited way, Britain and France unilaterally) have halved their deployed strategic nuclear forces in terms of nuclear re-entry vehicles (warheads) under the 1991 Strategic Arms Reduction Treaty (START I) and are expected to reduce them by another 60 per cent by 2012 under the 2002 Strategic Offensive Reduction Treaty (SORT) and a new START agreement. Combined with cuts in both sides' tactical nuclear arms, the reductions will apparently amount to more than 80 per cent over the 20-year period since the middle eighties.

But there is the other side of the coin. The decade and a half that has elapsed since the end of the Cold War has demonstrated at least three great paradoxes in regard to nuclear weapons.

The first is that mutual nuclear deterrence between the United States and the Soviet Union (and now Russia) has quietly outlived the two states' global rivalry and confrontation, with which it was closely associated from 1945 to 1991, and which continued in its self-perpetuating momentum even after the collapse of one of the main subjects of deterrence—the Soviet superpower. These inexorable dynamics of mutual nuclear deterrence have acquired a growing and negative "feedback effect" on political relations between former opponents, sustaining a muted, though multifarious, fear: of the supposed evil intentions of the "strategic partner;" of inadvertent or accidental nuclear attack; of possible loss of control over nuclear weapons leading to their acquisition by rebel groups or terrorists; of the one's plans to gain control over the other's nuclear weapons or to deliver a disarming strike against nuclear sites-all this in the absence of any real political basis for suspecting such horrific scenarios or actions.

The second paradox is that, with the removal of the fear of escalation of any nuclear weapon use to a global catastrophe, the United States, Russia, and some other nuclear weapon states have become much more casual about contemplating initiation of the actual combat use of nuclear weapons in service to specific military missions. Thus, the end of the Cold War has actually lowered, not raised, the nuclear threshold, to say nothing of not bringing an end to nuclear warfare planning altogether.

During the George W. Bush administration, Washington emphasized the right to launch pre-emptive selective nuclear strikes, thereby promoting a doctrine of actual nuclear warfare rather than of traditional nuclear deterrence. This example is being followed by Russia, although with some reservations and a variety of controversial official declarations. After a rather weak resistance, Moscow resigned herself to the Bush administration's lack of interest in arms control treaties. Instead, despite scarce funding, Russia unwisely is attempting to carry out a "balanced modernization" of all three legs of her nuclear triad (that is, air-, land- and sea-based systems); shrinking from discussing tactical nuclear weapons; and seeking to make up for her setbacks through the export of civilian nuclear technologies and materials, as well as massive arms sales abroad.

As early as 1993, democratic Russia officially repudiated the no-first-use commitment made by the totalitarian Soviet Union in 1982. During 2000 and 2001 Moscow reconfirmed that position, and it now says that nuclear weapons play a leading role in ensuring Russian national security. Moscow even acknowledges the possibility of "a selective and limited combat use" of strategic nuclear weapons in order to "de-escalate the aggression."[2] This implies accomplishing specific tasks involved in conducting and terminating nuclear warfare, rather than merely deterring aggression through the capacity to inflict "devastating retaliation;" as previously claimed by Soviet official military doctrine.

Not surprisingly, Great Britain, France and China are not going to undertake any limitations of their nuclear forces through arms control treaties, alleging that they lag far behind the two major nuclear powers. Indeed, all three are implementing planned long-term modernizations and, in some weapon systems, a build-up of nuclear arsenals. Besides, Britain and France are elaborating limited nuclear strike options of their own.

Now, as never before, nuclear deterrence looks like the factor most likely to remain a permanent part of international relations, at least until a more devastating or efficient weapon is invented. Moreover, this posture is taken not because of the colossal technical or political difficulties of achieving "general and complete nuclear disarmament;" but because of the presumably considerable "inherent advantages" of nuclear weapons as a means of sustaining national security and "civilizing" international relations.

Obviously, the Big Five (the United States, Russia, Great Britain, France and China) openly or tacitly treat nuclear deterrence as an indispensable and legitimate instrument of their security and military policies, even as they claim that other countries have no right to acquire nuclear weapons.

Smooth Theory and Harsh Reality of Nuclear Deterrence

It will never be proved with finality whether nuclear weapons and nuclear deterrence saved the world from a third world war during the Cold War decades. Fortunately history does not have a subjunctive mood. Certainly the Soviet Union and the United States conducted their foreign policy and military actions abroad with much greater caution than otherwise might be the case, and avoided direct armed conflict.

There was only one example when the great powers came to the brink of war and after some maneuvering stepped back out of the horror of possible nuclear conflagration. This was the Cuban missile crisis of October 1962. However, the irony of that case was that the crisis was provoked by the very nuclear deterrence that is now portrayed by many as an insurance against nuclear war. Moscow had decided to secretly deploy medium range missiles in Cuba to catch up with the US crash missile build-up of 1961–1964, which was provoked by Nikita Khrushchev's bluff of Soviet missile superiority after the triumph of Sputnik in 1957. Hence, the "remedy" (nuclear deterrence) barely saved the world from the catastrophe provoked by the application of that very "remedy" (arms race within the context of nuclear deterrence).

Be that as it may, the realities of nuclear deterrence are much more frightening and controversial than it seems in peacetime, when nuclear deterrence is no more than a theoretical notion buried in the deep background of day-to-day international affairs. If a new crisis happens, which cannot be excluded and which may be much more difficult to resolve in multipolar and uncontrolled international affairs, deterrence may once again move to the foreground of practical politics and fail, with catastrophic consequences. A miniature reminder of such a possibility occurred during the August 2008 conflict in the Caucasus.

It is commonly assumed that the sense of nuclear deterrence is making nuclear weapons not tools for conducting war, but a political instrument, which guarantees that nuclear weapons will not be used in practice—neither within the context of a premeditated attack nor as a result of the escalation of a non-nuclear conflict between nuclear nations. Now, in the sixth decade of the nuclear era, this circumstance is seen as being perfectly natural. Some even talk about the "civilizing" effect of nuclear weapons on politicians and the military and on international politics. However, the reality of nuclear deterrence is much more controversial, because there is no clear watershed between nuclear deterrence and nuclear warfighting.

Even the most destructive strategic nuclear forces carry out their political mission of deterrence specifically through their ability to carry out assigned combat missions—that is, destroy certain targets—and nothing else. These missions are embodied in operational plans, target lists and flight programmes loaded into ballistic and cruise missiles' on-board computers. These operational plans provide for the use of weapons with varying degrees of expected effectiveness in a first strike, a launch-on-warning (LOW) strike, a launch-under-attack (LUA) strike, or delayed retaliatory second strike. These options envision massive salvos, limited groupings, or even single missile nuclear strikes at various combinations of states and targets.

The "grey area" of no clear distinction between the concepts of deterring and waging nuclear war relates even more to operational-tactical and tactical nuclear systems (TNW) than is the case with strategic forces. Since TNW are viewed as means to promote success in a theatre or at the battlefield level more rapidly or to offset an enemy's superiority in conventional forces, it is nearly impossible to draw a distinction between deterrence and warfighting.

Moreover, the division of nuclear weapons into strategic and tactical categories is also quite conditional. For the Soviet Union/Russia, American forward-based systems in Europe have always been equated to strategic weapons, since from their forward bases they can reach deep into the territory of the Soviet Union/Russia. For Eastern Europe and Russian neighbours in Asia, in turn, Russia's tactical nuclear weapons are also seen as equivalent to strategic weapons, both in operational range and destructive consequences of their use.

Still more dangerous is the so-called hair-trigger nuclear posture, associated with weapons that must be launched quickly upon receiving information about an opponent's attack in order to avoid destruction on the ground. Altogether there are now probably about 2000 nuclear warheads on hair-trigger alert in line with LOW/LUA concepts and operational plans: most of the US and Russian intercontinental ballistic missiles (ICBMs), and the Russian submarine-launched-ballistic missiles (SLBMs) on submarines at bases.

With ICBM flight time being about 30 minutes and SLBM about 15–20 minutes, these concepts provide political leaders with a decision-making time of only 4–8 minutes (subtracting the time of missile attack detection and confirmation, and the time for the response launch sequence and fly-away). And this time would be available only if the leaders are safe and ready, and everything works perfectly according to planned procedures. Besides, the leaders will have to operate under enormous stress, receiving controversial intelligence information and the on-duty officers' interpretation of data from early warning systems. On top of all this, there is no difference between the upgrading of the alert status of strategic forces for the first (pre-emptive) or the second (retaliatory) strike. Operations officers would most probably interpret the actions of the other side in the most conservative way, and they would hardly have either the time or authority to educate political leaders on the details and possible misperceptions of strategic forces' crisis activities.

No doubt, maintaining several thousand nuclear warheads on hair-trigger alert is the ultimate absurdity of nuclear deterrence 20 years after the end of the Cold War, when political, economic and security relations, at least among the P5 (the five permanent members of the United Nations (UN) Security Council), render deliberate nuclear attack virtually unthinkable. Moreover, it was and remains extremely dangerous. During the Cold War years there were dozens of false alarms on both sides, and the fact that nuclear war did not erupt out of a technical malfunction or a decision-maker's miscalculation should be to an important degree attributed to sheer luck.

Nowadays a number of new dangers are contributing to such a catastrophic possibility. Proliferation of nuclear weapons and ballistic missiles to an expanding number of states with inadequate negative control makes unauthorized or deliberate provocative launch of a missile much more probable and threatening as a trigger of massive nuclear exchange. Of special concern is the proliferation of long-range cruise missiles and sea-based ballistic and cruise missiles, which are capable of delivering an anonymous strike. Some new nuclear states are politically unstable, and a launch of nuclear weapons may happen there as a result of civil war, putsch, or a contest among rival groups or between political and civilian leaders for control over nuclear weapons. As a result of nuclear proliferation there is a growing danger of terrorists getting access to nuclear materials or weapons. A terrorist nuclear explosion in one or several capitals might provoke a spontaneous nuclear exchange by great powers' forces on hair-trigger alert.

Disintegration of Arms Control

The third paradox is that with the end of the Cold War, the focus has been on doing away with nuclear arms limitations and reductions, transparency, and confidence building, rather than doing away with nuclear deterrence and eventually the nuclear weapons themselves. The victims of this process (primarily at the initiative of US policy makers of 2001–2008) already include the ABM treaty, START II, and the START III Framework Treaty, an Agreement on delineation between strategic and tactical BMD systems of 1997, as well as the entry into force of the Comprehensive Test Ban Treaty (CTBT), and negotiations on the Fissile Material Cut-off Treaty (FMCT). Emulating US policy, Russia suspended its adherence to the Treaty on Conventional Force Reduction in Europe (CFE) in 2007 and threatened to withdraw from the Intermediate Nuclear Forces and Short Range Nuclear Forces Elimination Treaty of 1987, if the US plan of deploying ballistic missile defence sites in Europe were to be implemented. Potentially, even the Nuclear Non-Proliferation Treaty (NPT) might fall apart—at least that is how it looked from the results of a disastrous NPT Review Conference of May 2005. The whole structure of nuclear arms control is collapsing, with most dire predictable consequences from the growth of new threats and risks.

The new beginning of a nuclear free world, started by the article of four respected American former state figures and later supported by the new US administration of Barack Obama, has been a promising change, but it is yet to be substantiated by practical policy.

How Relevant Is Nuclear Deterrence?

Of the main reasons why nuclear deterrence should be superseded by some type of constructive strategic relationship between the United States and Russia, and eventually among all nuclear weapon states, the first is nuclear deterrence's irrelevance to the real threats and challenges of the post-Cold War era.

Deterrence remains effective against the least probable or non-existent threats: nuclear or massive conventional attacks by great powers (and their alliances) against each other. But it does not work against the new real and present dangers, such as nuclear proliferation, international terrorism, ethnic and religious conflicts, drug and arms trafficking, trans-border crime and illegal migration, to say nothing of climate warming and world economic crisis. It has been a highly debatable point—whether nuclear disarmament could prevent nuclear proliferation in the past or might do it in the future. It is certain, however, that nuclear deterrence cannot stop proliferation, and it is quite probable that deterrence encourages further expansion of the "nuclear club."

The second reason for replacing deterrence with a new strategic relationship is that the relations involved in mutual nuclear deterrence place tangible limitations on the ability of great powers to cooperate genuinely in dealing with new threats and challenges. The degree of cooperation during Cold War times, when most arms control treaties, including the nuclear NPT, were concluded, is not enough for the new era. Such endeavours—cooperation between the leading states' secret services and special forces, joint counter-proliferation policies (for example, Russian participation in the Proliferation Security Initiative, and the envisioned actual US-Russian combat operations against terrorists and rogue and failed states), officially initiated joint early missile launch warning and BMD systems—virtually imply a common security alliance of a new type. Much stricter nuclear and missile export control regimes, greater emphasis on securing and accounting for nuclear warheads and nuclear materials (which implies broad transparency and access to each other's secret sites), verifiable cessation of production of weapons-grade nuclear materials throughout the world, common policy on internationalization of nuclear fuel cycle facilities, and ambitious Global Partnership projects—all these require a greater magnitude of trust and cooperation among partner states.

But all of these are impossible to imagine while the United States and Russia still aim thousands of nuclear warheads at each other, keep missiles on hair-trigger alert, and modernize nuclear forces to preserve devastating retaliatory capabilities against each other. Besides, as was mentioned above, the momentum of nuclear deterrence, in combination with new threats and missions and technological developments, may destabilize strategic relations among the great powers, further undercutting their ability to think and act together.

The current crisis over the Iranian nuclear programme, despite the apparent similarity of the US and Russian positions, provides a good illustration of this point. Neither the United States nor Russia wants Iran to have uranium enrichment or plutonium reprocessing capabilities, to say nothing of Teheran's potential acquisition of nuclear weapons. However, action in the form of UN Security Council sanctions against Iran, or a UN authorization of the use of military force, is where US-Russian unanimity stops. For the United States, the prospect of eventually being targeted by Iranian nuclear missiles is totally unbearable and warrants all means of prevention. For all conservative Russians, the dire political, economic, and security implications of supporting (even if passively) UN sanctions or US military action against Iran—Russia's long-standing partner—may be seen as too high a price to pay. After all, as hard-liners would point out, Russia is already targeted by thousands of US nuclear weapons, as well as by the nuclear weapons of American allies and partners (Britain, France, Israel and Pakistan). A nuclear-armed Iran would not add much to this picture, and it probably would target its missiles elsewhere anyway.

The prospect of Iranian nuclear materials or weapons being leaked to Islamic terrorist organizations is much more frightening. However, Russian hawks would claim that this is a hypothetical scenario, while actual transactions of that kind might have been already attempted or done through Abdul Qadeer Khan's black market connections with the Taliban, al Qaeda and Iran—without ensuing aggressive US attempts to investigate and prosecute the case. Apparently this benign position was motivated by Washington's desire not to destabilize her partner regime in Pakistan, which is important for other American interests. As for Russia, the majority of her political-strategic elite would not support sacrificing cooperation with its partner regime in Iran, which is likewise important for many Russian interests, in favour of cooperation with the United States, particularly in view of the state of US-Russian strategic relations.

The third reason for initiating a new strategic relationship in place of mutual nuclear deterrence is the problem of resource allocation. In annual defence budgets the allocations for nuclear weapons are relatively small (10–15 per cent).[3] But the costs of the whole 20- to 30-year service life cycle of strategic systems (research and development, procurement, deployment, maintenance, and eventual dismantling and utilization) are staggering. Sustaining nuclear deterrence at current levels, or even at reduced levels (such as the 1550 deployed warheads called for under the New START treaty), is an expensive luxury, given that the two biggest powers assign the bulk of these forces the mission of destroying each other, as well as serving "as a hedge against future uncertainty." This aimless "hedge" may be relatively inexpensive for the United States, which has the largest overall defence budget in the world (about as big as the sum of all military spending by the other major powers), and which fully modernized its strategic nuclear force during the eighties and nineties, investing in "strategic capital" that will last for decades into the future. Still, even for the United States it would be easy to find a much better allocation for these resources, whether within her defence budget or outside it, in particular in times of harsh economic crisis.

The burden of maintaining robust nuclear deterrence is much heavier for Russia, which is now implementing a "balanced modernization" of all elements of her strategic triad and planning to keep up with the New START treaty ceilings. Faced with a severe deficiency of appropriations for an expensive military reform, as well as modernization and restructuring of its conventional forces, Russia nonetheless has to spend huge sums on nuclear weapons. The budget share for nuclear deterrence is relatively still bigger for France, Britain and China.

By maintaining mutual nuclear deterrence, the great powers are wasting resources that otherwise could be applied to more appropriate military and security tasks and missions. Moreover, significant scientific and technical intellectual resources are tied up by nuclear deterrence. Powerful state, business, research and political organizations are locked into sustaining nuclear confrontation in economic, technical and mental respects, instead of addressing the more realistic and urgent needs of national and international security.

Transforming Nuclear Deterrence

There are three principal routes for doing away with mutual nuclear deterrence, at least as a principal operational mode of US-Russian military relationship.

The first of the three avenues towards the end of nuclear deterrence is to further reduce and "de-alert" Russian and American strategic nuclear forces. The second is to develop and deploy a joint ballistic missile early warning system and a missile proliferation monitoring system. The third is to develop and deploy joint BMD systems. Initially, the second and third avenues might be limited to nuclear and missile proliferation threats, but eventually—in parallel with transformation of the nuclear forces of both sides—they could embrace a growing part of the strategic assets of the two powers and their allies, and thus would transform their present mutual nuclear deterrence into a qualitatively new type of strategic relationship.

This new relationship could be called "strategic nuclear partnership," "cooperative nuclear weapons policies," "a common nuclear security framework," "a mutual nuclear insurance (or assurance) strategy," or any number of other names, depending on one's tastes and semantic skills. In any case, the main problem is not the term, but the substance.

The first two steps on this long way are the ratification of CTBT by the United States, China, and other nations, without which the treaty cannot enter force, and successful ratification of the New START treaty by the United States and Russia to replace START I, which expired in December 2009.

After a decade of mockery and neglect, nuclear disarmament has returned to US-Russian and other nuclear powers' official documents and political commitments. This was due to the famous article by four renowned American statesmen in favour of moving towards a nuclear-free world. The change of administration in Washington and President Barack Obama's commitment to this idea have made nuclear disarmament once again a subject of practical diplomacy.

The new US President promised to achieve the ratification of CTBT. In the spring of 2010, Washington and Moscow signed the New START treaty, which envisions reductions in the number of operationally deployed strategic delivery vehicles to 700 and of warheads to 1550 for each side. However, besides the problems of ratification, which the new treaty met in the US Senate, a number of issues remain disturbing for Russia, and they may reappear in the negotiations for a follow-on to

New START. One is the future of the US BMD deployment plan, in particular in Europe. The first stages of deployment on ships in the Mediterranean, Rumania, and Bulgaria pose no serious challenge to Russia's deterrence. However, there is no US commitment to limit or coordinate with Moscow further deployments in any clear way, while Russia has to plan its strategic forces and arms control policies ten to fifteen years in advance. Other problems are the US insistence on counting strategic warheads by their actual (operational) deployment, instead of maximum missile and bomber loadings, as in START I. Also, a growing concern of Moscow is the American deployment of thousands of strategic (long-range) air-launched and sea-launched cruise missiles and dozens of ballistic missiles with conventional precision-guided warheads on heavy bombers and submarines, capable of a counterforce strike at opponents' strategic forces.

In contrast to President Obama's call for a nuclear-free world, the practical policy of the United States for the first disarmament step is quite conservative, to say nothing of the open opposition of a part of the American political elite to New START. The Pentagon wishes to implement most of the New START reductions through removal of some multiple independently targetable re-entry vehicle (MIRV) warheads from missiles to storage, and by conversion of many strategic weapons for delivering conventional munitions. The first would leave the United States with a big reconstitution potential (the possibility of returning warheads from storage to missiles), and the second with a new conventional counterforce capability. Both are quite disturbing for Moscow.

Despite all general reservations about nuclear disarmament, Russia has been gradually becoming more receptive to the idea of further nuclear disarmament. On the other hand, Russia is reluctant to commit herself to much deeper reductions after the New START, in view of US/NATO advantages in BMD technologies and conventional weapon systems and forces, the potential threat from other nuclear weapon states (all eight of which have weapons that can reach Russian territory), and American space support and potential strike capabilities, as embodied in the Prompt Global Strike concept and systems.

If these obstacles are overcome, the process of nuclear disarmament will gain momentum and, eventually, involve other nuclear states and address the problems of tactical nuclear weapons, nuclear weapons in storage, fissile materials production and stocks, as well as numerous interacting military, political, and economic problems.

We may hope that, in the long run, nuclear disarmament will do away with nuclear deterrence, which would permit stopping and reversing nuclear proliferation. It goes without saying that this would be a long and difficult process. A future nuclear-free world cannot be just the present world minus nuclear weapons. It would have to be a very different world, not free only from nuclear weapons but also from weapons based on new physical principles, from large conventional wars, and from any arbitrary use of force by strong nations against weak ones. All in all it will have to be a world with a very different system of international security. But wouldn't it be better to strive for a new foundation of international security instead of the mutual capacity of states to kill in several hours dozens of millions of each other's citizens?

Notes

1. This essay draws on earlier work done with Vladimir Dvorkin: Alexei Arbatov and Vladimir Dvorkin, *Beyond Nuclear Deterrence* (Washington, DC: Carnegie Endowment for International Peace, 2006).
2. *Aktualnye zadachi razvitia Vooruzhennykh Sil Rossiyskoi Federatsii* [Urgent tasks of the armed forces of the Russian Federation] (Moscow, October 2003): 41–42.
3. *Updated Summary Tables: FY 2010 Budget of the United States* (Washington, DC: The White House, 2010). http://www.whitehouse.gov/omb/budget/fy2010/assets/summary.pdf.

Alexei G. Arbatov is a member of the Russian Academy of Sciences. He leads the Academy's Centre for International Security at the Institute of World Economy and International Relations. He is a former scholar in residence and the chair of the Carnegie Moscow Centre's Non-proliferation Program. Formerly, he was a member of the State Duma, vice chairman of the Russian United Democratic Party (*Yabloko*) and deputy chairman of the Duma Defence Committee.

CHAPTER 20

Nuclear Abolition or Nuclear Umbrella? Choices and Contradictions in US Proposals

Matthew Evangelista

Deterrence continues to be a relevant consideration for many states with regard to threats from other states. But reliance on nuclear weapons for this purpose is becoming increasingly hazardous and decreasingly effective.
—George P. Shultz, William J. Perry, Henry A. Kissinger, and Sam Nunn, "A World Free of Nuclear Weapons," Wall Street Journal, 4 January 2007

Originally published in Catherine McArdle Kelleher and Judith Reppy, eds., *Getting to Zero: The Path to Nuclear Disarmament* (Stanford CA: Stanford University Press, 2011): 296–316.

M. Evangelista (✉)
Cornell University, Ithaca, NY, USA

P. Foradori et al. (eds.), *Arms Control and Disarmament*,
https://doi.org/10.1007/978-3-319-62259-0_20

Although not suited for every twenty-first century challenge, nuclear weapons remain an essential element in modern strategy.
—National Security and Nuclear Weapons in the 21st Century, White Paper (Washington, DC: United States Departments of Defense and Energy, September 2008)

The subtitle to Stanley Kubrick's 1964 film classic, *Dr. Strangelove*, is "How I Learned to Stop Worrying and Love the Bomb." This black comedy, starring Peter Sellers in three different roles, depicted a world gone crazy with exaggerated fears and far-fetched strategies for coping with them—most notoriously the Doomsday Machine, a Soviet device to launch nuclear Armageddon automatically in the event of a perceived attack from the United States. The film's main conceit stems from the fact that the Doomsday Machine could not serve its purpose of deterring such an attack, because the Soviet leaders neglected to tell their US counterparts of its existence. In this respect *Dr. Strangelove* distilled an element of real-world nuclear politics: just two years earlier the actual Soviet leadership had kept secret the presence of Soviet nuclear weapons in Cuba, rendering them incapable of deterring US action, and instead triggering a serious international crisis. Kubrick's work, considered a masterpiece of comic invention, reflected reality in other respects. The script was in fact based on the musings of professional nuclear strategists, many housed at the RAND Corporation in Santa Monica, California. One may hesitate to claim that the RAND strategists *loved* the Bomb, but they certainly found it useful for the wide range of tasks, related to US foreign and security policy, about which they worried.[1] Others—members of the public, citizens and leaders of foreign countries, scholars, and intellectuals—worried more about the Bomb itself, in particular, the proliferation of nuclear weapons in the arsenals of the so-called superpowers and of the arcane strategies for their use. They advocated nuclear arms control and disarmament and changes in strategy, such as pledges of "no first use" and the de-alerting of nuclear systems, intended to forestall the doomsday scenario depicted in Kubrick's movie.

The current situation resembles the era of *Dr. Strangelove* in two regards. There are still people who express concern about the dangers of nuclear proliferation—mainly to countries without existing nuclear arsenals and to terrorist groups—and there are still people who harbour ambitious objectives for the use of nuclear weapons. And some people, including many proponents of the long-term goal of a nuclear-free world, appear to do both: they are worried about the Bomb, but, for certain purposes, they continue to love it.

In the first decade of the twenty-first century the call for nuclear disarmament and a move towards "nuclear zero" has come from an unlikely source: four former US officials—William Perry, Sam Nunn, George Shultz, and Henry Kissinger—with close ties to the Cold War nuclear strategists. Indeed Kissinger, the most famous member of the foursome, was a long-time consultant to RAND and the author of an early influential work of nuclear strategy, which advocated a prominent role for nuclear weapons in the defence of Western Europe.[2] One cannot say that such former US officials have become worried about nuclear weapons only recently, however. They always expressed concern about *other* countries' nuclear weapons—particularly those of the Soviet Union, sometimes those of China, and, however briefly, even those of France. But the policies they pursued in the furtherance of their understanding of US foreign and security interests suggested that they were not worried about US nuclear weapons. Rather they valued those nuclear weapons as central to resolving US security problems, and they developed strategies for nuclear use in a wide range of contingencies. Many of those contingencies—such as deterrence of an attack against European allies or discouraging China from settling her dispute with Taiwan by force—remain a part of US military policy, and, as such, they pose a major barrier to nuclear disarmament.

The key element that prompted the four former officials to launch their nuclear disarmament initiative was a worry that "the deadliest weapons ever invented could fall into dangerous hands," as they put it in the opening paragraph of their *Wall Street Journal* article of 15 January 2008, a year after they announced their original appeal in the same newspaper.[3] The implication is that nuclear weapons have been in safe hands since their invention in 1945, even though those hands dropped atomic bombs on Hiroshima and Nagasaki, killing tens of thousands of innocents; came close to nuclear war over Cuba in 1962; put nuclear weapons on high alert during the 1973 Middle East war; and blundered into numerous hair-raising accidents and mistakes in handling nuclear weapons, some as recently as 2007, others in the distant past—their occurrence discovered in formerly secret documents and through interviews.[4] Even aside from the military use and near-use of nuclear weapons since 1945, the process of mining and enriching uranium, creating plutonium, manufacturing the weapons components, and testing the weapons in the atmosphere and underground posed life-threatening risks to many thousands of people and devastated the natural environment. None of this seemed to worry the custodians of the nuclear arsenals, who gave every indication of having instead come to Love the Bomb—as long as it seemed to them to enhance US security.

Now they worry, because the Bomb could fall into dangerous hands and put US security at risk. As the second paragraph of the 2008 *Wall Street Journal* article explains, paraphrasing the one a year earlier, "with nuclear weapons more widely available, deterrence is decreasingly effective and increasingly hazardous."[5] In releasing his administration's *Nuclear Posture Review* in April 2010, President Barack Obama articulated the same concerns. The review, he argued, "recognizes that the greatest threat to U.S. and global security is no longer a nuclear exchange between nations, but nuclear terrorism by violent extremists and nuclear proliferation to an increasing number of states."[6] The main concern of the former officials and the current president is, not surprisingly, for the effectiveness of US deterrence. Rhetorically eliding "U.S. and global security," as the president did, will not convince everyone that they are the same thing. And as long as the initiative for a nuclear-free world gives priority to US security, it will not make much of an impression on states that aspire to obtain nuclear weapons, let alone on terrorist groups, which are presumably undeterred by the traditional remedy of nuclear retaliation.

In the pages that follow, I describe the barriers to achieving nuclear disarmament posed by US policy. They include: (1) that, despite changes announced by President Obama, the United States continues to depend on the threat of nuclear retaliation—often called "extended deterrence" or the quaintly nonsensical "nuclear umbrella"—to serve a wide range of security concerns; and (2) that the "nuclear zero" initiative provides no way of addressing the security concerns of the countries to which it is intended to deny nuclear weapons. Although the first steps towards nuclear zero seem promising—working with Russia and the other members of the nuclear club in the framework of traditional arms control to reduce their arsenals—even here there are serious difficulties. US policy towards ballistic missile defence and the expansion of the North Atlantic Treaty Organization (NATO) risk undermining cooperation with Russia and limit the possibilities for Russian reductions in nuclear weapons.

Even if these first steps are successful—convincing the acknowledged nuclear-weapon states to begin the process of disarmament—what would be the next steps in the absence of cooperation by the aspiring nuclear-weapons states, such as Iran and North Korea? Stopping those states from acquiring or accumulating nuclear weapons would then require coercive measures such as economic sanctions or military action. The nuclear-free initiative begins to look like a ploy by the dominant nuclear powers to rally support for such measures without seeming too hypocritical. The second part of the chapter takes up this issue by considering US objectives in pursuing nuclear zero.

The third part of the chapter draws on historical experiences to suggest some alternative means to move towards nuclear disarmament by connecting the nuclear predicament to the broader security environment. The end of the Cold War in Europe came when the rationale for deployment of nuclear weapons disappeared. Initiatives on the part of the Soviet Union—such as unilateral reductions in her offensively oriented armed forces and political liberalization within the Soviet bloc—removed the threat of major conventional war in central Europe. Thus it was no longer necessary for the United States to pose the risk of nuclear escalation to deter such a war. Scholars and activists had prepared the groundwork for the changes that ended the Cold War arms race by explicitly proposing initiatives that would take account of the links between conventional and nuclear war. Prospects for a successful "nuclear zero" initiative will also need to take account of the links between nuclear weapons and broader security concerns—not only for the nuclear "haves," but also the nuclear "have-nots" and the nuclear "wannabes."

The end of the nuclear arms race between the United States and the Soviet Union is owed not only to the material changes, such as withdrawal of Soviet armed forces from central Europe, but also to ideational ones. The top leadership in the two countries—most notably Ronald Reagan and Mikhail Gorbachev—endorsed concepts such as "common security," where one side's security does not come at the expense of the other's. They publicly expressed their antipathy to nuclear weapons. Their views reflected widespread public revulsion and alarm at the prospect of nuclear war. Nuclear disarmament will become easier to achieve if leaders of the countries that have deployed nuclear weapons would actively contribute to stigmatizing their possession and use—the focus of the final part of the chapter. To do so, they will have to face the contradictions in their existing strategies and make a choice between keeping their nuclear umbrellas and pursuing nuclear abolition.

The Purposes of Nuclear Weapons

Before considering how to get rid of nuclear weapons, we should review why they exist. The common answer is "deterrence," and the common follow-up question is "how much is enough?" to achieve it. When we think about nuclear weapons, however, the most important question should not be "how many?" but "what for?" We typically think of deterrence as the threat of retaliation with nuclear weapons by one country to forestall a nuclear attack by another. This is only one narrow category of deterrence, however, and it usually requires an adjective to specify its constrained role: *limited, minimum, finite*, or *existential* are the ones often attached. Beyond this limited purpose for nuclear weapons, the United States has pursued many far more ambitious ones. One of the most important purposes of US nuclear weapons has been to deter war against US allies or military conflict that implicates US interests abroad but does not pose a threat of nuclear attack against the United States itself. Such purposes typically fall under the designation *extended deterrence*. Without addressing head-on the purposes of US possession and planned use of nuclear weapons, many of the well-meaning discussions about how to reduce nuclear arsenals are beside the point. Experts who write about such topics as "redefining deterrence" without posing the question "deterrence of what?" are missing a key element of the picture.[7]

If we consider the history of nuclear weapons and US policy we can see that from the beginning the United States deployed nuclear weapons for purposes that went far beyond the deterrence of a nuclear attack against itself. The United States developed the first atomic bombs during World War II initially in response to fears that Nazi Germany would do so. In that respect the new weapons might have been considered a deterrent to Germany's prospective use. But Germany was defeated before she developed a nuclear weapons capability. Following the end of the European war, the US government used two atomic weapons against the Japanese cities of Hiroshima and Nagasaki in August 1945. Their purpose was not deterrence, but actual *use* in the service of what later became known as "compellence"—dropping the bombs was intended to compel the Japanese authorities to surrender quickly and perhaps to influence the policy of the Soviet Union.[8]

For the next four years, the United States held an atomic monopoly, and following the first Soviet atomic test in August 1949 she continued to maintain a monopoly on "deliverable" weapons into the early fifties. By definition US development, production, and deployment of atomic weapons during this period served not to deter the threat of nuclear attack by other countries—because no other countries could mount such an attack—but a variety of other purposes. The United States deployed "atomic-capable" B-29 bombers to Europe during the Berlin Crisis, for example, to signal resolve to the Soviet Union. She developed a worldwide system of air bases, planned already during the later stages of World War II (and without a specific post-war enemy in mind), and eventually used it to surround the Soviet Union with nuclear-armed aircraft.[9] One of the purposes was to deter the Soviet Union from starting a war. US leaders during the late forties justified monopoly possession of nuclear weapons as a counter to what they claimed was a Soviet superiority in conventionally armed forces poised to pour across the borders of the Soviet Union in pursuit of worldwide military conquest. The security of Western Europe was a particular focus of concern. Even though declassified documents have made it clear that US analysts and political figures overestimated the strength of Soviet forces in the immediate post-war era, perceptions of Soviet conventional superiority served as a justification for a US policy of extended nuclear deterrence—the threat of US nuclear attack against the Soviet Union to deter that country from invading Western Europe.[10]

Much of the subsequent development of US nuclear strategy and weapons was premised on the need to bolster the credibility of extended deterrence—for example, to allay doubts that the United States would risk global nuclear devastation to protect her European allies. To that end, the United States deployed thousands of so-called tactical nuclear weapons into Europe starting in the early fifties. But US nuclear weapons had roles to play beyond Europe as well. In the early eighties, for example, following the Iranian Revolution and the Soviet invasion of Afghanistan, the United States made increasingly explicit threats to escalate to the level of nuclear war in the event that a hostile power seeking militarily to deny United States access to oil supplies could not be defeated by conventional forces.[11] In the Pacific region, the United States has deployed nuclear weapons and employed nuclear threats for a variety of purposes over the decades: to influence China's policies towards Taiwan, to deter North Korean aggression against the South, and—however chimerical—to bolster the prospects for US success in her disastrous war against Vietnam.[12]

To the extent that the United States still relies on nuclear weapons for a range of purposes associated with extended deterrence, it will be harder to move towards a nuclear-free world. Historically the US Department of Defense has accorded nuclear weapons both a deterrent role and a "war-fighting" role in the event of military conflict. As late as May 2009 a doctrinal document from the US Air Force (USAF), for example, stressed the importance of using nuclear weapons as a deterrent against countries suspected of developing chemical or biological weapons.[13] In April 2010, President Obama rejected this particular role for nuclear deterrence, when he declared that "we will not use or threaten to use nuclear weapons against non-nuclear weapons states that are party to the Nuclear Non-Proliferation Treaty and in compliance with their nuclear non-proliferation obligations."[14] Even limited to nuclear-armed states, and countries such as Iran and North Korea, which the United States does not consider to be in compliance with the Non-Proliferation Treaty (NPT), US military policy for nuclear war is still ambitious. Perhaps battle plans will change in response to Obama's commitment "to reduce the role of nuclear weapons in our national security strategy." But as it stands, the Pentagon intends to use nuclear weapons not only at the "strategic" level against countries that have attacked the United States, but also "in support of theatre objectives" during an ongoing military conflict. Such use would have both military and political objectives, as the USAF's doctrinal statement describes: "While the use of nuclear weapons will affect an on-going engagement between friendly and enemy forces, their use should also be designed to help achieve the political goals of the operation." The document claims that "the law of armed conflict does not expressly prohibit the possession or use of nuclear weapons," and although "the destruction wrought by nuclear weapons can be immense," it can also "be and limited for a particular scenario."[15]

For many decades, such US plans for "tailored" and "limited," as well as massive, use of nuclear weapons have coincided with an inhibition on the part of US leaders actually to resort to nuclear war—something Nina Tannenwald has described as the "nuclear taboo."[16] The political reluctance to launch a nuclear attack rests uneasily with the detailed military planning for doing so. The political commitment to *use* nuclear weapons in defence of allies rests uneasily with the goal of a nuclear-free world, even if Obama limited that commitment to defence of allies menaced by a nuclear-armed state.

As former and current US leaders endorse "nuclear zero," they still maintained and frequently reiterated a commitment to deter attacks against US allies by threat of nuclear retaliation. That commitment is captured in the expression "nuclear umbrella," which so readily trips off the tongues of both supporters and opponents of moving towards a nuclear-free world. In a May 2009 article in the *Wall Street Journal*, for example, William Perry, a signatory of the original call for zero nuclear weapons, writes with two other former US officials that:

> an effective strategy to reduce nuclear dangers must build on five pillars: revitalizing strategic dialogue with nuclear-armed powers, particularly Russia and China; strengthening the international nuclear non-proliferation regime; reaffirming the protection of the U.S. nuclear umbrella to our allies; maintaining the credibility of the U.S. nuclear deterrent; and implementing best security practices for nuclear weapons and weapons-usable materials worldwide.[17]

In a response in the same newspaper the following month, Richard Perle, a former Defense Department official, and Republican Senator John Kyl criticized US President Barack Obama's commitment to a nuclear-free world for, among other things, its effect on "allies who may one day lose confidence in our nuclear umbrella."[18]

The criticism of Obama was misplaced. The US president has regularly reasserted his country's commitment to use nuclear weapons to deter attacks against various countries, and some of his fellow Democrats have proposed additional commitments. In June 2009, for example, Senator John Kerry, chair of the Senate Foreign Relations Committee, suggested in an interview with London's *Financial Times* that "Israel should be included under the U.S.'s nuclear umbrella."[19] That same month, at a summit meeting in Washington, DC, with South Korean President Lee Myung-bak, Obama reaffirmed that the US security commitment to the Republic of Korea included a nuclear component. As the Korean President put it at the joint press conference, "President Obama reaffirmed this firm commitment towards ensuring the security of South Korea through extended deterrence, which includes the nuclear umbrella." Both presidents vowed that North Korea should not be allowed to possess nuclear weapons.[20] To illustrate the possible consequences of the US commitment to use nuclear weapons in defence of South Korea, and to deny those weapons to its northern neighbour, the

Korea Times accompanied its article about the summit meeting with an illustration titled "Possible U.S. Nuclear Umbrella Scenario." It depicted a map of the Korean peninsula, with B-52 bombers, F-117 "stealth" bombers, carrier-based aircraft, tactical nuclear-armed missiles, and 155 mm artillery pieces all carrying out a nuclear attack on a spot on the map not too far north of the border between North and South Korea.[21] Whether or not the illustration reflected accurately what weapons the United States might use in a nuclear attack in the region, it represented a rare depiction of the meaning of the anodyne expression, "nuclear umbrella."

The dual approach of pursuing nuclear disarmament while issuing commitments of extended deterrence and expanding the "nuclear umbrella" constitutes official US policy. In his speech in Prague in April 2009, announcing his administration's support for a nuclear-free world, President Obama, for example, promised simultaneously to "reduce the role of nuclear weapons in our national security strategy" and extend the nuclear deterrent to the Czech Republic as a member of NATO: "Make no mistake," he cautioned. "As long as these weapons exist, the United States will maintain a safe, secure and effective arsenal to deter any adversary, and guarantee that defense to our allies—including the Czech Republic."[22] He repeated nearly the same words a year later when releasing the Nuclear Posture Review.

President Obama explicitly connected the US commitment to defend the Czech Republic, including with nuclear weapons, to the North Atlantic Treaty: "NATO's Article V states it clearly: An attack on one is an attack on all. That is a promise for our time, and for all time." During the last year of the Bush administration, especially following the military conflict between the Russian Federation and the Republic of Georgia in August 2008, the issue of extending NATO membership to Georgia and to Ukraine came to the fore. To the degree that the Obama administration pursues the expansion of NATO to those two countries, both of which border Russia, it will further complicate the prospects for nuclear disarmament. Would the United States and her allies be tempted to resurrect the arcane system of extended deterrence that characterized NATO's Cold War nuclear strategy, with its various "steps" along the "escalation ladder" to achieve "escalation dominance," and its thousands of "tactical" nuclear weapons deployed on European soil?

This is the sort of situation for which the term "Strangelovian" was coined—and today's situation is not so different, given the purposes that nuclear weapons continue to fulfil in US security policy, despite official endorsement of a nuclear-free future. Especially in the wake of Obama's speech in Prague, however, numerous observers began to call into question the role of tactical nuclear weapons stationed in Europe, and, in some cases, the merits of extended nuclear deterrence.[23] In announcing the Nuclear Posture Review in April 2010, Obama chose not to address the weapons in Europe, promising instead to consult with US allies about them.[24]

US Interest in a Nuclear-Free World

In Obama's Prague speech the President described US interest in preventing further countries from developing nuclear weapons and stressed the importance of the NPT. "The basic bargain," he claimed, "is sound: Countries with nuclear weapons will move towards disarmament, countries without nuclear weapons will not acquire them." But he stressed the need for better enforcement measures for countries in the latter category: "We need real and immediate consequences for countries caught breaking the rules or trying to leave the treaty without cause." Specifically, "violations must be punished." Even more specifically, "[We] must stand shoulder to shoulder to pressure the North Koreans to change course" and deal with the "real threat" posed by "Iran's nuclear and ballistic missile activity." Finally, he argued, "[We] must ensure that terrorists never acquire a nuclear weapon," a possibility that he described as "the most immediate and extreme threat to global security."

Thus, the Obama administration's motives for seeking a nuclear-free world echo those of the original proposal from Perry, Nunn, Shultz, and Kissinger. Unlike them—and in a rare, if not unprecedented, acknowledgement for a US president—Obama allowed that "as the only nuclear power to have used a nuclear weapon, the United States has a moral responsibility to act." Although Obama claimed, at another point in his speech, that "moral leadership is more powerful than any weapon," his administration remains concerned about nuclear weapons in dangerous hands. It is intent on denying such weapons to North Korea, Iran, and terrorist groups. But his critics have wondered how initiatives on the way to a nuclear-free world, such as US ratification of the Comprehensive Nuclear Test Ban Treaty (CTBT), would achieve that goal. Would US ratification of the test ban treaty make North Korea, Iran or al Qaeda any less interested in obtaining nuclear arms?[25]

A key argument that proponents of nuclear zero summon is that taking such near-term initiatives helps address the fact, as Perry, Nunn, Shultz, and Kissinger wrote, that "non-nuclear weapon states have grown increasingly skeptical of the sincerity of the nuclear powers" in their commitments under the NPT to move towards nuclear disarmament.[26] The problem is—as Kyl, Perle, and other critics point out—that professions of US sincerity are unlikely to dissuade Iran or North Korea (let alone al Qaeda) from their nuclear ambitions. By ratifying the CTBT and even negotiating deeper reductions with Russia, the United States could render her position less hypocritical on the matter of nuclear disarmament, but that will not be enough to achieve the main goal that motivates the nuclear zero initiative: the fear of nuclear weapons in the hands of "irresponsible" states or groups.[27] Moreover, to the extent that US security policy still relies on nuclear weapons for a variety of purposes—not least to threaten (deter) states such as North Korea and Iran—that policy will still be seen to embody a double standard. Critics will understand the US rhetorical commitment to a nuclear-free world, and even the intermediate steps in that direction, as a means to summon an international consensus to punish the states that refuse to go along.

The crux of the matter is that the United States pursues two contradictory paths to reducing the threat of nuclear weapons in "dangerous hands." The first is embracing nuclear disarmament as a long-term goal. The second is maintaining nuclear weapons and nuclear deterrence until that goal is achieved. In an interview with the *New York Times*, President Obama made the point with characteristic clarity: "We will retain our deterrent capacity as long as there is a country with nuclear weapons."[28] Thus, in Obama's lights, the last step to a nuclear-free world would be US monopoly possession of nuclear weapons. This, however, is a world we have already experienced. Monopoly possession of atomic weapons by the United States in the second half of the forties served as an inducement, not a deterrent, to other countries' acquisition of nuclear arsenals. The disarmament scheme premised on the US atomic monopoly—the Baruch Plan—demonstrably failed. Only when other states, most notably the Soviet Union, had built their own nuclear capabilities, were they willing to negotiate restrictions.

It is not difficult to draw the historical parallels to the era of US atomic monopoly and the arms control negotiations of the Cold War. Why would Iran or North Korea forgo the nuclear option given their current security predicaments? A look at the map shows Iran effectively surrounded by nuclear-armed states: Pakistan on its eastern border, India just beyond, Russia to the north, Israel to the west, and the United States with aircraft carriers and nuclear-armed submarines deployed in the Persian Gulf, Mediterranean Sea, and Indian Ocean, as well as the global reach of her land-based intercontinental missiles. The *Korean Times* article about the US nuclear umbrella presents a good approximation of the nuclear threat posed to North Korea, without even taking into account the nuclear arsenals of neighbouring Russia and China.

The security environment that confronts Iran and North Korea, and the incentives they face in making judgments about their nuclear options, are obvious to many well-informed international observers. In an interview with the British Broadcasting Corporation (BBC), for example, Mohammed El Baradei, then head of the International Atomic Energy Agency, "told the BBC that countries with nuclear weapons were treated differently to those without. He said North Korea, with a bomb, was invited to the conference table, while Saddam Hussein's Iraq, without one, was—as he put it—pulverised." Regarding Iranian motives, he said "[It] is my gut feeling that Iran would like to have the technology to enable it to have nuclear weapons, if it decides to do so." The Iranians, in his view, "want to send a message to their neighbours, to the rest of the world, don't mess with us." Even recognizing the incentives for Iran to obtain a nuclear weapons capability, El Baradei offered as solution that the nuclear "haves" take the initiative: "The only safe future, he said, was widespread nuclear disarmament led by the existing nuclear powers."[29] As long as Iran and North Korea perceive those existing nuclear powers—the United States, in particular—as hostile, they are unlikely to disarm first on the promise of future disarmament by their adversaries. But that seems to be the only plan on offer by the Americans.

Relating Nuclear Disarmament to Broader Security Concerns

Previous efforts at nuclear disarmament suggest that states achieve progress only when they take other states' security concerns into account. During the Cold War that meant US acknowledgement that the Soviet Union would insist on achieving "parity" with the United States before she would pursue mutual reductions. A more significant reduction of the nuclear threat came with the end of the Cold War and the East-West conflict, when the governments on both sides—spurred by nongovernmental organizations and popular movements—recognized the interconnection of security issues with political concerns, including human rights.[30] A key turning point came when the reformist Soviet leadership of Mikhail Gorbachev acknowledged the effect on US and NATO nuclear policy of the large Soviet conventional armed presence in central Europe and the role that the Soviet Army had played in bolstering the rule of communist parties in the region. Gorbachev's unilateral initiatives—such as deep reductions and restructuring of conventional forces, a halt to Soviet nuclear testing, and an acknowledgement of "freedom of choice" in the domestic political order of the East European states—paved the way to the end of the Cold War and the prospect for further reductions in nuclear weapons. Although Ronald Reagan was sympathetic to the goal of abolishing nuclear weapons, his successors were more sceptical and squandered an opportunity to achieve more dramatic progress. Nevertheless, that the United States and Russia ceased to consider each other mortal enemies in a nuclear stand-off constituted a major achievement.

Picking up where Gorbachev and Reagan left off—an explicit goal of Shultz, Perry, Kissinger, and Nunn, articulated in their first *Wall Street Journal* article—is a sensible first step towards achieving nuclear disarmament. Yet the fact that Russia and the United States are no longer the leaders of rival military alliances does not mean there are no security concerns that could complicate the prospects for disarmament. Indeed, the Russian government has made clear two of those concerns: the continued expansion of NATO, and US plans, promoted by the administration of George W. Bush, to deploy components of a missile defence system in Eastern Europe, ostensibly to defeat an attack from Iran. Russian officials expressed concern about the system and doubts about the rationale.

Regardless of what one thinks of NATO and of the merits of extending security guarantees to countries along Russia's border with which Moscow has conflictual relations, no one can plausibly argue that NATO expansion enhances the prospects for nuclear disarmament. The Russian armed forces have to plan for the contingency of war, and if they contemplate a war with states allied to a nuclear-armed United States, they must also contemplate the use of nuclear weapons. The continued expansion of NATO—both the eastward expansion and the expansion into a worldwide military force for "out of area" missions—will hinder nuclear disarmament, much as the deployment of large, offensively oriented conventional forces did during the Cold War. Few countries, Russia included, are going to be willing to give up their nuclear weapons if the United States continues to flaunt her global dominance in conventional military forces.

On missile defence, Moscow still sees the connection between nuclear defence and disarmament in pretty much the traditional way it was understood during the Cold War. One side's deployment of defences undermines the deterrent effects of the other side's retaliatory nuclear offensive forces. The Russian public interpretation of US plans for missile defence in Eastern Europe held that they were directed against Russia and, in former President Putin's words, would "upset the balance" of nuclear forces. Russian analysts suggested the new deployments were intended to neutralize Russia's capability to launch a retaliatory nuclear attack against Europe if Russia faced a nuclear attack from the West. Secretary of State Condoleezza Rice dismissed these concerns as "ludicrous." In fact, the proposed deployments in the Czech Republic and Poland were not well suited for their stated mission of defending Europe from an Iranian attack; if upgraded, they would have been more effective in hindering a Russian missile attack. In response to the announcement of US deployment plans, Putin proposed to the Bush administration more cooperative means of dealing with a potential Iranian threat, including sharing data from Russian radar systems and even joint operation of early-warning centres in Moscow and Brussels.[31] The Bush administration preferred its own plan. In November 2008, following the election of Barack Obama, Defense Secretary Robert Gates dismissed Putin's offer out of hand rather than let the new administration make its own decision.[32]

The incoming Democratic administration did, in fact, revise the Republican decision—substituting for it one that it described as more plausible from a security standpoint. In the process it opened the possibility for new sites for deployment in Eastern Europe and a redistribution of the contracts to a new set of military industrial firms within the United States.[33]

Although the change in policy on missile defence offered an opportunity to try to address Russian security concerns, Obama's advisers seemed to go out of their way not to do so, even in the interest of negotiating reductions in nuclear arms. As Michael McFaul of the National Security Council staff insisted in July 2009, "[We're] not going to reassure or give or trade anything with the Russians regarding NATO expansion or missile defence."[34] The proposal offered at NATO's Lisbon summit in November 2010 to cooperate with Russia on missile defence suggests a departure from such an obstinate approach. Russia is still, however, likely to find the prospect of missile-defence components in Romania and Poland troubling.

Russia is the easy problem, compared to addressing the security concerns of Iran and North Korea. Resolution of the conflict with those two countries may have to await changes in their internal policies, much as the Soviet reforms of the *perestroika* era combined domestic and international change. But as in the Soviet case, the United States must be prepared to recognize genuine initiatives towards reconciliation when they appear. In Prague, President Obama expressed such readiness when he stated that his "administration will seek engagement with Iran based on mutual interests and mutual respect. We believe in dialogue." That line received warm applause compared to his earlier threat to punish violators of the non-proliferation regime. His Czech audience also reacted positively to the prospect that "if the Iranian threat is eliminated, we will have a stronger basis for security, and the driving force for missile defense construction in Europe will be removed." That prospect, however distant, would simultaneously remove one of the barriers to nuclear reductions with Russia. In his June 2009 press conference with the South Korean president, Obama indicated that there "is another path available to North Korea" as well,

> a path that leads to peace and economic opportunity for the people of North Korea, including full integration into the community of nations. That destination can only be reached through peaceful negotiations that achieve the full and verifiable denuclearization of the Korean peninsula.[35]

The United States is presumably not prepared to put her own nuclear arsenal on the table during such negotiations. Under the circumstances, one may doubt whether the North Korean leadership sees the negotiations as fully "peaceful," when they are conducted under the shadow of the "nuclear umbrella" that the United States extends to her South Korean ally.

The end of the Cold War and the East-West arms race took many observers by surprise. But the proposals that led to the withdrawal of hundreds of thousands of troops from central Europe and substantial reductions in nuclear weapons were many decades in the making. Peace researchers and activists in the United States and Europe worked with their counterparts in the Soviet Union to promote initiatives for restructuring and reduction of military forces and respect for human rights.[36]

Sympathetic diplomats and political leaders incorporated many of the ideas into their policies. Today's security challenges seem no less daunting, and the solutions will have to be equally creative and bold.

The Normative Context for Nuclear Disarmament

Many observers have noted the antipathy towards nuclear weapons that Ronald Reagan shared with Mikhail Gorbachev. These weapons, in order to achieve their deterrent effect, must inspire terror, and they did so by threat of mass destruction of innocent life—even if civilians were not deliberate targets. More than a half century ago a prominent strategist of the RAND Corporation referred to the US-Soviet nuclear stand-off as a "delicate balance of terror."[37] A quarter century after that, US and Soviet leaders, fearing that the balance might not be adequate to prevent a nuclear war, thought about the dreadful consequences. In the case of Reagan and Gorbachev that thinking spurred them to action in the cause of nuclear disarmament. They were undoubtedly influenced by the fact that people worldwide shared their fear, and at certain points in history many of them protested against nuclear arms in large numbers. Research has suggested that a prerequisite for meaningful limitations on nuclear weapons has been popular mobilization, and that fear and a sense that political leaders are acting irresponsibly tended to fuel that mobilization.[38] Public reaction to the threat of radioactive fallout from the massive nuclear tests in the fifties and the near catastrophe of the Cuban Missile Crisis, for example, contributed towards the first achievements in arms control, such as the Limited Test Ban Treaty. Loose talk by members of the Reagan administration about fighting and winning a nuclear war, combined with a worsening of US-Soviet relations in the eighties, spurred a worldwide peace movement that inspired Gorbachev's initiatives and gave him some hope that they would be well received.[39] The overall normative context included a stigmatization of nuclear weapons as dangerous and potentially genocidal, no matter who owned them.

A deliberate and forthright condemnation of nuclear weapons by the leaders of the nuclear-armed states could make an important contribution to the prospects for global nuclear disarmament. President Obama made a gesture in that direction by acknowledging US moral responsibility for the atomic bombings of Hiroshima and Nagasaki—implicitly recognizing as morally dubious the disproportionate killing of innocents even in the service of a just cause. Many observers have likened the process of nuclear disarmament, or the more ambitious goal of abolishing war, to the abolition of slavery.[40] Tannenwald has argued that the use of nuclear weapons has already attained the status of a taboo, but that no such stigma attaches to the possession of nuclear weapons, nor to the planning for their use. She argues that the taboo against use could be strengthened by what she calls "virtual abolition schemes" entailing changes in "habit, attitude, norms, law." These might include international pledges of "no first use" of nuclear weapons and criminalization of the use of nuclear weapons, with threats of war-crimes trials of leaders who violate the prohibition.[41] President Obama's commitment of April 2010 not to use US nuclear weapons against (most) non-nuclear states, however welcome, does not constitute a "virtual abolition scheme" in Tannenwald's sense.

The logic of such proposals assumes that the more a practice is considered morally abhorrent the less likely it is to recur. The process of nuclear disarmament would be bolstered if the leaders of states that possess nuclear weapons not only rejected their use, but apologized for their acquisition and possession. As it stands, their current position—advocating disarmament but retaining the threat of nuclear annihilation—undermines the goal of a nuclear-free world. Imagine if leaders instead emphasized only the negative side of nuclear weapons: the tremendous economic and environmental costs they have imposed over the decades, and even on future generations (if one considers the legacy of genetic damage caused by radioactive fallout); the near misses from accidents and during Cold War crises, and what the consequences might have been had those crises triggered a war; and what the consequences will be if even a small fraction of the world's current nuclear arsenal is used. Consider again the parallel to the abolition of slavery. No one doubts that any country that harbours slave-traders, or individuals found to enslave others, would be subject to unqualified condemnation. No one would argue that one should balance such condemnation against the economic or psychological benefits that accrue to slave-holders. Why should nuclear weapons be treated any differently?

If the possession of nuclear weapons were universally stigmatized, many of the problems that arise in discussions of the merits of nuclear zero would diminish in significance. Consider the problem of verification, and the concern—often associated with Jerome Wiesner, professor at the Massachusetts Institute of Technology and science adviser to President John F. Kennedy—that as fewer nuclear weapons exist the ones that remain become more significant. This formulation applies when countries treat nuclear weapons as valuable additions to their arsenals. But what if they treat them rather as stigmatized instruments of genocide, which call forth universal condemnation? On the topic of verifying "nuclear zero," Andrew Mack has pointed to the potential role of whistle-blowers as a way of preventing cheating on a disarmament agreement: "No state that contemplated reneging on its disarmament commitments could be certain that its transgression would not be revealed *from within*. If just one individual refused to go along with the deception, all would be revealed." He points to several familiar cases: "Israel's nuclear-weapons programme had its Mordechai Vanunu, Russia's chemical-weapons plans had their Vil S. Mirzayanov, and Saddam Hussein had his defector son-in-law, Hussein Kamel Hassan."[42] One could imagine that the prospect of whistle-blowing would be even more threatening to potential violators if the weapons in question were publicly condemned by all of the world's leaders and held in revulsion by all of the world's citizens.

The stigmatization of nuclear weapons to the extent that no country would admit to desiring them may seem an unrealistic goal. The history of prior abolition movements, as well as the evidence from moderate successes in arms control during the Cold War, suggest that a prerequisite for such a significant change may be popular mass mobilization contributing to a gradual evolution in the normative context. Unless people experience the right combination of fear and hope, they may not be willing to act. If instead, along with *Dr. Strangelove*, we stop worrying and learn again to love the Bomb, our prospects for achieving nuclear disarmament will diminish.

Notes

1. Fred Kaplan, *The Wizards of Armageddon* (New York: Simon and Schuster, 1983).
2. Henry A. Kissinger, *Nuclear Weapons and Foreign Policy* (New York: Harper, 1957).
3. George P. Shultz et al., "Toward a Nuclear-Free World," *Wall Street Journal*, 15 January 2008.
4. In August 2007, six cruise missiles armed with nuclear warheads went missing when USAF personnel mistakenly loaded them onto a plane and flew them across the country. For an analysis of previous accidents and near-misses, see the masterful study by Scott D. Sagan, *The Limits of Safety: Organizations, Accidents, and Nuclear Weapons* (Princeton, NJ: Princeton University Press, 1993).

5. Shultz et al., "Toward a Nuclear-Free World."
6. The White House, Office of the Press Secretary, "Statement by President Barack Obama on the Release of Nuclear Posture Review," 6 April 2010.
7. Yousaf Butt, "Redefining Deterrence: Is RRW Detrimental to U.S. Security Calculus?," *Bulletin of the Atomic Scientists* 64 (2 December 2008): 15.
8. Tsuyoshi Hasegawa, *Racing the Enemy: Stalin, Truman, and the Surrender of Japan*, (Cambridge, MA: Harvard University Press, 2005); Sean L. Malloy, *Atomic Tragedy: Henry L. Stimson and the Decision to Use the Bomb against Japan* (Ithaca, NY: Cornell University Press, 2008).
9. Melvyn P. Leffler, "The American Conception of National Security and the Beginnings of the Cold War, 1945–1948," *The American Historical Review* 89 (April 1984): 346–381.
10. Matthew Evangelista, "Stalin's Postwar Army Reappraised," *International Security* 7 (Winter 1982–1983): 110–138; Phillip A. Karber and Jerald A. Combs, "The United States, NATO and the Soviet Threat to Western Europe: Military Estimates and Policy Options, 1945–1963," *Diplomatic History* 22 (Summer 1998): 399–429; and Matthew Evangelista, "The 'Soviet Threat': Intentions, Capabilities, and Context," *Diplomatic History* 22 (Summer 1998): 439–449.
11. Christopher Paine, "On the Beach: The Rapid Deployment Force and the Nuclear Arms Race," *Middle East Research and Information Project (MERIP) Reports* 111 (January 1983), reprinted in *Peace Studies: Critical Concepts in Political Science*, vol. 2., ed. Matthew Evangelista (London: Routledge, 2005).
12. Nina Tannenwald, *The Nuclear Taboo: The United States and the Non-Use of Nuclear Weapons since 1945* (New York: Cambridge University Press, 2007) ch. 4–6.
13. USAF Chief of Staff, *Nuclear Operations*, Air Force Doctrine Document 2–12 (Washington, DC: USAF Chief of Staff, 7 May 2009).
14. The White House, "Statement by President Barack Obama."
15. USAF Chief of Staff, *Nuclear Operations*, 4, 8.
16. Tannenwald, *Nuclear Taboo*.
17. William J. Perry, Brent Scowcroft, and Charles D. Ferguson, "How to Reduce the Nuclear Threat," *Wall Street Journal*, 28 May 2009.
18. John Kyl and Richard Perle, "Our Decaying Nuclear Deterrent," *Wall Street Journal*, 30 June 2009.
19. Daniel Dombey and Tobias Buck, "Moment of Truth Looms over Impasse on Enrichment," *Financial Times*, 11 June 2009, 4.
20. "President Obama Held a News Conference with South Korean President Lee Myung-bak," 16 June 2009; transcript available at http://www.washingtonpost.com/wp-dyn/content/article/2009/06/16/AR2009061601627.html.

21. Jung Sung-ki, "US Nuclear Umbrella: Double-Edged Sword for S. Korea," *The Korea Times*, 24 June 2009.
22. The White House Office of the Press Secretary, "Remarks by the President Barack Obama in Prague as Delivered," Hradcany Square, Prague, Czech Republic, 5 April 2009, http://prague.usembassy.gov/obama.html.
23. Carl Bildt and Radek Sikorski, "Next, the Tactical Nukes," *New York Times*, 2 February 2010; Agence France Presse, "Allied Bid for Obama to Remove US European Nuclear Stockpile," 20 February 2010, at http://nz.news.yahoo.com/a/-/world/6830367/allied-bid-for-obama-to-remove-us-european-nuclear-stockpile/; Massimo D'Alema, et al., "Per un mondo senza armi nucleari," *Corriere della Sera*, 24 July 2008; "New Think and Old Weapons," *New York Times* (editorial), 27 February 2010.
24. David E. Sanger and Peter Baker, "Obama Limits When U.S. Would Use Nuclear Arms," *New York Times*, 5 April 2010.
25. Kyl and Perle, "Our Decaying Nuclear Deterrent."
26. George P. Shultz et al., "A World Free of Nuclear Weapons," *Wall Street Journal*, 4 January 2007.
27. The term appears frequently in connection with the nuclear ambitions of North Korea and Iran. See, for example, "Blair Condemns 'Completely Irresponsible Act," *Independent* (UK) 9 October 2006, on the reaction to a North Korean nuclear test; and Agence France Presse, "US Military Chief Slams Iran's 'Irresponsible Influence'," *France 24 International News*, 2 May 2008.
28. William J. Broad and David E. Sanger, "Obama's Youth Shaped His Nuclear-Free Vision," *New York Times*, 5 July 2009.
29. "Iran 'Would like Nuclear Option'," *BBC*, 17 June 2009, http://news.bbc.co.uk/2/hi/middle_east/8104388.stm.
30. This was the explicit program of the European Nuclear Disarmament movement. See, for example, Edward P. Thompson, *Protest and Survive* (London: Campaign for Nuclear Disarmament, 1980); Edward P. Thompson, *Beyond the Cold War* (New York: Pantheon, 1982); Edward P. Thompson, *The Heavy Dancers* (New York: Pantheon, 1985); and Jean Stead and Danielle Grünberg, *Moscow Independent Peace Group* (London: Merlin Press, 1982). For a scholarly account of the transnational human-rights efforts, see Daniel C. Thomas, *The Helsinki Effect: International Norms, Human Rights, and the Demise of Communism* (Princeton, NJ: Princeton University Press, 2001).
31. George N. Lewis and Theodore A. Postol, "European Missile Defense: The Technological Basis of Russian Concerns," *Arms Control Today*, 1 October 2007.
32. Tom Parfitt and Ian Traynor, "US Rejects Kremlin's Call to Scrap Missile Shield," *The Guardian*, 14 November 2008.

33. On Romania's interest in hosting a system, see Mihaela Iordache, "Après la 'Guerre des étoiles:' la Roumanie au centre du nouveau système de défense américain?" *Europolitik*, 15 October 2009, http://www.newropeansmagazine.orgfcontent/view/IoI72/I/. On the redistribution of contracts, see Roxanna Tiron, "Missile Defense Shift Redirects Billions in Government Contracts," *The Hill*, 20 September 2009, http://thehill.com/business-a-lobbying/59449-missile-defense-shift-redirects-billions-in-contracts.
34. Stefan Wagstyl and Edward Luce, "US and Russia Square up over Missile Shield," *Financial Times*, 3 July 2009.
35. Transcript of press conference, 16 June 2009.
36. The work of Randall Forsberg was particularly important in this regard. See, for example, Randall Forsberg, "The Freeze and Beyond: Confining the Military to Defense as a Route to Disarmament," *World Policy Journal* 1, no. 2 (1984): 287–318; and "Parallel Cuts in Nuclear and Conventional Forces," *Bulletin of the Atomic Scientists* 41 (August 1985): 152–156.
37. Albert Wohlstetter, "The Delicate Balance of Terror," *Foreign Affairs* 37 (January 1959): 211–234.
38. Jeffrey W. Knopf, *Domestic Society and International Cooperation: The Impact of Protest on United States Arms Control Policy* (New York: Cambridge University Press, 1998); David S. Meyer, *A Winter of Discontent: The Nuclear Freeze and American Politics* (Boulder, CO: Praeger, 1990); and David Cortright, *Peace Works: The Citizen's Role in Ending the Cold War* (Boulder, CO: Westview, 1993).
39. Robert Scheer, *With Enough Shovels: Reagan, Bush and Nuclear War* (New York: Random House, 1982); for a contemporary analysis of Gorbachev's proposal for a nuclear-free world, see Matthew Evangelista, "The New Soviet Approach to Security," *World Policy Journal* 3 (Fall 1986): 561–599.
40. James Lee Ray, "The Abolition of Slavery and the End of International War," *International Organization* 43 (Summer 1989): 405–439; Randall Forsberg, "Toward a Theory of Peace: The Role of Moral Beliefs" (PhD diss., Massachusetts Institute of Technology, June 1997); a summary version is available as "Socially-Sanctioned and non-Sanctioned Violence: On the Role of Moral Beliefs in Causing and Preventing War and Other Forms of Large-Group Violence," in *Peace Studies*, vol. 1, ch. 5.
41. Tannenwald, *Nuclear Taboo*, ch. 10.
42. Andrew Mack, "America, Russia, and a Nuclear-Free World," *openDemocracy.net*, 7 June 2009, http://www.opendemocracy.net/article/america-russia-and-a-nuclear-free-world, original emphasis.

Matthew Evangelista is Professor of History and Political Science in the Department of Government at Cornell University. He has served as Director of Cornell's Judith Reppy Institute for Peace and Conflict Studies and as Chair of the Department of Government. He taught at the University of Michigan and has been a visiting scholar at Harvard, Stanford, the Brookings Institution, and several Italian universities.

Conclusion: ISODARCO as an Epistemic Community

Paolo Foradori, Giampiero Giacomello, and Alessandro Pascolini

This collection of contributions by some of the most influential experts in nuclear weapons has addressed numerous issues related to the threat of nuclear weapons and the risk of their proliferation. Despite the diversity in their backgrounds, origins, areas of expertise, political orientations and time of writing, the various authors share some common assumptions about nuclear weapons and policies that can be summarized in the following four main propositions.

First, wide consensus exists among them that nuclear weapons are a very special category of weapons, radically different from conventional and also other non-conventional ones (i.e., chemical and biological weap-

P. Foradori
School of International Studies
University of Trento
Trento, Italy

G. Giacomello
Department of Political and Social Sciences
University of Bologna
Bologna, Italy

A. Pascolini
Department of Physics and Astronomy "Galileo Galilei"
University of Padua
Padua, Italy

P. Foradori et al. (eds.), *Arms Control and Disarmament*,
https://doi.org/10.1007/978-3-319-62259-0

ons). Nuclear weapons are truly weapons of mass destruction (WMDs) that can produce catastrophic consequences in indiscriminate and disproportionate ways. All authors, regardless of the time of their writing, are fully aware of these weapons' exceptionality, their revolutionary impact on international relations and the inherent risks that their very presence has for the future of humanity. Despite differences in language, the current notion of the "catastrophic humanitarian consequences" of nuclear weapons—which has animated the debate on nuclear disarmament over the last few years and was featured in the failed 2015 Review Conference of the Non-Proliferation Treaty (NPT)—is clearly anticipated in many chapters of this volume written decades ago with remarkable vision. Given their extraordinary destructive power, nuclear weapons must be carefully controlled and, in an ideal world, eliminated, although many of the authors betray serious doubts about the viability of full-scale disarmament (see also below), or "nuclear zero," as we call it today.

Second, none of the authors downplays the fact that nuclear weapons strategies inherently imply a very dangerous "gamble." If disarmament is, in the best possible scenario, a difficult and long-term objective, *Dr. Strangelove*'s "stop worrying and love the Bomb" attitude cannot be an option. Despite the best intentions and calculations, human beings make mistakes and things go wrong for a variety of unintentional and unplanned reasons. Deterrence is not foolproof and failure is always possible. There is no "fail safe" device. The (tragic) trouble is that if something goes wrong in the nuclear realm, consequences can be catastrophic, and there is no going back.

Third, even if the world has profoundly changed during the 50-year period covered in this volume, the threat of nuclear weapons has not waned. While the spectre of a nuclear holocaust obviously surfaces as a major concern in the chapters of the first two parts of the book dedicated to the years of East-West nuclear confrontation, the third part of the post-Cold War era appears to be equally frightening. Contrary to expectations, the "nuclear peace-dividend" at the end of the bipolar system turned out to be very meagre and nuclear threats have persisted. Despite significant reductions in nuclear arsenals since the height of the superpower rivalry in the mid-eighties, it is not possible to lower our guard. As a reading of the preceding chapters clearly shows, the "second-nuclear age" is as dangerous as the first.

If this is the case, the fourth and final teaching that can be inferred from reading this volume is that only complete and irreversible disarmament represents a definitive solution to the dangers posed by nuclear weapons.

However, there is no consensus among the contributors regarding the feasibility of nuclear disarmament, and for many this goal remains mere wishful thinking given current and near-future circumstances. However, all contributors consider it to be of the utmost importance to halt the further spread of nuclear weapons. Favouring Scott Sagan's arguments to Kenneth Waltz in the famous debate between nuclear pessimists and optimists, the volume's contributors clearly subscribe to the view that "more can only be worse."

While waiting and actively working for the conditions of nuclear disarmament to become ripe, international efforts towards non-proliferation must be strongly supported. As the reading of this book suggests, and the history of ISODARCO shows, this objective can be promoted by being an active member of an epistemic community and by engaging in disarmament and non-proliferation education.

An Actual Epistemic Community

Long before Francis Bacon's "knowledge is power" aphorism, control over knowledge and information was a key ingredient of power. In fact, the emergence and diffusion of new ideas and data may lead to new patterns of behaviour that disrupt the status quo. However, if fear of altering the present order is overcome, these patterns may prove to be a game changer for international policy coordination in the most diverse circumstances, including arms control and nuclear proliferation, as the chapters in this volume demonstrate.

Equally essential for appreciating the value of the contributions presented here is the notion of epistemic communities—networks of knowledge-based experts—that will be briefly discussed in this section. Strictly speaking, this collection of contributions by prominent researchers is not about these informal groups of professionals with authoritative and policy-relevant expertise. Nonetheless, the chapters represent the efforts of many scientists and scholars to expose their ideas and policy solutions to the appraisal of their peers. Their ultimate goal was to develop viable solutions to various nuclear problems that could be adopted by adversarial governments (first and foremost, the United States and Russia) and increase confidence, reduce tension and tame risks of nuclear escalation. In so doing, these scientists and scholars created a close and vibrant epistemic community. This outcome was unplanned; it was a by-product of many interactions and international meetings, including the ISODARCO courses. Incidentally, "the quality and frequency of meetings point toward the nature of interaction among members of an epistemic community,"[1] which is also one of the indicators of success; and, in this respect, the record of ISODARCO is remarkable.

Epistemic communities have become a recent and successful concept in International Relations theory over the last 20 years. The most quoted reference is the 1992 special issue of *International Organization* (IO) entitled *Knowledge, Power, and International Policy Coordination.* In that issue, Peter Haas defined epistemic communities according to four criteria that are shared amongst members[2]: (1) normative and principled values; (2) causal beliefs; (3) tools to validate or refute causal claims and (4) the tools (or practices) utilized to prescribe policy solutions. Restricted by international and national constraints, Haas continues,[3] epistemic communities are called upon by decision makers, who are being confronted with public policy issues of increasing complexity, to help reduce the uncertainty and provide clarification. It is the prestige, training and reputation of the community that breed the type of expertise valued by political elites.

In the same issue of IO, an article by Emanuel Adler[4] recalls a national group of experts selected by the American government as the basis for negotiations with the Soviets, ultimately becoming the seed of the anti-ballistic missile (ABM) regime. Paraphrasing Adler, an American epistemic community played a key role in creating the international shared understanding and practice of nuclear arms control. In the absence of nuclear war, leaders' expectations of nuclear war and of its control were affected by causal theories, abstract propositions and models that, given their "scientific" and technical nature, were developed by an epistemic community. The theoretical and practical ideas from the arms control epistemic community turned into political expectations, which were also shared by Soviet leadership. This commonality of interests then became embodied in the 1972 ABM treaty. The very same process of generating workable policy ideas that subsequently percolated into the policy-making arena is exemplified in many of the contributions included in the present volume.

Why is the relationship between international networks of experts and policy-makers so fundamental? Because how decision makers define state interests and formulate policies to address complex and technical issues can be a function of the manner in which the issues are presented by those specialists to whom they turn for advice in the face of uncertainty. This is where epistemic communities thrive: identifying the cause-and-effect relationships of complex problems; helping governments define their interests; framing the issues for collective debate; and proposing specific policies or identifying salient points for negotiation.

A policy problem of a technical nature is an essential ingredient to appreciate the level of influence an epistemic community is likely to have.[5] When a policy problem is highly technical, experts may dominate the policy process, thus limiting other actors' ability to influence politics.[6] Natural scientists have a tendency to adopt a far more technocratic and science-based view of government, which is also evident in some of the contributions in this volume. Nonetheless, as policy issues "are seldom purely technical or purely political," as Giandomenico Majone puts it,[7] scientists cannot really avoid a certain degree of politicization of science. Social scientists, by training and nature, are more amenable to the latter phenomenon, and this too is palpable in other chapters of this work.[8]

In addition to Adler, other scholars (most notably Matthew Evangelista[9] but also Neil Mitchell et al.[10]) have applied the concept of epistemic communities to explain why superpowers cooperated on nuclear issues during the Cold War. Indeed, what is surprising about this state of affairs is not the modest results, compared to the enormity of a potential nuclear Armageddon, but, given the animosity and the leeway of the competing ideologies, that there were any results at all. Cooperation was not "by default" and networks of nuclear scientists did contribute to "the long peace."[11] This volume traces the development of that process of influence and indicates how many political obstacles (most notably mutual suspicion) had to be overcome.

Today, epistemic communities are recognized tools of international governance.[12] They not only advise states but also non-state actors (NSAs) with decision-making powers.[13] As a testimony of the concept's success, epistemic communities have been at the centre of a variety of studies, from water management,[14] to control of the bureaucracy,[15] to pension reform.[16] Their most remarkable feature, however, as confirmed by this volume, is that they have been crucial in the analysis of nuclear proliferation and disarmament, topics strongly associated with the realm of the "high politics" dear to realists. In that domain, realists find transnational cooperation to be an oxymoron, as states tend to be quite distrustful of each other. Robert Jervis concluded that, short of an individualistic security policy leading to disaster, "it may be doubted whether there will ever be strong political pressure in favour of a [security] regime."[17] However, with nuclear proliferation, that is exactly what happened. Most importantly, it was not an isolated case, as it also happened, for example, with chemical weapons[18] and other security matters.[19]

There is a final point worth considering. The literature on epistemic communities is quite clear in admitting one fundamental failure, namely the dissemination/diffusion of knowledge. At times, their access to high-level knowledge and information may make scientists and scholars akin to "mandarins," who have no incentives to communicate their knowledge to the public. Indeed, if everyone knew about something, how important would the experts be? In a sense, we are back to Bacon's "knowledge is power:" if information is controlled by, or restricted to, even well-meaning epistemic communities, there will always be substantial room for distortion and manipulation. If, however, information is shared, distributed, and passed on to the next generations, the influential power of epistemic communities, almost paradoxically, will be extended thanks to the enlarged number of informed citizens.

This is why education, in this case disarmament and non-proliferation education, and teaching by networks of professionals are essential. The chapters in this volume were all delivered as lectures to students and fellow experts, and this is the main mission of ISODARCO, as illustrated in the next section.

Contributing to Disarmament and Non-proliferation Education

Indeed, the core mission of ISODARCO has always been in the realm of disarmament and non-proliferation education (DNPE). If, as recalled, "knowledge is power," then—in the words of former United Nations (UN) Secretary General Kofi Annan—"education is quite simply, peace building by another name."[20]

The relevance of DNPE as a fundamental tool in curbing the spread of nuclear weapons, with a view to their elimination, has been recognized since the first special session of the UN General Assembly (UNGA) on disarmament issues in 1978. Since then, the sense of urgency has been increasing, prompting the UNGA in resolution 55/33/E of 20 November 2000 to request that the UN Secretary General prepare, with the assistance of a group of qualified experts, a detailed study on the matter.

After two years of preparation, the *United Nations Study on Disarmament and Non-Proliferation Education*[21] was submitted to the First Committee of the General Assembly at its 57th session on 9 October 2002, and endorsed in General Assembly Resolution 57/60. The Study builds upon and seeks to revitalize past efforts concerned directly with disarmament education, which it considers to be an integral part of *peace education*. It starts from the recognition that the need for DNPE has never been greater and urges new thinking to pursue and achieve disarmament and non-proliferation goals. The overall purpose of DNPE is described as, "impart[ing] knowledge and skills to empower individuals to make their contribution, as national and world citizens, to the achievement of general and complete disarmament under effective international control."[22]

The Study emphasizes the continued and growing significance of the non-proliferation of WMDs and the need for new thinking to address the security challenges of the post-Cold War environment. It tackles new elements such as the great potential of innovative pedagogical methods in connection with new information and communication technologies, particularly the Internet; and the importance of introducing DNPE into post-conflict situations as a contribution to peace-building; the great need to introduce gender perspectives into security issues, and specifically, into non-proliferation. Most importantly, the Study contains 34 specific recommendations for action to be undertaken by governments, regional organizations, the UN, and other international organizations as well as municipal and religious leaders.[23] It also seeks to establish close collaboration between the experts and civil society, practitioners, and scholars, including educators and academic institutions primarily at the secondary and tertiary levels of education.[24]

While the UN initiative has received wide support its implementation has lagged behind, with DNPE continuing to be a largely underutilized (and underfinanced) tool for promoting *peace*, *disarmament* and *non-proliferation*.[25] The trouble is, in the words of William C. Potter—founder and director of the Center for Non-Proliferation Studies, one of the world's leading institutions engaged in non-proliferation and disarmament education and training—that the two emerging post-Cold War disarmament challenges are *ignorance* and *complacency*. With the end of the Cold War and the diminution of the traditional danger of superpower nuclear conflict, there is the widespread (mistaken) perception that there are no longer any real nuclear dangers.[26]

This reading of the current security situation is dangerously flawed. Although we have likely escaped the danger of a nuclear annihilation, the presence and proliferation of WMDs continue to pose a serious threat to global security today. Currently, WMDs and their delivery vehicles are still perceived by many countries as necessary or highly desirable weapons for a variety of reasons, including serving as a deterrent to security threats, as a force equalizer complementing conventional military capabilities, as prestige and status enhancers, or for bolstering regime security.[27]

The fight against nuclear proliferation is still long and uncertain, and disarmament remains a very long-term objective. Meanwhile, the risk of actual use continues to be high. Notwithstanding deep cuts in the US and

Russian (former Soviet) arsenals since the end of the Cold War, the 16,000 nuclear weapons still in existence, some of which are mounted on ballistic missiles ready to be launched on very short notice, are a potent reminder of the immediate ability to enact world annihilation. Despite rhetorical commitment to "a world free of nuclear weapons," hardly any devaluation of those weapons has occurred within nuclear weapons countries, while sophisticated and extremely expensive modernization programs are ongoing to maintain the efficiency and reliability of these weapons for the foreseeable future.

In the meantime, the nuclear club has welcomed three new members with the nuclear tests of India and Pakistan in 1998 and North Korea in 2006. If, with a bit of luck, we survived the Cold War and deterrence helped to save the world from total destruction, there is no guarantee that the same doctrines will perform at all in a multipolar nuclear world, which is likely to grow more nervous and unstable over time. Moreover, the scenarios in which NSAs acquire nuclear capabilities and mount a catastrophic attack are real and urgent, especially since 9/11. In this respect, securing nuclear materials has become a top priority for many governments and non-governmental institutions alike. The legitimate spread of nuclear energy in troubled regions, particularly the Middle East, might represent a profound proliferation challenge, as the Iranian case shows. Verifying compliance in a changing technological setting will become an ever more daunting task.

In this volatile and evolving "nuclear context," combating ignorance, complacency, and a culture of violence through DNPE continues to be a task of critical relevance. As the UN Secretary-General has rightly highlighted, "what we know little about, we care little to do anything about."[28]

The very limited awareness of the risks of nuclear proliferation applies to not only the general public but also otherwise well-educated citizens. If DNPE concerns all stages of education, ISODARCO's specific target lies at the university, specifically the postgraduate, level. Experts in arms control and disarmament are in short supply in both developing and developed countries, including nuclear weapons states. The earlier generation of scholars and practitioners increasingly complains that very few younger experts are available to replace them.[29] An entire new generation is growing up without a clear perception of the risks posed by nuclear weapons, wrongly perceived as relics of the Cold War, while non-proliferation studies have become a marginal and neglected field of research and teaching in most colleges and universities. With each passing year, there are also fewer opportunities to access the living memory of the *hibakusha* (the surviving victims of the 1945 atomic bombings of Hiroshima and Nagasaki), who are able to provide fundamental testimony of the catastrophic and inhumane consequences of nuclear weapon usage. The scant interest in disarmament also concerns policy makers, an overwhelming majority of whom are not only relatively uninterested in international matters as such but also woefully uneducated about non-proliferation and disarmament issues. To aggravate the matter, looking for quick solutions to immediate crises, governments tend not to invest adequately in long-term training programs.[30]

DNPE is not only about educating and cultivating critical thinking on nuclear weapons-related issues, it is also a form of advocacy: a way of building a social movement of action towards achieving enhanced national and international security with lower levels of arms and ultimately general and complete disarmament under effective international control. As the 2002 UN Study notes, development of global DNPE and culture cannot be accomplished easily, cheaply or quickly, and hence a major effort is required to build communities of independent disarmament and non-proliferation specialists.[31] As highlighted, to be effective, community-building must be sustained over an extended period of time. The 50 year-long engagement of ISODARCO in DNPE and training represents, as the outstanding authors of this book have shown, a very significant contribution to this long-term objective.

Paolo Foradori is an Associate Professor at the University of Trento, School of International Studies. From 2009 to 2011 he was Marie Curie Fellow at the Center for Non-proliferation Studies, Monterey Institute of International Studies, California, and until 2014 an associate at the Managing the Atom Program, Belfer Center for Science and International Affairs at Harvard University's Kennedy School of Government.

Giampiero Giacomello is Associate Professor of Political Science in the Department of Social and Political Sciences, University of Bologna, where he has taught Strategic Studies for several years. His research interests include strategic theory, cyber-security and war-gaming and simulation. He has published extensively on these themes and has been a visiting professor at several European and American universities. His association with ISODARCO dates back to 1988.

Alessandro Pascolini is a theoretical physicist and senior scholar at the University of Padua, where he also taught courses on arms and disarmament and science communication. He is the vice-president of ISODARCO.

Notes

1. Mai'a K. D. Cross, "Rethinking Epistemic Communities Twenty Years Later," *Review of International Studies* 39, no. 1 (2013): 137–160.
2. Peter M. Haas, "Introduction: Epistemic Communities and International Policy Coordination," *International Organization* 46, no. 1 (1992): 1–35, esp. 3.
3. Ibid., 17.
4. Emanuel Adler, "The Emergence of Cooperation: National Epistemic Communities and the International Evolution of the Idea of Nuclear Arms Control," *International Organization* 46, no. 1 (1992): 101–145.
5. John G. Ruggie, "International Responses to Technology: Concepts and Trends," *International Organization* 29, no. 3 (1975): 557–583.
6. B. Guy Peters, "The Problem of Policy Problems," *Journal of Comparative Policy Analysis* 7, no. 4 (2005): 349–370, esp. 360.
7. Giandomenico Majone, *Evidence, Argument, and Persuasion in the Policy Process* (New Haven, CT: Yale University Press, 1989), 3.
8. Patrik Marier, "Empowering Epistemic Communities: Specialised Politicians, Policy Experts and Policy Reform," *West European Politics* 31, no. 3 (2008): 513–533.
9. Matthew Evangelista, "The Paradox of State Strength: Transnational Relations, Domestic Structures, and Security Policy in Russia and the Soviet Union," *International Organization* 49, no. 1 (1995): 1–38.
10. Neil J. Mitchell et al., "Elite Beliefs, Epistemic Communities and the Atlantic Divide: Scientists' Nuclear Policy Preferences in the United States and European Union," *British Journal of Political Science* 37, no. 4 (2007): 753–764.
11. "The long peace" comes from the well-known book by John L. Gaddis, *The Long Peace: Inquiries into the History of the Cold War* (Oxford: Oxford University Press, 1987).
12. Elias G. Carayannis, Ali Pirzadeh, and Denisa Popescu, *Institutional Learning and Knowledge Transfer Across Epistemic Communities, New Tools of Global Governance* (New York, Dordrecht, Heidelberg, London: Springer, 2012).
13. Cross, "Rethinking Epistemic Communities".
14. Farhad Mukhtarov and Andrea K. Gerlak, "Epistemic Forms of Integrated Water Resources Management: Towards Knowledge Versatility," *Policy Sciences* 47, no. 2 (2014): 101–120.
15. Stuart Shapiro and David Guston, "Procedural Control of the Bureaucracy, Peer Review, and Epistemic Drift," *Journal of Public Administration Research and Theory* 17, no. 4 (2007): 535–551.
16. Marier, "Empowering Epistemic Communities".

17. Robert Jervis, "Security Regimes," *International Organization* 36, no. 2 (1982): 357–378, esp. 378.
18. Richard M. Price, *The Chemical Weapons Taboo* (Ithaca, NY: Cornell University Press, 1997).
19. See, for example, Chap. 2 in Giampiero Giacomello, *National Governments and Control of the Internet: A Digital Challenge* (London and New York: Routledge, 2005), or Mikael Karlsson, "Epistemic Communities and Cooperative Security: The Case of Communicable Disease Control in the Baltic Sea Region," *Journal of International and Area Studies* 11, no. 1 (2004): 79–100.
20. Quoted in UN Secretary-General, "Address" to *Learning Never Ends Colloquium, Calls Education Investment Which Yields Highest Profit*, press release SG/SM/7125, 10 September 1999, 2.
21. United Nations, *United Nations Study on Disarmament and Non-Proliferation Education* (New York: UN General Assembly Report A/57/124, 30 August 2002).
22. Ibid., point 6.
23. To review the implementation of the recommendations by states, international organizations and NGOs, the Secretary-General produces biennial reports. The reports are available at: http://www.un.org/disarmament/education/sg-reports.html.
24. Action Item 22 of the 2010 Review Conference of the NPT acknowledges the key role of NPDE in achieving a world without nuclear weapons and calls on the member states to fully implement the recommendations of the Study.
25. See Elena Sokova, "Disarmament and Non-Proliferation Education. Recent Developments and the Way Forward," *CTBTO Spectrum* 19 (September 2012): 16–18.
26. William C. Potter, "A New Agenda for Disarmament and Non-Proliferation Education," *Disarmament Forum* 3 (2001): 5–12.
27. Paolo Foradori, "Introduction," in *Still the Century of Overkill? Strengthening the Control of Weapons of Mass Destruction*, ed. Paolo Foradori (Baden-Baden: Nomos, 2014): 13–25.
28. United Nations, *Study on Disarmament*, 4.
29. Sokova, "Disarmament and Non-Proliferation Education", 18.
30. Potter, "A New Agenda for Disarmament", 5.
31. United Nations, *Study on Disarmament*, point 10.

Afterword: Historical Notes on ISODARCO

Carlo Schaerf

The idea to organize residential courses on the topics of disarmament and arms control was born during a conversation that professor Edoardo Amaldi and I had in the summer of 1962 at Villa Monastero in Varenna (on the Como lake) during a residential course for physicists organized by the Italian Physical Society. For a period of one-to-two weeks some 30–40 junior researchers live together with 10–15 senior scholars, listen to their presentations and have the very useful opportunity to interact with the lecturers and other participants.

At that time, Amaldi was the most influential scientist in Italy.[32] A former member of the Rome Fermi group, after the Second World War he had contributed to the renaissance of Italian science, the establishment of the European Organization for Nuclear Research (CERN) and of the European Space Agency (ESA). He was very worried by the development of the nuclear arms race and was a member of the Continuing Committee of the Pugwash Conferences on Science and World Affairs.

Between 1960 and 1963, I was working at Stanford University in the laboratory directed by Wolfgang K. H. Panofsky, founder and first director of the Stanford Linear Accelerator Center. He was also involved in prob-

C. Schaerf
ISODARCO, Rome, Italy

P. Foradori et al. (eds.), *Arms Control and Disarmament*,
https://doi.org/10.1007/978-3-319-62259-0

lems of international security as a member of the President's (Kennedy) Science Advisory Committee.

During those years the United States was approaching her largest arsenal of nuclear weapons and the Soviet Union was rapidly expanding her own (in 1963 the United States had approximately 29,000 nukes and the Soviet Union roughly 4,200).[33] Those were the years of the Berlin and Cuban crises, and the possibility of a nuclear confrontation between the countries of the North Atlantic Treaty Organization (NATO) and the Warsaw Pact was a practical concern (around Stanford some people were building fall-out shelters in their backyards). Some experts, including within the Kennedy administration, started worrying about this trend and some open discussions on the development of the nuclear arms race were encouraged, even outside the closed circles of expert officials. At Stanford, there was a high concentration of knowledgeable people and a Faculty Forum on Arms Control started to meet once a month. They were introduced by a presentation by some expert, from a university, a think tank, the administration or the military, and were followed by open discussions. The meetings were public and I attended most of them and became interested in the topic.

During the conversation with Amaldi, I reported my experience at Stanford and mentioned the idea of organizing a residential course on the scientific and technical questions connected with the problems of international security, the nuclear arms race and the possible use of the extensive nuclear arsenals in war. While it was clear that the questions of international security, the arms race and military confrontations pertain to the political domain, we were at the same time convinced that, in this technological and nuclear age, their understanding requires a wealth of scientific and technical knowledge often unfamiliar not only to the general public but also to most politicians, journalists and academics. Amaldi expressed interest in the idea and after my return to Italy in November 1963 we started to work on it.

The first course was organized at Villa Falconieri in Frascati (Rome) on 13–25 June 1966. It had 9 lecturers (including Rolf Bjornerstedt, William Epstein, Bernard Feld, Mihailo Markovic, Bernard Röling and Paolo Sylos-Labini), 3 visiting officials, 23 participants and 7 observers. It was made possible by a grant from the Geneva office of the Carnegie Endowment for International Peace, the use of Villa Falconieri provided by the Italian Ministry of Public Education, and grants from the office of the Italian Prime Minister and the Minister of Foreign Affairs. Amaldi opened the meeting with a speech illustrating the initiative, its goals and its role as a substantial and effective educational complement to the activities of the Pugwash Conferences. The main topics discussed included the effects of nuclear weapons and nuclear war, nuclear strategy, armaments and world security, technological and political problems on the road to disarmament, the economic aspects of disarmament and the prospects for peaceful coexistence.

To the best of our knowledge, this was the world's first advanced course devoted to the study of the scientific, technological, economic and political problems related to the quest for the reduction and final elimination of nuclear weapons. This experience showed that politically sensitive arguments regarding the nuclear arms race and the prospects of nuclear war could be presented and discussed in a relaxed academic atmosphere with participants from many different countries, ideological backgrounds and from both sides of the Iron Curtain.

Encouraged by its success, a second course was organized in 1968 at Collegio Ghislieri, Pavia. Particular attention was given to international law, the role of the United Nations (UN), and the control of disarmament agreements. Lectures were given by, among others, Francesco Calogero, Vasily Emelyanov, Jules Moch and Louis Sohn. Emelyanov, the former chairman of the Soviet Union Atomic Energy Commission, illustrated the result of the UN study on the effects of nuclear war that he had co-authored. He must have made a positive report back home since after Pavia we had a constant participation of leading personalities from Russia, including Oleg Reutov, Georgy Arbatov, Vassili Goldanskii, Mikhail Milstein, Sergei Kapitza, Alexei Arbatov, Nadezhda Arbatova, Alexei Vasilyev, Dmitry Chereshkin, Vitaly Tsygichko, Sergei Batsanov, Eugene Miasnikov, Vladimir Orlov, Ruslan Khasbulatov and Alexander Nikitin.

The third course was held in 1970 at the magnificent Duino Castle (Trieste) with the participation of, among other experts, Mary Kaldor, Anatol Rapoport, Frank Barnaby, William Gutteridge, Kosta Tsipis and Julian Perry Robinson. New themes discussed included chemical and biological warfare and disarmament, security problems of developing countries, and the application of game theory and simulations to international policy.

The first three courses of ISODARCO (at that time called ISSODAC) were organized without a formal structure, all the work being essentially done personally by Amaldi, myself and some colleagues. During the preparation of the fourth course, mostly in order to apply for grants at the national and international level, we established a non-profit association formally registered in Rome on 18 January 1972. The main points of its statute are:

> The principal aim of the Association is to foster interest for scientific problems related to disarmament and peace with all available means. More specifically it organizes, participates in and contributes to meetings, seminars, conferences and professional and cultural refresher courses. It collects documentation, promotes and carries out research on these problems. The Association is non-profit and not aligned to any party or political ideology. All proceeds will be used for its institutional goals.

Amaldi was the first President and myself the Director; when Amaldi passed away, I became President. The statute of the association underwent two minor modifications in 1991 and 2015 to make it more compatible with Italian and international regulations for non-profit institutions. ISODARCO membership has always remained limited, composed mostly of Italian scientists. A part-time secretary and the assistance of a financial administrator and treasurer were the only help during the periods of maximum activity, with some local support during the courses.

The fourth course (Padua, 1972) considered new technologies, their implications for political and strategic doctrines, and their impact on the perspectives of disarmament, with the participation, among others lecturers, of David Carlton, Jack Ruina, Hans Morgenthau, Thomas Schelling, Francesco Cavalletti, Alan Dowty, Joseph Goldblat, George Ratjens and Herbert York. David Carlton served as co-editor of the proceedings of the course and continued his collaboration for several years as course director and editor.

These first ISODARCO courses did confirm our founding assumption: there was a substantial need to provide serious basic scientific and technical information to people interested in international politics and security and those trained in political, social and human sciences, as well as to people involved in the peace movement willing to play a more informed and competent role in their activity. It also confirmed the basic tenet that the problems of war and peace, arms and disarmament, and international security are political and social, and that they involve a large spectrum of disciplines including economics and international law. Therefore, the topics of ISODARCO courses from the very beginning included lectures on international politics and relations, international law, the economics of the arms race, among other topics. Moreover, if our initial attention was to the Cold War and the East-West conflict between the superpowers and their allies, very soon we had to add to our attention other forms of conflicts, among smaller states and asymmetric forms of warfare.

The basic structure of ISODARCO courses had now stabilized: 10–15 lecturers are invited by the course directors to present one or two lectures each, and to take part in all meetings, contributing to the discussions. Participants apply and are selected on the basis of their qualifications; they are provided with accommodation and full board for a nominal admission fee that covers only a fraction of the cost of the course and of their hospitality. Lecturers receive no honorarium. Normally they are offered hospitality for themselves and a companion for the entire duration of the course. The home institutions of the lecturers contributed substantial indirect support to ISODARCO by providing them with the cost of travel. This economically unattractive offer to our lecturers, imposed by our limited financial resources, did not prevent the participation of very eminent scholars, motivated by the level of the presentations and the opportunity to become involved in intellectually engaging discussions in a very friendly atmosphere, with a group of people very heterogeneous not only from the national and ideological points of view but also from different academic disciplines and work experiences. In this way, a signature style of the ISODARCO approach to disarmament and non-proliferation education (DNPE) was forged: several lecturers and participants attend a number of courses, creating in this way a world-wide epistemic community of thought and culture.

By this time, we had learned that the best part of ISODARCO were the discussions that followed each lecture and the opportunity for lecturers and participants to interact with each other out of the sessions. Each session is organized as a 45-minute presentation followed by 45 minutes of open discussion; and, most of the time, we limited the official working hours to four sessions a day to leave free time for informal interactions among all participants, seminars offered by lecturers and participants, and spontaneous working groups and round tables. In order to make the discussions free and effective, we keep them private, asking all participants not to attribute specific opinions to anybody. This made it easier for younger participants to ask candid questions and for everybody to express personal opinions even if different from the official positions of the political authorities of their countries. Session chairpersons are instructed to give priority, during the discussion time, to younger participants and non-lecturers.

It is also clear that the major asset of ISODARCO is its capacity to attract participants from many different disciplines, countries, working experiences and ages, host them for 7–14 days in the same place, and provide a series of high quality presentations on some topic related to international security. Courses are generally hosted in a college or student residence that could provide hospitality (room and board, a class room and an office for the secretary) for all our participants (lecturers and students). Keeping all our participants sleeping, eating, and attending sessions in the same place produces immediately a very lively intellectual community that may easily overcome ideological and national barriers in the quest for a better understanding of the problems facing them. A catalyst factor is the presence of several young participants in their twenties: undergraduate, graduate and PhD students, young research associates and faculty members, and other young people at the start of their career in diplomacy, the administration, the military, the police, to name a few.

After four courses devoted to the arms race and the prospects for arms control, the next course (Urbino, 1974) was devoted to international terrorism. At that time terrorism in Europe was dominated by groups engaged in continuing domestic campaigns of violence, such as the Provisional Irish Republican Army, Spain's Basque separatists (ETA), German Rote Armee Fraktion (RAF) and Italian Brigate Rosse. The macabre events of the 1972 Munich Olympics exemplified the potential of terrorist actions to attract very wide media attention and be a powerful tool of psychological warfare in asymmetric conflicts. It further illustrated the ability of terrorist acts committed by individuals and group to capture the attention of states, which would not yield in showing their resolve in the face of violent attacks such as those witnessed at Munich and others that were to follow. It was our impression that the media success of this operation would encourage similar actions in the future, enlarging the number of perpetrators and possible targets, and making these operations much more diffuse as an emerging threat to international security. Of course, we could not envision at that time the 2996 victims of the 9/11 attacks on the Twin Towers in New York, or the 77 lives taken by one deranged individual in Norway on 22 July 2011, or the recent 133 victims of the Paris attacks in November 2015, or the present daily toll of tens of lives in many countries, due to terrorist violence.

The Urbino course attracted 60 lecturers and participants from 20 different nationalities and a very wide spectrum of professional experiences. Among the lecturers we can mention Pierre Hassner, Brian Jenkins, Victor Gilinsky, Ciro Zoppo, Gaston Bouthoul and John Bowyer Bell. The papers presented at this course produced an influential book, *International Terrorism and World Security*.[34] In 2015, this book was reprinted by Routledge Library Editions in their series on Terrorism and Insurgency.

The next course on international terrorism was held in 1978, at Ariccia (Roma), and included lectures by Yonah Alexander, Frank Wright and Alessandro Silj. The book that emerged from its proceedings has also been recently reprinted.[35] The problems of international terrorism and anti-terrorism were again on ISODARCO's agenda after 9/11 (courses of 2006, 2007 and 2008), with particular attention to the political and military reactions and their impact on human rights; new lecturers on the subject included Matthew Evangelista and Giancarlo Tenaglia (directors of several courses), Steve Wright, George Joffé, Sandra Ionno Butcher, Paul Ingram, Steve Miller, Amos Nadan, Dennis Gormley, Juergen Altmann, Malcolm Dando, Zia Mian, Takao Takahara, Luigi Caligaris, Maati Monjib, Laura Reed and Virginie Guiraudon. For us, the main result of these meetings was the possibility of presenting and discussing, in an academic atmosphere, scholarly analyses on highly emotional political problems with a highly variegated audience including Israelis, Palestinians, Iranians, Africans, and so on.

The impact on arms control of technological innovations in nuclear and conventional armaments, including space systems, was the subject of several courses, namely those at Nemi (Rome, 1976), Venice (1980, 1984 and 1988), Verona (1982), San Miniato (1986) and L'Aquila (1990). New experts on these subjects joined our faculty, including Herbert Scoville, Enid Schoettle, Jorma Miettinen, Pierre Lellouche, Lawrence Freedman, Alexander De Volpi, Olga Šuković, HylkeTromp, Jane Sharp, Dietrich Schroeer, Sverre Lodgaard, Bhupendra Jasani, Robert Neild, Catherine Kelleher (director of several courses), Shu Yuan Hsieh and Frank von Hippel.

At Ariccia (1978), attention was also paid to the problems related to the control of energy and strategic raw materials, with lectures by Joseph Rotblat, Harold Feiveson and Robert Williams; the matter was reconsidered at Candriai (Trento, 2001), with attention also given to climate change by new experts, including Venance Journé and Mirco Elena (director of several courses).

The year 1988 marked a watershed moment in the life of ISODARCO. A major event was the convening in China of the first ISODARCO Beijing Seminar on Arms Control, the fruit of collaboration with the Institute of Applied Physics and Computational Mathematics (IAPCM) and the China Institute of Contemporary International Relations (CICIR). The seeds of this important development were planted at the 1986 San Miniato course, which was attended by Hu Side—who at that time was director of the China Academy of Engineering Physics—along with Hua Xinsheng and Chen

Xueyin. During an informal conversation, Hu Side proposed to explore the possibility of organizing a similar meeting in China in order to allow Chinese scholars of different disciplines and institutions to discuss openly the problems of international peace and security with scholars from foreign countries. At the first ISODARCO Beijing seminar, there were 8 western scholars, including Richard L. Garwin and Frank von Hippel, and 45 Chinese scholars. They represented a wide spectrum of disciplines, from the social and political sciences to physics, and came from several different institutions including the military. Despite the language barrier, which limited direct contacts, the meeting was very fruitful and it was decided to have a similar seminar every two years. After 2004 it was renamed as PIIC Beijing Seminar to include in its acronym the names of all the main sponsoring organizations, having the Program for Science and National Security Studies (PSNSS) joined ISODARCO, IAPCM and CICIR. In his book,[36] Evan S. Medeiros discusses the role of ISODARCO Seminars in the development of Chinese strategic thinking. The collaboration initiated in 1988 continues to this day with the regular participation of Chinese scholars and younger researchers at the ISODARCO courses in Italy and the collaboration of ISODARCO people to the seminars in China.

As a second novelty, in 1988, we organized our first ISODARCO Winter Course that proved to be successful; in these courses, the sessions are scheduled in the early morning, late afternoon and after dinner. When the sun is high we leave people free to organize spontaneous activities like extra seminars offered by the participants, working groups, showing of films, informal discussions or a walk on the snow or a run on the easy ski slopes of the Paganella mountain. For several years we had annual winter and summer courses in addition to bi-annual seminars in China and occasional special seminars in Taipei, Amman and Venice; when, in 2005, financial constraints forced us to scale back our activities we decided to continue with the yearly winter courses sacrificing the summer courses.

After the end of the Cold War the field of arms control presented new dimensions; in particular, the international transfer of military technologies became critical. These problems were considered at Folgaria (1993), L'Aquila (1993), Pontignano (Siena, 1996), Rovereto (2000), with the participation of Ruth Adams (who directed three courses), Cui Liru, Richard Ullman, Karlheinz Lohs, Gary Chapman, Jean Pascal Zanders, Bruce Larkin, Judith Reppy (director of several courses), George Bunn, Virginia Gamba, Götz Neuneck and Patricia Lewis.

The political changes in Russia and Eastern Europe created a new landscape; this both ushered in new and in some cases daunting crises and conflicts, as well as cultivated novel perspectives for the whole European and sub-regional security. To these themes many courses from 1993 to 2005 were dedicated, with the participation of many newcomers including Lamberto Zannier, Jan Prawitz and Ioan Mircea Pascu.

Military applications in cyberspace and autonomous systems were considered in Rovereto (1999), Trento (2002) and Andalo (2012 and 2013) with Giampiero Giacomello, Chunmei Kang, Gian Piero Siroli, Herb Lin, Li Hua, Denise Garcia, Peter Dombrowski and Carlo Trezza.

The 2009 speech in Prague by President Barack Obama and the climate of collaboration between Russia and the United States enhanced the perspective of the final elimination of nuclear weapons. ISODARCO devoted the 2009, 2010 and 2011 courses to explore the possibility and consequences of a world without nuclear weapons; several previous lecturers took part in these courses, with new experts including James Acton, Avner Cohen, David Holloway, Jeffrey Lewis, Tariq Rauf, Paolo Cotta Ramusino, Giorgio Franceschini, Rebecca Johnson, Harald Müller and Tom Sauer.

The problems and perspectives of nuclear governance were the subjects of the most recent courses (2014, 2015, and 2016), with lectures by Paolo Foradori, Laura Rockwood, Mark Suh, Hamad Alkaabi, Jacek Bylica, Mark Fitzpatrick, Alexander Kmentt, Hannu Kyröläinen, Grégoire Mallard, Benoît Pelopidas, Asghar Soltanieh, Tibor Tóth, Jiang Yimin, Martin Malin, Joseph Pilat and Jenni Rissanen.

In its first 50 years, ISODARCO has produced 52 courses (51 in Italy and one in Germany), two seminars in Taipei, one each in Amman and Venice, and 14 seminars in China in cooperation with our Chinese colleagues. In total, about 3000 interested and active lecturers and participants from some 80 different countries have attended ISODARCO meetings. With the broadening of ISODARCO interests, it became necessary to involve people with different expertise in the organization of its courses. A total of 27 different course Directors from Italy, the United States, the United Kingdom, Germany and Canada have collaborated in the organization of the courses. The ISODARCO meetings have resulted in the publication of 29 books by such British and American houses as Macmillan, John Wiley, St. Martin's Press, Dartmouth-Ashgate, Palgrave, Stanford University Press and Bloomsbury.[37] ISODARCO has also carried out and published an extensive research on political violence in Italy, supported financially by the Italian National Research Council.[38]

As an educational institution, ISODARCO has never issued any political statement or endorsed any particular political position and all presentations made to ISODARCO are made under the personal responsibility of the speaker and not of ISODARCO or the institutions to which the speaker belonged. Published reports about ISODARCO meetings present the topic discussed and the opinions expressed during the discussions but without attribution of any specific opinion to any participant.

Fifty years of ISODARCO continuous activity have been made possible by the financial support provided by major foundations, mainly the John D. and Catherine T. MacArthur Foundation, the Ford Foundation and the Volkswagen Foundation, universities and research institutions, national and local authorities, and some individual donors. Among those deserving of special mention are the Carnegie Endowment for International Peace and the Carnegie Corporation of New York, the Italian National Research Council (CNR), the universities of Roma "Tor Vergata" and Trento and several local institutions in the Trentino region. Substantial indirect support has been provided by the institutions that have paid for the travel costs of course directors and lecturers. The several people working for ISODARCO as volunteers provided an essential support without which we would have not survived several difficult times.

Notes

1. Carlo Rubbia, *Edoardo Amaldi: Scientific Statesman* (Geneva: CERN, 1991).
2. Robert S. Norris, and Hans M. Kristensen, "Global Nuclear Weapons Inventories, 1945–2010," *The Bulletin of the Atomic Scientists* 66 (July/August 2010): 77–83.
3. David Carlton and Carlo Schaerf, eds., *International Terrorism and World Security* (London: Croom Helm, 1975; Abingdon: Routledge Library Editions, 2015).
4. David Carlton and Carlo Schaerf, eds., *Contemporary Terror: Studies in Sub-State Violence* (London: Macmillan, 1981; Abingdon: Routledge Library Editions, 2015).
5. Evan S. Medeiros, *Reluctant Restraint: The evolution of China's Non-Proliferation Policies and Practices, 1980–2004* (Stanford, CA: Stanford University Press, 2007).
6. A complete list of ISODARCO's publications is available on ISODARCO's web site www.isodarco.it.
7. Carlo Schaerf et al., *Venti anni di violenza politica in Italia, 1969–1988. Cronologia ed analisi statistica* (Rome: Università degli Studi di Roma "La Sapienza", Centro Stampa d'Ateneo, 1992).

Selected Bibliography

Adler, Emanuel. 1992. The Emergence of Cooperation: National Epistemic Communities and the International Evolution of the Idea of Nuclear Arms Control. *International Organization* 46 (1): 101–145.

Amaldi, Edoardo, and Carlo Schaerf, eds. 1967. *International School on Disarmament and Arms Control: Proceedings of the First Course*. Rome: ISODARCO.

———, eds. 1969. *International School on Disarmament and Arms Control: Proceedings of the Second Course*. Rome: ISODARCO.

Arbatov, Alexei, and Vladimir Dvorkin. 2006. *Beyond Nuclear Deterrence*. Washington, DC: Carnegie Endowment for International Peace.

Ball, Desmond. 1986. The Development of the SIOP, 1960–1983. In *Strategic Targeting*, ed. Ball and Richelson, 57–83.

Ball, Desmond, and Jeffrey Richelson, eds. 1986. *Strategic Nuclear Targeting*. Ithaca, NY: Cornell University Press.

Barnaby, Frank, and Carlo Schaerf, eds. 1972. *Disarmament and Arms Control: Proceedings of the Third Course Given by the International School on Disarmament and Arms Control*. New York: Gordon and Breach.

Barton, John H. 1977. The Proscription of Nuclear Weapons: A Third Nuclear Regime. In *Nuclear Weapons and World Politics*, ed. David C. Gompert, Michael Mandelbaum, Richard L. Garwin, and John H. Barton, 151–211. New York: McGraw-Hill.

Bildt, Carl, and Radek Sikorski. 2010. Next, the Tactical Nukes. *New York Times*, February 2.

Blechman, Barry M., and Cathleen Fischer, eds. 1988. *The Silent Partner: West Germany and Arms Control*. Cambridge, MA: Ballinger.

P. Foradori et al. (eds.), *Arms Control and Disarmament*,
https://doi.org/10.1007/978-3-319-62259-0

Brennan, Donald G., ed. 1961. *Arms Control, Disarmament, and National Security*. New York: G. Braziller.

Broad, William J. 1995. Quietly, U.S. Converts Uranium into Fuel for Civilian Reactors. *New York Times*, June 19.

Broad, William J., and David E. Sanger. 2009. Obama's Youth Shaped His Nuclear-Free Vision. *New York Times*, July 5.

Brown, Harold. 1980. *Department of Defense Annual Report, Fiscal Year 1981*. Washington, DC: US Department of Defense.

Bundy, McGeorge. 1969–1970. To Cap the Volcano. *Foreign Affairs* 48 (1): 1–20.

Burck, Gordon M. 1992. The Chemical Weapons Convention Negotiations. In *Verification Report 1992*, ed. John B. Poole and Richard Guthrie, 126–128. London: VERTIC.

Butt, Yousaf. 2008. Redefining Deterrence: Is RRW Detrimental to U.S. Security Calculus. *Bulletin of the Atomic Scientists* 64 (December): 15.

Carayannis, Elias G., Ali Pirzadeh, and Denisa Popescu. 2012. *Institutional Learning and Knowledge Transfer Across Epistemic Communities, New Tools of Global Governance*. New York: Springer.

Carlton, David, Klaus Gottstein, Mirco Elena, and Paul Ingram, eds. 1995. *Controlling the International Transfer of Weaponry and Related Technology*. Aldershot: Dartmouth.

Carlton, David, and Carlo Schaerf, eds. 1977. *Arms Control and Technological Innovation*. London: Croom Helm.

———, eds. 1982a. *The Arms Race in the 1980s*. London; New York: Macmillan; St. Martin's Press.

———, eds. 1991. *The Arms Race in an Era of Negotiations*. London; New York: Macmillan; St. Martin's Press.

———, eds. 1981. *Contemporary Terror: Studies in Sub-State Violence*. London: Macmillan. Reprinted. Abingdon: Routledge Library Editions, 2015.

———, eds. 1975a. *The Dynamics of the Arms Race*. London: Croom Helm.

———, eds. 1982b. *The Hazards of the International Energy Crisis: Studies of the Coming Struggle for Energy and Strategic Raw Materials*. London; New York: Macmillan; St. Martin's Press.

———, eds. 1975b. *International Terrorism and World Security*. London: Croom Helm. Reprinted. Abingdon: Routledge Library Editions, 2015.

———, eds. 1984. *Reassessing Arms Control*. London; New York: Macmillan; St. Martin's Press.

Cohen, Bernard L. 1977. The Disposal of Radioactive Wastes from Fission Reactors. *Scientific American* 236 (June): 21–31.

Commission on Integrated Long-Term Strategy. 1988. *Discriminate Deterrence*. Washington, DC: US Government Printing Office.

Committee on International Security and Arms Control, National Academy of Sciences. 1991. *The Future of the U.S.-Soviet Nuclear Relationship*. Washington, DC: National Academy of Sciences.
———. 1994. *Management and Disposition of Excess Weapons Plutonium*. Washington, DC: National Academy Press.
———. 1995. *Management and Disposition of Excess Weapons Plutonium: Reactor-Related Options*. Washington, DC: National Academy Press.
Cortright, David. 1993. *Peace Works: The Citizen's Role in Ending the Cold War*. Boulder, CO: Westview.
Cross, Mai'a K.D. 2013. Rethinking Epistemic Communities Twenty Years Later. *Review of International Studies* 39 (1): 137–160.
D'Alema, Massimo, Gianfranco Fini, Giorgio La Malfa, Arturo Parisi, and Francesco Calogero. 2008. Per un mondo senza armi nucleari. *Corriere della Sera*, July 24.
Daugherty, William, Barbara Levi, and Frank von Hippel. 1985–1986. The Consequences of 'Limited' Nuclear Attacks on the United States. *International Security* 10 (4): 3–45.
Deutsch, Karl W. 1963. *The Nerves of Government*. New York: The Free Press of Glencoe.
De Vries, Klaas G. 1979. Responding to the SS-20: An Alternative Approach. *Survival* 21 (6): 251–255.
Dombey, Daniel, and Tobias Buck. 2009. Moment of Truth Looms over Impasse on Enrichment. *Financial Times*, June 11.
Dulles, John Foster. 1954. Policy for Security and Peace. *Foreign Affairs* 32 (April): 353–364.
Dunn, Lewis A., and Herman Kahn. 1976. *Trends in Nuclear Proliferation, 1975–1995: Projections, Problems, and Policy Options*. Croton-on-Hudson, NY: The Hudson Institute.
Ehmke, Horst. 1987. A Second Phase of Détente. *World Policy Journal* 4 (Summer): 363–382.
Eichholz, Geoffrey G. 1976. *Environmental Aspects of Nuclear Power*. Ann Arbor, MI: Ann Arbor Science Publishers.
Evangelista, Matthew. 1982–1983. Stalin's Postwar Army Reappraised. *International Security* 7 (Winter): 110–138.
———. 1986. The New Soviet Approach to Security. *World Policy Journal* 3 (Fall): 561–599.
———. 1995. The Paradox of State Strength: Transnational Relations, Domestic Structures, and Security Policy in Russia and the Soviet Union. *International Organization* 49 (1): 1–38.
———. 1998. The 'Soviet Threat': Intentions, Capabilities, and Context. *Diplomatic History* 22 (Summer): 439–449.

———. 1999. *Unarmed Forces: The Transnational Movement to End the Cold War.* Ithaca, NY: Cornell University Press.

Feiveson, Harold A., Richard H. Ullman, and Frank von Hippel. 1985. Reducing U.S. and Soviet Nuclear Arsenals. *Bulletin of the Atomic Scientists* 47 (August): 144–151.

Fetter, Steve, Trevor Findlay, Jozef Goldblat, Jan Prawitz, Frode Ringdal, and Joseph Rotblat. 1992. Expert Study on Questions Related to a Comprehensive Test Ban Treaty. In *Towards a Comprehensive Test Ban Treaty.* Oslo: Royal Norwegian Ministry of Foreign Affairs.

Fischer, David, and Harald Müller. 1991. *A Treaty in Trouble: Europe and the NPT After the Fourth Review Conference.* Frankfurt: PRIF.

Fisher, Roger, and William L. Ury. 1981. *Getting to Yes: Negotiating Agreement Without Giving In.* London: Penguin Group.

Foradori, Paolo. 2014. Introduction. In *Still the Century of Overkill? Strengthening the Control of Weapons of Mass Destruction*, ed. Paolo Foradori, 13–25. Baden-Baden: Nomos.

Forbes, Henry W. 1962. *The Strategy of Disarmament.* Washington, DC: Public Affairs Press.

Forsberg, Randall. 1984. The Freeze and Beyond: Confining the Military to Defense as a Route to Disarmament. *World Policy Journal* 1 (2): 287–318.

———. 1985. Parallel Cuts in Nuclear and Conventional Forces. *Bulletin of the Atomic Scientists* 41 (August): 152–156.

———. 1997. Toward a Theory of Peace: The Role of Moral Beliefs. PhD diss., Massachusetts Institute of Technology.

Franck, Thomas M. 1990. *The Power of Legitimacy Among Nations.* New York: Oxford University Press.

Freedman, Lawrence. 1986. British Nuclear Targeting. In *Strategic Targeting*, ed. Ball and Richelson, 109–126.

———. 1977. *US Intelligence and the Soviet Strategic Threat.* London: Wiley.

———. 1979. Time for a Reappraisal. *Survival* 21 (5): 198–201.

———. 1982. *The Evolution of Nuclear Strategy.* London: Palgrave Macmillan.

Gaddis, John L. 1987. *The Long Peace: Inquiries into the History of the Cold War.* Oxford: Oxford University Press.

Garwin, Richard L. 1976. Effective Military Technology for the 1980s. *International Security* 1 (Fall): 50–77.

———. 1977. Reducing Dependence on Nuclear Weapons: A Second Nuclear Regime. In *Nuclear Weapons and World Politics: Alternatives for the Future (1980s Project/Council on Foreign Relations)*, ed. David C. Gompert, Michael Mandelbaum, Richard L. Garwin, and John H. Barton. New York: McGraw-Hill.

———. 1985. How Many Orbiting Lasers for Boost-Phase Intercept. *Nature* 315 (23 May): 286–290.

Garwin, Richard L., and Hans A. Bethe. 1968. Anti-Ballistic-Missile System. *Scientific American* 218 (March): 21–31.

Garwin, Richard L., and Roal'd Z. Sagdeev. 1991. Verification of Compliance with the ABM Treaty and with Limits on Space Weapons. In *Verification – Monitoring Disarmament*, ed. Francesco Calogero, Marvin L. Goldberger, and Sergei P. Kapitza, 23–44. Boulder: Westview Press.

Garwin, Richard L., John Tike, and Yevgeny P. Velikhov. 1984. Space Weapons. *Bulletin of the Atomic Scientists* 40 (5): 1S–15S.

Gasteyger, Curt. 1991. The Remaking of Eastern Europe's Security. *Survival* 33 (March/April): 111–124.

Giacomello, Giampiero. 2005. *National Governments and Control of the Internet: A Digital Challenge*. London: Routledge.

Glasstone, Samuel. 1964. *The Effects of Nuclear Weapons*. Washington, DC: US Department of Defense.

Haas, Peter M. 1992. Introduction: Epistemic Communities and International Policy Coordination. *International Organization* 46 (1): 1–35.

Halperin, Morton H. 1987. *Nuclear Fallacy: Dispelling the Myth of Nuclear Strategy*. Cambridge, MA: Ballinger.

Hasegawa, Tsuyoshi. 2005. *Racing the Enemy: Stalin, Truman, and the Surrender of Japan*. Cambridge, MA: Harvard University Press.

Hebel, L. Charles, Eldon L. Christensen, Fred A. Donath, Warren E. Falconer, Leon J. Lidofsky, Ernest J. Moniz, Thomas H. Moss, et al. 1978. Report to the American Physical Society by the Study Group on Nuclear Fuel Cycles and Waste Management. *Reviews of Modern Physics* 50 (1): S139.

Heisbourg, Francois. 1987. Can the Atlantic Alliance Outlast the Century? *International Affairs* 63 (Summer): 413–423.

Herman, Michael. 1991. Intelligence and Arms Control Verification. In *Verification Report 1991*, ed. John B. Poole. New York: The Apex Press for VERTIC.

Hoehn, William E. 1979. *Outlasting SALT II and Preparing for SALT III*. Santa Monica, CA: RAND.

International Commission on Radiological Protection (ICPR). 1977. Recommendations of the ICRP. ICRP Publication 26, *Annals of the ICRP* 1, n. 3.

IISS. 1980. *The Military Balance 1980–1981*. London: International Institute for Strategic Studies.

Iordache, Mihaela. 2009. Après la 'Guerre des étoiles:' la Roumanie au centre du nouveau système de défense américain? *Europolitik*, October 15.

Ireland, Timothy. 1978. The Year of Departure: 1950. *The Fletcher Forum of World Affairs* 2 (Summer): 114–138.

Jervis, Robert. 1982. Security Regimes. *International Organization* 36 (2): 357–378.

Johnson, Lyndon B. 1968. Annual Budget Message to the Congress, Fiscal Year 1968, January 24, 1967. In *Public Papers of the Presidents of the United States, 1967.* Book 1. Washington, DC: US Government Printing Office.

Kahn, Herman. 1960. *On Thermonuclear War.* Princeton, NJ: Princeton University Press.

———. 1965. *On Escalation: Metaphors and Scenarios.* New York: Praeger.

Kaiser, Karl, Georg Leber, Alois Mertes, and Franz-Josef Schulze. 1981–1982. Nuclear Weapons and the Preservation of Peace. *Foreign Affairs* 60 (5): 1157–1170.

Kamp, Karl-Heinz. 1991. *Die Sicherheit der Sowjetischen Atomwaffen.* St. Augustin: Konrad-Adenauer-Stiftung.

Kaplan, Fred. 1983. *The Wizards of Armageddon.* New York: Simon and Schuster.

Karber, Phillip A., and Jerald A. Combs. 1998. The United States, NATO and the Soviet Threat to Western Europe: Military Estimates and Policy Options, 1945–1963. *Diplomatic History* 22 (Summer): 399–429.

Karlsson, Mikael. 2004. Epistemic Communities and Cooperative Security: The Case of Communicable Disease Control in the Baltic Sea Region. *Journal of International and Area Studies* 11 (1): 79–100.

Kemp, Geoffrey. 1991. *The Control of the Middle East Arms Race.* Washington, DC: Brookings Institution Press.

Kennan, George F. 1982. *The Nuclear Delusion: Soviet-American Relations in the Atomic Age.* New York: Pantheon Books.

Keohane, Robert O. 1989. *International Institutions and State Power.* Boulder, CO: Sage Publications.

Kidder, Ray E. 1992. How Much More Nuclear Testing Do We Need? *Arms Control Today* 22 (September): 11–14.

Kissinger, Henry A. 1957. *Nuclear Weapons and Foreign Policy.* New York: Harper.

———. 1961. *The Necessity for Choice: Prospects of American Foreign Policy.* New York: Harper.

———. 1979. NATO: The Next Thirty Years. *Survival* 21 (November-December): 264–268.

Knopf, Jeffrey W. 1998. *Domestic Society and International Cooperation: The Impact of Protest on US Arms Control Policy.* New York: Cambridge University Press.

Kokoshin, Andrei. 1988. A Soviet View on Radical Weapons Cuts. *Bulletin of the Atomic Scientists* 44 (March): 14–17.

Kötter, Wolfgang, and Harald Müller. 1990. *Germany and the Bomb: Nuclear Policies in the Two German States and the United Germany's Nonproliferation Commitments.* Frankfurt: PRIF.

Krauss, Melvyn B. 1986. *How NATO Weakens the West.* New York: Simon and Schuster.

Küntzel, Matthias. 1991. Die Bundesrepublik Deutschland zwischen Nuklearambition und Atomwaffenverzicht: Eine Untersuchung der Kontroverse um den Beitritt zum Atomwaffen-Sperrvertrag. PhD diss., University of Hamburg.

Kyl, John, and Richard Perle. 2009. Our Decaying Nuclear Deterrent. *Wall Street Journal*, June 30.

Labrie, Roger P., ed. 1979. *SALT Handbook: Key Documents and Issues, 1972–1979.* Washington, DC: American Enterprise Institute for Public Policy Research.

Lanchbery, John. 1995. Reviewing the Implementation of Biodiversity Agreements. In *Verification 1995: Arms Control, Peacekeeping and the Environment*, ed. John B. Poole and Richard Guthrie. Boulder, CO: Westview Press for VERTIC.

Lee, William T. 1986. Soviet Nuclear Targeting Strategy. In *Strategic Targeting*, ed. Ball and Richelson, 84–108.

Leffler, Melvyn P. 1984. The American Conception of National Security and the Beginnings of the Cold War, 1945–48. *The American Historical Review* 89 (April): 346–381.

Leitenberg, Milton. 1980. NATO and WTO Long-Range Theatre Nuclear Forces. In *Arms Control in Europe; Problems and Prospects*, ed. Karl E. Birnbaum, 146–149. Laxenburg: Austrian Institute for International Affairs.

Lenin, Nikolaj. 1964. The Disarmament. In *The Collected Works of Vladimir I. Lenin*, vol. 19, 2nd Russian ed. Moscow: Progress Publishers.

Levi, Barbara G., and Tony Rothman. 1985. Nuclear Winter: A Matter of Degrees. *Physics Today* 38 (September): 58–65.

Lewis, Harold W., Robert J. Budnitz, Albert W. Castleman, David E. Dorfan, Fred C. Finlayson, Richard L. Garwin, L. Charles Hebel, et al. 1975. Report to the American Physical Society by the Study Group on Light-Water Reactor Safety. *Reviews of Modern Physics* 47 (Supplement 1).

Lewis, George N., and Theodore A. Postol. 2007. European Missile Defense: The Technological Basis of Russian Concerns. *Arms Control Today*, October 1.

Lincoln, Abraham. Letter to Albert G. Hodges, 4 April 1864. In *The Collected Works of Abraham Lincoln*, edited by Roy P. Basler, Marion Dolores Pratt, and Lloyd A. Dunlap, 88-93. New Brunswick, NJ: Rutgers University Press, 1953.

Luttwak, Edward N. 1987. Soviet Military Strategy in the Emerging Postnuclear Era. In *National Security Issues of the USSR*, ed. Murray Feshbach, 277–289. Dordrecht: Nijhoff.

Mahnke, Dieter. 1972. *Nukleare Mitwirkung: Die Bundesrepublik Deutschland in der Atlantischen Allianz, 1954–1970.* Berlin: De Gruyter.

Majone, Giandomenico. 1989. *Evidence, Argument, and Persuasion in the Policy Process.* New Haven, CT: Yale University Press.

Malashenko, Igor. 1991. Behind the Arms Cuts: A Jousting of Republics. *International Herald Tribune*, October 22.

Malloy, Sean L. 2008. *Atomic Tragedy: Henry L. Stimson and the Decision to Use the Bomb against Japan.* Ithaca, NY: Cornell University Press.

Mandelbaum, Michael, Richard L. Garwin, David C. Gompert, and John H. Barton. 1977. *Nuclear Weapons and World Politics.* New York: McGraw-Hill.

Marier, Patrik. 2008. Empowering Epistemic Communities: Specialised Politicians, Policy Experts and Policy Reform. *West European Politics* 31 (3): 513–533.

Maull, Hans W. 1990/1991. Germany and Japan: The New Civilian Powers. *Foreign Affairs* 69 (Winter): 91–106.

May, Michael M., George F. Bing, and John D. Steinbruner. 1988–1989. Strategic Arsenals After START: The Implications of Deep Cuts. *International Security* 13 (1): 90–133.

McArdle Kelleher, Catherine, and Judith Reppy, eds. 2011. *Getting to Zero: The Path to Nuclear Disarmament.* Stanford, CA: Stanford University Press.

McCarthy, Eugene. 1969. *The Year of the People.* Garden City, NY: Doubleday and Company.

McNamara, Robert S. 1962. Defense Arrangements of the North Atlantic Community. *Department of State Bulletin* 47 (9 July): 64–70.

———. 1967. The Dynamics of Nuclear Strategy. *Department of State Bulletin* 57 (9 October): 443–451.

———. 1968. *The Essence of Security: Reflections in Office.* New York: Harper and Row.

———. 1986. *Blundering Into Disaster: Surviving the First Century of the Nuclear Age.* New York: Pantheon Books.

Mearsheimer, John. 1990. Back to the Future: Instability in Europe after the Cold War. *International Security* 15 (Summer): 5–56.

Medeiros, Evan S. 2007. *Reluctant Restraint: The Evolution of China's Non-Proliferation Policies and Practices, 1980–2004.* Stanford, CA: Stanford University Press.

Menshikov, Valeri. 1995. On the Situation with the Storage of Plutonium and Enriched Uranium in Tomsk-7. *Yaderny Control* (February): 3–4.

Meyer, Stephen M. 1987. Soviet Nuclear Operations. In *Managing Nuclear Operations*, ed. Aston B. Carter, John D. Steinbruner, and Charles A. Zraket, 495–512. Washington, DC: The Brookings Institution.

Meyer, David S. 1990. *A Winter of Discontent: The Nuclear Freeze and American Politics.* Boulder, CO: Praeger.

Mikhailov, Victor, Valeri Bogdan, Victor Murogov, Vladimir Kagramanian, Nickolai Rabotnov, Valentina Rudneva, and Mikhail Troyanov. 1994. Utilization of Plutonium in Russia's Nuclear Power Industry. *Post-Soviet Nuclear Complex Monitor* (March): 9–17.

Mitchell, Neil J., Kerry G. Herron, Hank C. Jenkins-Smith, and Guy D. Whitten. 2007. Elite Beliefs, Epistemic Communities and the Atlantic Divide: Scientists' Nuclear Policy Preferences in the United States and European Union. *British Journal of Political Science* 37 (4): 753–764.

Mukhtarov, Farhad, and Andrea K. Gerlak. 2014. Epistemic Forms of Integrated Water Resources Management: Towards Knowledge Versatility. *Policy Sciences* 47 (2): 101–120.

Müller, Harald. 1990. No Need to Fear United Germany. *Bulletin of the Atomic Scientists* 46 (April): 14–15.

———. 1991. A United Nations of Europe and North America. *Arms Control Today* 21 (January-February): 3–8.

———. 1992. The Politics of Technology Transfer. In *Germany and the Middle East; Patterns and Prospects*, ed. Shahram Chubin, 154–176. London: Pinter.

Nerlich, Uwe. 1973. *Der NV-Vertrag in der Politik der Bundesrepublik Deutschland: Zur Struktur eines aussenpolitischen Prioritätenkonfliktes.* Ebenhausen: Stiftung Wissenschaft und Politik.

———. 1976. *The Alliance and Europe: Part V. Nuclear Weapons and East-West Negotiations.* London: International Institute for Strategic Studies.

Norris, Robert S., and Hans M. Kristensen. 2010. Global Nuclear Weapons Inventories, 1945–2010. *The Bulletin of the Atomic Scientists* 66 (July/August): 77–83.

Nye, Joseph S. 1989. Nuclear Learning and the Evolution of U.S.-Soviet Security Competition. In *Windows of Opportunity: From Cold War to Peaceful Competition in U.S.-Soviet Relations*, ed. Graham Allison and William L. Ury, 131–162. Cambridge, MA: Ballinger.

Osgood, Robert E. 1957. *Limited War: The Challenge to American Strategy.* Chicago, IL: University of Chicago Press.

Paine, Christopher. 2005. On the Beach: The Rapid Deployment Force and the Nuclear Arms Race. In *Peace Studies: Critical Concepts in Political Science*, ed. Matthew Evangelista, vol. 2. London: Routledge.

Panel on Nuclear Weapons Safety of the U.S. House Committee on Armed Services. 1990. *Nuclear Weapons Safety.* Washington, DC: US Government Printing Office.

Parfitt, Tom, and Ian Traynor. 2008. US Rejects Kremlin's Call to Scrap Missile Shield. *The Guardian*, November 14.

Pascolini, Alessandro, and Dietrich Schroeer, eds. 1997. *The Weapons Legacy of the Cold War.* Aldershot: Ashgate-Dartmouth.

Perry, William J., Brent Scowcroft, and Charles D. Ferguson. 2009. How to Reduce the Nuclear Threat. *Wall Street Journal*, May 28.

Peters, B. Guy. 2005. The Problem of Policy Problems. *Journal of Comparative Policy Analysis* 7 (4): 349–370.

Potter, William C. 1991. "What if Armenia Joins Nuclear Club?" *Los Angeles Times*, October 25.

———. 2001. A New Agenda for Disarmament and Non-Proliferation Education. *Disarmament Forum* 3: 5–12.

Price, Richard M. 1997. *The Chemical Weapons Taboo.* Ithaca, NY: Cornell University Press.

Ramberg, Bennett. 1986. Targeting Nuclear Energy. In *Strategic Targeting*, ed. Ball and Richelson, 250–266.

Ray, James Lee. 1989. The Abolition of Slavery and the End of International War. *International Organization* 43 (Summer): 405–439.
Risse-Kappen, Thomas. 1988. *The Zero Option: INF, West Germany and Arms Control.* Boulder, CO: Westview Press.
Rittberger, Volker, ed. 1990. *International Regimes in East–West Politics.* London: Pinter.
Rosecrance, Richard. 1987. *Der neue Handelsstaat: Herausforderungen für Politik und Wirtschaft.* Frankfurt: Campus Verlag.
Rosenberg, David A. 1986. U.S. Nuclear War Planning, 1945–1960. In *Strategic Targeting*, ed. Ball and Richelson, 5–56.
Rotblat, Joseph. 1978. The Risks for Radiation Workers. *The Bulletin of the Atomic Scientists* 34 (September): 41–46.
———, ed. 1982. *Scientists, the Arms Race and Disarmament.* London: Taylor and Francis.
Royal Commission on Environmental Pollution. 1976. *Sixth Report: Nuclear Power and the Environment.* London: Her Majesty's Stationery Office.
Rubbia, Carlo. 1991. *Edoardo Amaldi: Scientific Statesman.* Geneva: CERN.
Ruggie, John G. 1975. International Responses to Technology: Concepts and Trends. *International Organization* 29 (3): 557–583.
Sagan, Scott D. 1993. *The Limits of Safety: Organizations, Accidents, and Nuclear Weapons.* Princeton, NJ: Princeton University Press.
Sammut, Dennis. 1995. The CSCE and Russian Peacekeeping. In *Verification 1995: Arms Control, Peacekeeping and the Environment*, ed. John B. Poole and Richard Guthrie. Boulder, CO: Westview Press for VERTIC.
Sanger, David E., and Peter Baker. 2010. Obama Limits When U.S. Would Use Nuclear Arms. *New York Times,* April 5.
Schaerf, Carlo, Giuseppe De Lutiis, Alessandro Silj, Francesco Carlucci, Emilio Bellucci, and Stefania Argentini. 1992. *Venti anni di violenza politica in Italia, 1969–1988. Cronologia ed analisi statistica.* Rome: Università degli Studi di Roma "La Sapienza", Centro Stampa d'Ateneo.
Scheer, Robert. 1982. *With Enough Shovels: Reagan, Bush and Nuclear War.* New York: Random House.
Schelling, Thomas C. 1960. *The Strategy of Conflict.* Cambridge, MA: Harvard University.
Schlesinger, James. 1974. Strategic Forces. *Department of Defense Annual Report to the President and the Congress.* Washington, DC: US Government Printing Office, March 4.
Schmidt, Helmut. 1962. *Defence or Retaliation.* Edinburgh: Oliver and Boyd.
Scribner, Richard A., Theodore J. Ralston, and William D. Metz. 1985. *The Verification Challenge: Promise and Problems of Strategic Nuclear Arms Control Verification.* Boston, MA: Birkhauser.
Senghaas, Dieter. 1991. Friedliche Streitbeilegung und kollektive Sicherheit im neuen Europa. *Europa-Archiv* 10: 311–317.

Shapiro, Stuart, and David Guston. 2007. Procedural Control of the Bureaucracy, Peer Review, and Epistemic Drift. *Journal of Public Administration Research and Theory* 17 (4): 535–551.

Sharp, Jane M.O. 1987. After Reykjavik: Arms Control and the Allies. *International Affairs* 63 (Spring): 239–358.

Shultz, George P., William J. Perry, Henry A. Kissinger, and Sam Nunn. 2007. A World Free of Nuclear Weapons. *Wall Street Journal*, January 4.

———. 2008. Toward a Nuclear-Free World. *Wall Street Journal*, January 15.

Sloan, Stanley R. 1988. *Defense Burdensharing: US Relations with the NATO Allies and Japan*. Washington, DC: Library of Congress, Congressional Research Service.

Smith, Jessica, David Laibman, and Marilyn Bechtel, eds. 1977. *Building a New Society: The 25th Congress of the Communist Party of the Soviet Union*. New York: New World Review Publications.

Snyder, Glenn. 1984. The Security Dilemma in Atlantic Politics. *World Politics* 36 (July): 461–496.

Sokova, Elena. 2012. Disarmament and Non-Proliferation Education. Recent Developments and the Way Forward. *CTBTO Spectrum* 19 (September): 16–18.

Spector, Leonard S. 1987. *Going Nuclear*. Cambridge, MA: Ballinger.

Stead, Jean, and Danielle Grünberg. 1982. *Moscow Independent Peace Group*. London: Merlin Press.

Sung-ki, Jung. 2009. US Nuclear Umbrella: Double-Edged Sword for S. Korea. *The Korea Times*, June 24.

Talbott, Stobe. 1979. *Endgame: The Inside Story of SALT II*. New York: Harper and Row.

Tannenwald, Nina. 2007. *The Nuclear Taboo: The United States and the Non-use of Nuclear Weapons Since 1945*. New York: Cambridge University Press.

Thomas, Daniel C. 2001. *The Helsinki Effect: International Norms, Human Rights, and the Demise of Communism*. Princeton, NJ: Princeton University Press.

Thompson, Edward P. 1980. *Protest and Survive*. London: Campaign for Nuclear Disarmament.

———. 1982. *Beyond the Cold War*. New York: Pantheon.

———. 1985. *The Heavy Dancers*. New York: Pantheon.

Tiron, Roxanna. 2009. Missile Defense Shift Redirects Billions in Government Contracts. *The Hill*, September 20.

Tuchman, Barbara. 1982. *The Alternative to Arms Control*. Los Angeles: Center for International and Strategic Studies University of California.

Ullman, Richard H. 1988. Ending the Cold War. *Foreign Policy* 72 (Fall): 135–139.

———. 1991. *Securing Europe*. Princeton, NJ: Princeton University Press.

United Nations. 1990. *Verification in All Its Aspects: Study on the Role of the UN in the Field of Verification*. New York: UN General Assembly A/45/372, August 28.

———. 2002. *United Nations Study on Disarmament and Non-Proliferation Education*. New York: UN General Assembly Report A/57/124, August 30.

United Nations Scientific Committee on the Effects of Atomic Radiation (UNSCEAR). 1977. *Sources and Effects of Ionising Radiation*. New York: UNSCEAR.

US Arms Control and Disarmament Agency. 1964. *Documents on Disarmament 1964*. Washington, DC: US Government Printing Office.

US Atomic Energy Commission. 1975. *Reactor Safely Study: An Assessment of Accident Risks in US Commercial Power Plants*. Washington, DC: US Nuclear Regulatory Commission.

US Department of Defense. 1976. *Report of Secretary of Defense Donald H. Rumsfeld to the Congress on the FY1977 Budget and its Implications for the FY1978 Authorization Request and the FY1977–1981 Defense Programs*. Washington, DC: US Government Printing Office, January 27.

US Department of Energy. 1996a. *Plutonium: The First 50 Years: United States Plutonium Production, Acquisition and Utilization from 1944 Through 1994*. Washington, DC: US Department of Energy.

———. 1996b. *Stopping Weapons-Grade Plutonium Production in Russia*. Washington, DC: US Department of Energy.

US National Academy of Sciences. 1972. *The Effects on Population of Exposure to Low Levels of Ionising Radiation, 1972*. Washington, DC: The National Academies of Sciences.

US-Russian Commission on Economic and Technological Cooperation. 1996. *Report on the Nuclear Energy Committee*. Washington, DC: Library of Congress, Congressional Research Service.

Wagstyl, Stefan, and Edward Luce. 2009. US and Russia Square Up Over Missile Shield. *Financial Times*, July 3.

Waltz, Kenneth N. 1979. *Theory of International Relations*. New York: McGraw-Hill.

Weinberg, Alvin M. 1967. *Reflections on Big Science*. Cambridge, MA: The MIT Press.

Wohlstetter, Albert. 1959. The Delicate Balance of Terror. *Foreign Affairs* 37 (January): 211–234.

Yost, David S. 1986. French Nuclear Targeting. In *Strategic Targeting*, ed. Ball and Richelson, 127–156.

Young, Oran R. 1989. *International Cooperation: Building Regimes for Natural Resources and the Environment*. New York: Cornell University Press.

Index[1]

[1]Note: Page numbers followed by 'n' refers to notes.

P. Foradori et al. (eds.), *Arms Control and Disarmament*,
https://doi.org/10.1007/978-3-319-62259-0

The manufacturer's authorised representative in the EU is Springer Nature Customer Service Centre GmbH, Europaplatz 3, 69115 Heidelberg, Germany. If you have any concerns regarding our products, please contact ProductSafety@springernature.com

Printed and bound by CPI Group (UK) Ltd, Croydon, CR0 4YY
25/09/2026
02222881-0002